WATER QUALITY AND AVAILABILITY

A Reference Handbook

WATER QUALITY AND AVAILABILITY

A Reference Handbook

E. Willard Miller
Department of Geography

Ruby M. Miller
Pattee Library

The Pennsylvania State University

CONTEMPORARY WORLD ISSUES

ABC-CLIO

Santa Barbara, California
Denver, Colorado
Oxford, England

Library of Congress Cataloging-in-Publication Data

Miller, E. Willard (Eugene Willard), 1915–
Water quality and availability : a reference handbook / E. Willard Miller, Ruby M. Miller.
p. cm.— (Contemporary world issues)
Includes bibliographical references (p. 165) and index.
1. Water quality—United States. 2. Water-supply—United States.
I. Miller, Ruby M. II. Title. III. Series.
TD223.M53 1992 333.91'0973—dc20 92-33057

ISBN 0-87436-647-X (alk. paper)

99 98 97 96 95 94 93 10 9 8 7 6 5 4 3 2

ABC-CLIO, Inc.
130 Cremona Drive, P.O. Box 1911
Santa Barbara, California 93116-1911

This book is printed on acid-free paper ♾.
Manufactured in the United States of America

Contents

Preface

Water is our most fundamental resource. It serves in many ways to maintain life, health, vigor, and social stability. Like air, water is bound up with human evolution—and doubtless human destiny—in many ways. One of the basic conditions for life on our planet is water. Modern civilization, however, imposes heavy demands on water, and the steady rise in consumption of water in our industrial society explains why we now regard our water supplies with great concern. Water shortages are already occurring, particularly in the drier regions of our nation. Pollution of our water sources is now a universal problem. Towns, cities, and industries have expanded so much that available water is diminishing to unsafe levels. To meet these difficulties, more thought needs to be given to planning for water availability in the future, with the ultimate goal of adopting a water policy that will insure the necessary quantity and quality of water needed by our industrial society in the future.

This volume begins with an introductory chapter that provides a background for sources of water, distribution and variability of precipitation, water supply, and means of increasing the water supply. Because of differences in the availability of water in the United States, water use doctrines have evolved that reflect the differing needs for water between dry regions and humid regions. The contamination of water began with the earliest human settlements. As industrial society evolved, the problems of contamination increased and today are immensely complex and wide-ranging. A major section of chapter 1 is devoted to modern-day contaminants.

As population increased in the western United States, the problems of providing an adequate supply of water to these dry regions have grown. Several sections of chapter 1 are devoted to the use of water in irrigated agriculture and the rising urban water consumption. The chapter concludes with a discussion of water management issues and strategies.

Because of the importance of water, laws and regulations have evolved at all levels of government from local to federal. The early laws and regulations focused on water for irrigation in the West, flood control, and power. More recent laws have stressed environmental protection, water quality, recreation, and ocean pollution. There is a growing recognition that water is one of our most precious resources and its availability and quality must be assured.

Other chapters provide information on a variety of topics related to water. A chronology lists some of the critical dates in water development. Many organizations have been established to study or regulate the uses of water and an overview of these appears in the organizations chapter. In recent years there has been a massive increase in the volume of literature on water. The selected bibliography, including annotations of about 100 books as well as several hundred journal articles and government documents, reflects this increase and is organized under a number of headings to facilitate use. At the end of the bibliography there is a list of selected journals that publish articles on water. The volume concludes with an annotated list of films dealing with many aspects of water, a glossary, an appendix of useful maps, and an index.

E. Willard Miller
Ruby M. Miller
The Pennsylvania State University

1

Water: A Fundamental Resource

OF ALL THE NATURAL RESOURCES ON THE EARTH necessary for life, water is the most important. Without it no life could exist on this planet. Many people in the United States consider an unlimited supply of inexpensive safe drinking water as an inherent right, but as the population continues to grow, it is becoming abundantly evident everywhere that water is a limited resource. To satisfy the water needs of urban areas it is now frequently necessary to transport water hundreds of miles, and with the coming of the age of environmental awareness and concern, issues of water quantity and quality are becoming a major public policy frontier in the 1990s.

In the nineteenth century, water treatment techniques reduced the occurrence of waterborne diseases to a very low level. Nonetheless, even though the possibility of epidemics of cholera and typhoid fever have virtually disappeared in the United States, there is a growing concern about contamination of water resources by such pollutants as pesticides, chemicals, and heavy metals. Municipalities now commonly report problems with water taste and odor and occasionally are beset with outbreaks of gastrointestinal diseases that may be traceable to a water source. This problem of impure water is evidenced by water use restrictions that are imposed on towns. Even more important are cases in which entire water systems have become contaminated, leaving towns without a water supply for extended periods of time.

To solve the growing water problem an informed and committed leadership will be needed to assist the public in making decisions about future water allocations. Only with an informed public and increased insights and understanding of the interrelated aspects of providing clean water will correct decisions be made to assure water for future generations. The technology is available and the societal framework of laws and regulations is now being developed to solve the water problems.

Physical Characteristics and Sources of Water

Water is unique among chemical substances. It is universally present and has remained unchanged in amount and physical character for millions of years. Of all the substances naturally occurring on the face of the earth, water comes closest to being the universal solvent. About half of all chemical elements are dissolved into natural water, some only in traces, some in abundance. Water has two important heat qualities. First, it has a very great ability to absorb heat. Heat capacity is expressed in terms of the amount of heat required to raise a given quantity of a substance by a given number of degrees. An enormous amount of heat is needed to warm water. For example, an iron bar will heat nearly ten times faster than water—it requires that much less heat to raise an iron bar to the temperature of water. Second, when ice melts it absorbs a certain amount of heat without an increase in temperature until it is entirely melted. This is known as latent heat, as it produces no change in the temperature of the ice—the heat goes entirely into changing the form of the substance. Because of the latent heat of water, the range of temperatures on the earth is smaller than it would be on a waterless planet.

Water can exist in all three matter states—solid, liquid, and gas—at a single location. Changes in form are controlled by energy levels that determine the availability of specific and latent heat energy in any mass of water. In changing its form from a liquid to a solid, water has the unique characteristic of contracting rather than expanding. This is why lakes and rivers have open water beneath the ice, making it possible for fish to survive even very cold winters.

In recent years there have been attempts to calculate the amount of water on the Earth (see Table 1.1). The most recent calculations indicate that the global water volume is 1.386×10^9 km^3, of which approximately 97.4 percent is ocean water and 2.6 percent fresh water.

TABLE 1.1

Global Water Volume

Volume	Km^3	%
Oceans	1,348,000,000	97.39
Polar ice caps, glaciers, icebergs	27,820,000	2.01
Groundwater, soil moisture	8,062,000	0.58
Lakes and rivers	225,000	0.02
Atmosphere	13,000	0.001

Source: A. Baumgartner and E. Reichel, *The World Water Balance* (Munich: Elsevier, 1975), p. 179.

It has been estimated that the global supply of water is about ten times the demand. It thus appears that no water shortage should exist in the foreseeable future. This is a very misleading picture, however. It is common knowledge that the quantity of water is unequally distributed over the earth. Some places have too much water, while other places have too little. Although the water supply is vast, we are tapping and using and wasting the easily obtained portions. Water that is easily obtained and cheap is becoming scarce, even as demand is rising.

Forms of Precipitation

The form of precipitation that reaches the ground depends on atmospheric conditions after the raindrops or ice crystals develop. Of these conditions, temperature is most important. If the entire troposphere is below freezing, ice crystals grow into snow crystals of legendary variety: hexagonal plates, starlike crystals, needles, or granular pellets. The form of these snow crystals depends on both temperature and relative humidity.

Other forms of precipitation occur when layers of air that are above 32 degrees Fahrenheit alternate with layers of air that are below 32 degrees Fahrenheit. When a shallow layer of below-freezing air occurs at ground level and a layer of above-freezing air lies on top of it, rain falling from the warmer upper layer crystalizes, turning to ice as it passes through the colder air. The accumulation of ice on trees, vines, and buildings under these conditions can be devastating.

The placement and patterns of freezing rain in rugged topography can vary widely. If the coldest air occurs in the valleys, ice will accumulate in the lower elevations. On occasion, the tops of mountain peaks may penetrate warmer air layers and escape the freezing rain altogether. The reverse can also occur. It is even possible for a cold upper-air layer and a cold air layer at ground level to be separated by a warmer, above-freezing layer of air in the middle. Under these conditions freezing rain occurs only at higher elevations and valleys receive only rain. Sleet or ice pellets reach the ground when there is a deep layer of below-freezing air near the ground and warmer air aloft. The rain that falls through the cold layer freezes into ice. Hail is associated with extreme weather such as thunderstorms and, on rare occasions, tornadoes. In these convectional storms, a hailstone may start to fall and then be carried back up several times before it reaches the ground. The hailstone may "grow" each time it is carried upward and receives another coating of water. Hailstones may thus, though only rarely, accumulate several layers and be as large as baseballs.

Hydrologic Cycle

Each year about 335,000 cubic miles of water are evaporated from the land and water surface of the Earth and approximately the same amount is precipitated back onto the surface. In this closed system the total water volume has remained constant for millions of years. This water cycle is powered by the energy from the sun and, because the amount of solar energy remains essentially the same from year to year, the total volume of water involved in the cycle is essentially constant.

The hydrologic cycle consists of two subsystems. The first is the precipitation and evaporation that occurs over the oceans, and the second is the precipitation and evapotranspiration that occurs over land surfaces. These two systems are connected by horizontal movement in the atmosphere, known as advection, and by runoff of water from land surfaces.

Within the hydrologic system, the water precipitated into and evaporated from the oceans is most important. Although the oceans cover only 71 percent of the Earth's surface, it is estimated that 79 percent of the precipitation and 84 percent of the Earth's total evaporation occur there. Seven percent of the evaporation of the ocean is carried onto the land by advection. The large amounts of moisture and energy available in the lower latitudes make the oceans proportionally more significant than land as a source of water vapors.

The land surface of the Earth receives about 21 percent of the Earth's total precipitation. Of this amount about 58 percent returns to the atmosphere by evapotranspiration and about 42 percent runs off into the oceans. Although land covers about 29 percent of the Earth's surface, it receives proportionally less precipitation because large areas of the land are semiarid and arid. Further, the greatest land masses are in the middle and higher latitudes where it is cooler and precipitation is less than in the warmer, tropical areas of the world.

Distribution and Variability of Precipitation

In the United States, precipitation varies from about 80 inches annually on the windward side of the mountains in the northwestern United States to less than 10 inches in the deserts of Arizona and Nevada. More significant, however, is the 20-inch isohyet. An isohyet is a line drawn on a map connecting points receiving equal rainfall. The 20-inch isohyet approximately parallels the hundredth meridian across the Great Plains, and separates the East from the arid West. East of the 20-inch isohyet most places receive 30 to 50 inches of precipitation annually, but much of the West receives less than 20 inches annually, with the exception of the Northwest. The people of most of the western United States have long recognized the problems of water availability. In the eastern United States there are sufficient quantities of water, if supplies are managed properly. In recent years, as metropolitan areas have grown, water availability has become a critical issue everywhere (see precipitation map, page 403).

It must be remembered that average precipitation figures for a month or year do not reveal the actual precipitation received at a place over the long term. In a short period, precipitation may vary sharply from the average. Although California had an average annual precipitation of 23.9 inches over a 52-year period, actual annual amounts ranged from 14.1 inches in the driest year to 42.1 inches in the wettest year. When considered on a monthly basis, the extremes are even greater.

In general, an inverse relationship exists between precipitation amount and precipitation variability, with the greatest fluctuations occurring in the arid and subarid regions. There are places in the desert of the southwestern United States that may receive no precipitation for years, and then receive as much as 20 inches during thunderstorms occurring within a few weeks. In the United States the Northeast has the least annual variability in precipitation, with the extremes departing only about 10 percent from normal.

Dams and Water Availability

Since precipitation is intermittent, and to overcome variations in demand as well as supply, dams are built to create reservoirs that provide a steady supply of water to urban areas. Some of the greatest engineering projects in recent years have been for water storage. Huge dams have created reservoirs sufficiently large enough to be called lakes. Many of these have multiple purposes, including power generation, irrigation, flood control, and recreation. The dams of the West are thus of major importance in developing the economy of the region.

The Fort Peck Dam, on the Missouri River in Montana, is one of the largest of all dams in volume. It is of the earth-fill type, about a mile thick at the base and 100 feet wide at the top, with a height of 250 feet. With 128 million cubic yards of material, it contains four times as much material as any previously completed dam. It was built for flood control, power development, and irrigation. The lake it created is 189 miles long with a shoreline of 1,600 miles, and when full it holds 19,417,000 acre-feet of water. Its annual power production is about 215 million kilowatt hours.

In the southwestern United States the Hoover Dam, built in the Black Canyon of the Colorado River between Arizona and Nevada, is the tallest concrete dam in the world with a height of 726 feet. Lake Mead, 119 miles long and covering an area of 246 square miles, is one of the largest artificial lakes in the world. It holds 32,359,000 acre-feet of water when full. The dam was built to provide irrigation water and electric power. With a capacity of over 1,300,000 kilowatts, it is operated by the city of Los Angeles and the Southern California Edison Company as agents for the United States government.

In the northwestern United States the Grand Coulee Dam on the Columbia River has one of the largest generating capacities with annual output of 1,975,000 kilowatts. In addition the dam provides irrigation water for a million acres. Its lake is 146 miles long and has an area of 127 square miles.

All dams are temporary. Rivers carry loads of eroded materials, causing the reservoirs behind dams to slowly fill with silt. The length of time that a dam is functional depends upon the size of the dam and the sediment load carried by the rivers. For example, the vast Roosevelt Reservoir of the Salt River of Arizona is rapidly filling with silt. Large portions of the river basin have been stripped of grasses and brush cover by natural forces and the soil is easily eroded during the desert thunderstorms. As the dam fills, the surface water gradually disappears and a swamp develops. Gradually the water of the swamp disppears and more sediment enters the region. In the final stage, dry land develops. This is a phenomenon that will be common in the twenty-first century. The use of the land will then depend upon the natural condition of the region. This type of land may become of use for agricultural purposes in the future.

Flood Control

After heavy periods of precipitation the ground becomes saturated and cannot absorb further moisture; a large percentage of additional water runs off into streams and rivers. River channels may be filled above capacity and flood. These floodwaters are not only destructive, but the water is lost for useful purposes.

In the past there was little effort expended to conserve floodwaters. Sometimes flooding was even perceived as being desirable. For example, the flooding of the Nile River in Egypt renewed the soil in the flood plain. With the construction of the Aswân Dam on the Nile, floods are now controlled.

In order to control the floodwaters, one or more dams are built, frequently at the headwaters of a river system, to create great storage reservoirs. The water from these reservoirs is released gradually, eliminating floods. The most important aspect of these reservoirs is that water is available for many users.

Surface Water and Groundwater

During a heavy rain and shortly thereafter, water can be seen flowing over the surface of the ground to the nearest creek or river. If the slope is steep, the runoff may be a sheet of water. A significant percentage of precipitation is carried away immediately but, during the initial runoff, some of the water may sink into the ground. The surface of the soil has often been compared to a blotter.

The amount of water found in soils depends on a number of factors. If the soil is sandy and gravelly, it will absorb a large proportion of the water. But at the same time, water flows through sandy and gravelly soils rapidly, and is retained in the soil for only a short while. Clay or fine-grained soil has less capacity to absorb water, as a consequence the soils are soon saturated and more water is lost through immediate runoff. On these soils, the infiltration rate is increased if there is a large amount of organic material, which is a high absorber of water, in the upper layer of soil.

Infiltration is the process by which water sinks into the soil. Infiltration rate refers to the inches of water that pass into the ground surface per unit of time. If rain falls at a rate of one inch per hour and the infiltration rate is 1/4 inch per hour, the remaining 3/4 inch per hour runs off the surface because it has not been infiltrated.

As the soil becomes saturated, the water in it moves downward through pores, holes, cracks, and joints to become groundwater. If the water fills all the air spaces in the soil, the saturated zone lies on the surface. If it fills only part of the space, the saturated zone will be below the surface. The top of the saturated zone is known as the water table.

Groundwater consists of the water that remains in underground reservoirs called aquifers. The quantity of underground water in the United States is enormous. It is estimated that the amount is at least six times greater than all the water stored in our surface lakes and reservoirs. The amount within 2,500 feet of the surface is equal to about 35 years worth of all surface water runoff, or about 400 times annual consumption.

Groundwater is a major source of water in the United States:

1. Groundwater provides drinking water for about 50 percent of all residences and about 97 percent of rural residences.
2. Municipal water service depends for about 35 percent of its supply from groundwater.
3. Groundwater accounts for some 40 percent of all agricultural irrigation water.
4. Manufacturing secures about 25 percent of its water from ground sources.
5. Groundwater is a major water source in many ecosystems valued for fish production, wildlife habitat, recreation, and other uses.
6. In periods of drought, groundwater frequently becomes a major source of water to maintain streams, freshwater lakes, and wetlands.

Water Budget

When the water supplies of an area are to be determined, a first step is to construct a water budget. This accounts for the water a region receives, the disbursement of the water, and the amount of water remaining. Although data are incomplete, there is sufficient information to estimate the annual water budget for the United States as a whole.

Each year the atmosphere brings 150 inches of moisture from the oceans to the land and returns 141 inches. The cycle appears to be out of balance. The difference is made up in precipitation and runoff of water. In a year's time, the atmosphere releases an average of 30 inches of precipitation to the land, but it receives only 21 inches back in the form of vapor caused by evapotranspiration. The remaining 9 inches of precipitation is discharged into the oceans, which returns to the atmosphere through evaporation to maintain the necessary balance. It is estimated that less than 0.1 inch of water annually moves by deep seepage of groundwater to the oceans.

There are few data available on the capital stock of water in lakes, soil moisture, and groundwater. Most studies have emphasized the change in the capital stock of water rather than total amounts. The average seasonal change in soil moisture and groundwater is estimated to be about 4 inches. Seasonal fluctuations in groundwater level are not very large.

It must be emphasized that regional water budgets vary greatly from the eastern to the western United States. To be of value, a water budget must be prepared for each climatic area. To illustrate, the water budget for the desert Southwest is considerably different from that of the rain-drenched southern Appalachians. The water budget provides a primary planning tool for all regions, but especially for the dry regions of the United States, where water supplies are most limited.

U.S. Water Consumption

The United States receives about 1,300,000 million gallons of water per day, of which about 240,000 million gallons per day are used for different purposes. This is about 7,000 gallons each day for every person in the nation. Of the water users, irrigation consumes about 110,000 million gallons per day or about 46 percent of the total. Of the irrigation water, 81,000 million gallons per day is used by plants and 29,000 million gallons per day is lost from the irrigation canals. Urban areas use about 17,000 million gallons per day and rural areas total about 3,000 million gallons per day.

Water use varies by season and in different sections of the nation. In summer when lawns and gardens are watered, water use goes up. In water-short areas such as the Southwest, water is conserved due to its scarcity. Although water supplies appear to be adequate overall, there can be local or even regional shortages resulting from uneven distribution of precipitation or excessive use of available supplies. In any given region other availability problems may arise due to seasonal and annual variations in precipitation or from the chance occurrence of a series of dry years.

Increasing the Water Supply

As demand approaches available supply, there are a number of actions that can be taken to increase the supply of water. Water resource development becomes increasingly important at this point.

Recycling

Recycling of water has been practiced since the beginning of civilization. The successive use of water by towns located along the course of a stream has increased with rising population. The planned reuse of unpotable water has been recognized by many as a viable alternative to new sources. The advantages include the reduction in both wastewater withdrawals and waste-water discharge.

Recycling systems have been divided into two general categories: those designed for indirect and those designed for direct reuse. Indirect reuse involves the discharge of wastewater into a surface or groundwater storage place and then reusing the water after removal of some of the objectional substances. Recycled water may be used directly in irrigation, in industry, and it even has some residential uses. A number of questions have been raised concerning the reuse of wastewater. The American Waterworks Association and the Water Pollution Control Federation recognize the potential use of wastewater, but caution that a health hazard may exist and further research is needed.

Control of Urban Runoff

As more and more land is paved in urban areas, the runoff of precipitation increases. The control of this runoff can add to the urban water supply. To illustrate, impervious areas such as parking lots, malls, tennis courts, and the like can be constructed so that runoff is stored in holding pools for later use. This water could be used to water lawns, golf courses, planted medians, and the like. It is estimated that judicious use of runoff water could supply one-fourth to one-half of the public water demand of small cities.

Reuse of Industrial Water

In the past most industries did not consider the amount of water used in plant operations. The cost of water compared to that of other inputs in production had been so low that little economic incentive existed to conserve water. With growing water shortages, the cost of water in many manufacturing processes has increased.

As water supplies become limited, there is growing economic incentive to conserve water by reusing it. Once the decision is made to reuse water, the results can be quite satisfactory. Water can be reused within a plant or recycled within various processes. Closed systems can decrease use and eliminate most discharge. Industrial conservation of water does incur expenses and may best be introduced during plant remodeling or modernization. For example, it is estimated that water consumption for the production of oil from oil shale could be reduced from five barrels to two or three barrels of water per barrel of oil, but at a significant rise in cost.

Water Conservation

There are many controls that result in users conserving water. The metering of water use is one of the most effective. Metering discourages less essential uses of water, such as lawn watering and car washing. It also encourages the repairing of leaks and better maintenance of plumbing fixtures. In-house use of water may also be reduced. Evidence indicates that metered irrigation water is used much more efficiently than when the water supply is unlimited.

Water-saving household devices cover a wide range of plumbing fixtures and household appliances. Many are cost-effective, easy to install and maintain, and require no change in existing use habits. These include plastic bottles or bricks in the toilet water closet to reduce water usage per flush, low-flow showerheads, and faucet aerators. The reduction of water use in toilets can save from 4,000 to 6,000 gallons per year per household; showerheads can save as much as 12,000 gallons per year, and aerators about 3,000 gallons. All of these devices are cost-effective, even at very low water prices. To be effective for a water system, however, they must be installed in a large percentage of households. It is estimated that if 5,000 households in a community installed such devices the savings would equal 100 million gallons of water per year.

Water Use Restrictions

Water use restrictions are enforced only in times of critical water shortages. When water supplies fall below a critical level and officials project that a water shortage may occur, voluntary restrictions are usually instituted. If the supply continues to decrease, the restrictions may be made mandatory. These restrictions usually apply to such uses as watering lawns and washing cars. As these restrictions inconvenience the user directly, they are normally removed as soon as possible.

Desalination

Desalination of salt water has begun in the desert regions of the world such as the Middle East and is now being developed in California.

Advances in technology over the past few years now make seawater a competitive source of drinking water. As California's population growth continues to outstrip its natural water supply, communities are looking beyond the present drought and investigating desalination as a long-term solution.

There are two basic techniques developed to desalt water. In the Middle East, most countries use distillation methods that heat the salt water and recapture the purified water as condensed steam. Although this process is costly, it can be efficient when coupled with a power plant that utilizes the waste heat.

In the United States, the most commonly considered method is a form of reverse osmosis, an ultrafiltration technology first developed in the late 1960s to supply superpure water for use in manufacturing computer chips. In reverse osmosis the salt water is pumped along a porous cylindrical membrane at high pressure. The process is continuous, and because the water is flowing along the membrane rather than perpendicular to it, the flow tends to flush the very small pores, so only occasional cleaning is needed. Water molecules pass through the membrane, but salt molecules, bacteria, and pollutants, which are larger, are left behind. Depending on the saline content, it takes two to three gallons of salt water to make one gallon of fresh water. Solid particles are collected and disposed of at a landfill; the concentrated brine is often blended with treated wastewater, which, being nonsalty, reduces the overall salinity to levels suitable for ocean discharge.

A pilot plant in the Marin Municipal Water District north of San Francisco is currently in operation. By reverse osmosis the plant reduces the solids from between 35,000 and 40,000 parts per million to less than 500, a level considered safe to drink. To improve the taste the plant runs the water over limestone. The pilot plant has been successful and the water district plans to build a permanent 5-million-gallon-per-day plant to supply 25 percent of its needs. It is estimated that the average monthly water bill for residences will rise from $25 to $31.

Other California communities considering or using desalination include Santa Barbara—which has built a $25 million plant to supply one-third of the city's needs—Monterey, San Diego, Morro Bay, Ventura, Oxnard, and Los Angeles. No other states are planning desalination, although reverse osmosis is commonly used in Florida to purify brackish water for drinking.

There are many advantages to developing desalination plants to secure fresh, clean water. Technology has advanced rapidly so that it is commercially feasible. Compared to alternatives, such as interbasin water transfers by pipelines or tanker ships, desalination plants offer communities the advantage of an unlimited water supply. The plant can be placed anywhere on the coast it is needed. At present, costs are two to three times the cost of water from natural sources, but in areas where water supplies are drying up or polluted, it is a viable alternative. With development, costs will certainly decrease.

Drought

A drought occurs when evapotranspiration exceeds precipitation over a considerable period of time. In a time of water shortage, soil moisture and groundwater are both depleted. There are basically two types of drought. The first type includes those droughts that are a fundamental part of a climatic region. The second type of droughts are those that are unpredictable and not associated with a normal precipitation regime.

The desert and semiarid regions of the world are characterized by permanent drought. In the southwestern United States, evapotranspiration rates generally exceed precipitation. The sparse vegetative landscape reflects the lack of water. Agriculture survives only with irrigation, and water supplies must be imported from great distances to serve metropolitan areas. In many of these areas groundwater is being depleted faster than it is replenished (see groundwater overdraft map on page 404).

Droughts that cannot be predicted occur throughout the United States. They may last for a few weeks to a number of years. The Great Plains suffered one of the longest and most severe droughts in U.S. history from 1931 to 1938. The effects were felt throughout the nation in increased food prices. Thousands of families in the Dust Bowl area abandoned their farms and migrated to other areas. A second major drought occurred in the southern Great Plains and southwestern United States from 1950 to 1954. As the pastures dried up cattle ranchers had to send their cattle to other regions. To ease the hardships, the federal government offered farmers emergency credit and seed grain at low prices.

Droughts have occurred in other areas of the United States. In the early 1960s, a drought affected the northeastern United States for several years, and from 1975 to 1977 a lack of winter snowfall resulted in severe drought conditions for areas in the western United States.

In the late 1980s and early 1990s, a severe drought occurred in California.

During the summer in the normally humid eastern United States, a drought of several weeks duration may occur. The high temperatures induce rapid rates of evapotranspiration and even short periods of rain may not restore the water lost. The result is a near water deficiency that decreases crop yields, and water may be scarce for unessential uses for a short period.

Evolution of Water Quality Issues

Water quality has been a concern in the United States since the earliest settlements. The first legislation was enacted in 1647 when the Massachusetts colony legislature passed a law designed to prevent the pollution of the Boston harbor. This act was not to control disease, but to improve navigation. It was not until the nineteenth century that polluted water was recognized as a potential source of disease.

By the middle of the nineteenth century, before pathogenic microorganisms had yet been isolated, John Snow, a physician in England, demonstrated the causative relationship between a polluted water supply and disease transmission. By empirical evidence he showed that people who drank polluted water from the Southwash and Vauxhall Water Companies had a higher incidence of cholera than other Londoners. At the same period, physicians in the United States and Germany began campaigns for clean water supplies.

As urban populations grew, larger quantities of water were needed. This required the development of water systems that tapped water supplies across large areas. If the catchment basins were located in protected, upland, relatively uninhabited areas, the water was of high quality. If, however, the catchment basins were in lowlands in areas where sewage had been deposited in the stream, the water quality was poor. In such situations outbreaks of disease were common.

It was soon recognized that water from many catchment basins needed treatment. Early treatment facilities, such as sand filtration, produced significant decreases in waterborne diseases. Chemical treatment to purify water added an additional factor of safety. A major advance occurred in 1870 with the introduction of water chlorination. The "multiple barrier" concept was then born, including four stages of protection that began with a pure catchment basin, a large storage basin, filtration, and finally chlorination.

The next step in water purification came with the growth of economic activity early in the twentieth century, which also caused a growth in the volume and complexity of waste-water effluents. As a result of deposition of waste materials in streams and lakes, aquatic ecosystems were polluted with large quantities of effluents that could not be removed quickly. In the process of decay, bacteria removed the dissolved oxygen from the water. This process, called eutrophication, caused a major problem for the plants and fish that required oxygen. The scientific examination of eutrophication began in the mid-1920s. Since then, a means to control pollution has evolved.

In the 1950s and 1960s, attention focused on toxic materials and particularly pesticides in water. Many of these pollutants are not removed by common treatment processes and accumulate in body tissues. Often a link is suspected between the ingestion of traces of these compounds over protracted periods of time and the incidence of teratogenic and mutagenic effects in human beings. Consequently, the presence of these substances is a significant cause of alarm.

In recent years, epidemiological studies have examined "natural" dissolved constituents of water, which would not normally be defined as pollutants, but appear to have effects on health. For example, studies have linked nitrate concentrations in water with infant methemoglobin and gastric cancer. Another concern has been the adding of substances to the water supply to support good health. One of the most common and controversial substances in this regard is fluoride, which is added to the water supply to reduce tooth decay.

The most publicized water quality issue in the last decade has been the problem of acid precipitation and the effect on aquatic and terrestrial ecosystems. Acid precipitation can cover hundreds of thousands of square miles and is international in its distribution. The full magnitude of this water pollutant is only now becoming evident.

Water Use Doctrines

Rules about water use were among the first policies developed by settlers as they migrated westward across the North American continent. Essentially two policies developed—the riparian doctrine in the east and appropriation doctrine in the west—reflecting the availability of water. In essence, there were no restrictions on water use in areas using the riparian doctrine. In contrast, in the areas applying the appropriation doctrine, the use of water had to be controlled because of its scarcity. Legislation to date concerning water rights has largely centered on water use in the driest regions of the western United States.

Evolution of Water Use Doctrines

The laws and regulations affecting the use of water have evolved over time, reflecting changes in demand and quantity available. With population increase and economic growth, there is a growing call to conserve water. The problems are complex and there are disagreements as to the potential for real water savings depending upon such factors as evaporation losses, reuse of runoff, and amounts of reusable groundwater. Since farmers are major consumers of water, comprehensive water conservation efforts must consider agricultural practices. Some examples of farm conservation techniques include: improved management of irrigation water, improvements in the distribution system, and changes in irrigation application methods, such as changing from flood irrigation to drip irrigation.

Traditionally, surface and groundwater were treated separately, but it is now recognized that these sources of water are physically intertwined and must be managed as a single system.

Water quality concerns are of recent origin. Water pollution was given little attention in the formulation of early water laws and regulations. Water users are now cognizant of the problems of pollution. For example, agricultural users are becoming concerned when their water supplies have been used upstream by cities and industries.

As the demand for water increases in semiarid and arid regions, water management has become increasingly comprehensive and complex. Water can no longer be taken from streams and used in unlimited quantities. There is a growing public awareness that water regulation must protect the entire river basin. Sufficient water must be left upstream to serve agricultural, recreational, and ecological needs downstream. In semiarid and arid regions, these uses often compete for scarce resources. Several major rivers in the West are multistate or international. As a result, water policies must be coordinated among the states and nations sharing a single water basin.

Surface Water Doctrines

Riparian Doctrine

In the eastern United States, where water is quite plentiful, a body of law developed called the riparian doctrine. This doctrine was based on the rule of reasonable use, which stated:

> One who owned lands touching upon a stream was entitled to have the full flow of the stream coming by his place undiminished in quantity and unimpaired in quality, except that each landowner was entitled to make reasonable use of the water upon his own lands, provided: (1) that he returned the stream to its natural channel before it left his lands, so as not to deprive succeeding lower riparian owners of their rights, and (2) that use on his own land was reasonable in respect to the corresponding rights of other riparian owners lying below him. (C. Busly, "Water Rights and Our Expanding Economy," Journal of Soil and Water Conservation, 9 [March 1984]: 68.)

States east of the Great Plains use this riparian doctrine exclusively. In the riparian doctrine, the right to use water is considered real property, but the water itself is not the property of the landowner, and it cannot be transferred, sold, or granted to another person separate from the land. The rights of individual owners are also subordinate to public interest, and the state can subject water use to reasonable regulation.

Reasonable use of the water has been determined to be subject to appropriation by the state, as in the case of water on public lands and excess private water. A strict set of rules has been developed for riparian owners who must adhere to the philosophy of reasonable use.

In Texas, the riparian doctrine usually prevails in lawsuits. The riparian water user gains a vested right over the amount of water beneficially used, but is not entitled to water in excess of that quantity. The riparian doctrine was confirmed by the Texas courts as early as 1863 and as late as 1982. Oklahoma uses both systems, each operating in different areas of the state, depending on water availability.

Appropriation Doctrine

In the semiarid and arid regions of the western United States, the riparian doctrine, which was based on there being plentiful supplies of water, was inadequate. With initial settlement the eastern riparian laws were applied, but these were soon changed. With the discovery of gold in California, for example, an unwritten law developed that a mining claim could be worked using all the water required to secure the metal. This came to be known as the appropriation doctrine. It was gradually modified to fit the environmental circumstances.

The best modern definition of the doctrine is given in a 1950 report of the President's Water Resources Policy Commission:

> The appropriation doctrine . . . rests on the proposition that beneficial use of water is the basis of the appropriative right. The first in time is prior in right. Perfected only by use the. . .right is lost by abandonment. . . . An appropriative water right is not identified by ownership of riparian lands. . . . Its existence and relationship to other rights on the same stream are identified in terms of time of initiation of the right by start of the work to direct water coupled with an intent to make beneficial use of it and the diligence with which the appropriator completes the diversion of water for beneficial uses.

Application of the Doctrines

In the arid Southwest all states have statutory water laws. In four states—Arizona, Colorado, Nevada, and New Mexico—the appropriative doctrine is exclusive, whereas in three states—California, Oklahoma, and Texas—the appropriation and riparian doctrines coexist to different degrees.

Arizona, one of the driest regions of the nation with an average precipitation of less than 15 inches per year, has developed one of the most rigid systems to control the use of water resources. The state constitution states that riparian rights do not apply in Arizona. Instead a comprehensive water code has been developed which specifies that all sources of water are subject to appropriation and grants preference to domestic, municipal, and irrigation uses in that order. The appropriation laws of Colorado, Nevada, and New Mexico are similar to those of Arizona.

In states where the appropriation and riparian laws coexist, there has been an adaptation to particular conditions. In California a system of limitations is imposed on each type of law. For example, a grant of water rights does not confer ownership of the water, but only the right to use the water for beneficial purposes. In addition, only certain areas, primarily in southern California, may use water that is necessary for the maintenance of life and enjoyment of the property. The courts have developed a hierarchy of uses when water becomes scarce. The first priority of use is given to water used for drinking, cooking, and stock watering; second priority is to industry and irrigation; and third priority is to water that is removed from the stream and transported off the property. It has also been determined by the courts that in times of water scarcity the water must be equally available to all landowners.

A set of regulations has developed to define the responsibilities of the owners of a stream. An owner may not obstruct a stream, and the stream must not be polluted by those using the water upstream. Regrettably, the riparian doctrine was formulated at a time when there was little or no need for stream management. With no restrictions on the quantity of water that can be used, the riparian doctrine tends to encourage excess use or even waste of the water available. In essence, a state or regional water management system cannot efficiently evolve from a common law system that originated in another era.

Groundwater Allocation Doctrines

Groundwater allocation doctrines can be common law or statutory. Common law doctrines include the overlying rights of absolute ownership, reasonable use, and correlative rights. Historically, common law gave the landowner absolute ownership of all groundwater underlying a property. This absolute ownership doctrine, found in most states, meant a landowner could use the water without legal liability to neighboring owners.

A second groundwater rule in common law is that of reasonable use. This doctrine holds that groundwater may be used without waste or if water occurs on nonoverlying land. Overlying land is interpreted as land from which water is pumped. In Arizona this is known as the American Rule.

The California rule of correlative rights extends the reasonable use doctrine to permit groundwater use by nonoverlying users. This doctrine includes pro rata sharing during times of shortage and allows rights to be established for water stored underground, or recharged groundwater.

Sources of Water Contamination

In our modern industrial society there are hundreds of possible pollutants in a local water supply. For example, water runoff may pick up wastes, landfills may contain toxic materials, and injection wells may pump impurities into aquifers.

Natural Contamination

One of the most widespread forms of water pollution comes from sodium from natural salt deposits and, in coastal areas, from ocean sprays. Although a certain amount of sodium is essential to good health, the consumption of high levels of sodium has long been known to cause high blood pressure in humans. It is estimated that sodium concentrations in nearly half the nation's cities is now sufficiently high that it is difficult to observe a strict low-sodium diet. Many bottled waters, especially mineral water, contain high concentrations of sodium.

There are also natural deposits of sulfur that enter water systems and give the water a distinctly distasteful odor. The same is true of iron. Some mineral-rich waters have a therapeutic effect, and in a number of places health spas have been developed to utilize these waters.

Human-Generated Contamination

In our modern industrial society there are many human generated sources of water pollution (see Table 1.2 on page 40).

Agricultural Contamination

The major types of water contamination from agriculture are from fertilizers, wastes from livestock operations, and pesticides. The extent of the contamination depends upon amounts of chemicals used, type of precipitation, temperature, groundwater penetration, surface evaporation, soil maintenance, and farming practices.

FERTILIZERS The three basic fertilizers applied to crops are nitrates, phosphates, and potash. Since phosphates and potash are not readily soluble, they are minor contaminants of water. Nitrates, in contrast, are easily dissolved in water. As a result, when nitrate fertilizers increase the nitrates in the soil, these nitrates are free to move into the groundwater. The use of nitrates in the United States has increased from 2.5 million metric tons in 1960 to over 10 million metric tons in 1990. By tonnage, the states with the largest usage are Illinois, Iowa, Nebraska, Kansas, Texas, and California.

Many studies have been made of the contamination of groundwater by nitrates. Halberg found the groundwater contamination of the Big Spring Basin in Iowa was due primarily to increased use of nitrate fertilizer ("Nitrates in Groundwater in Iowa," Proceedings of Nitrogen and Groundwater Conference, Chemical and Fertilizer Association, Ames, Iowa, 1982). Another study of 650 square miles in central Wisconsin found nitrate-nitrogen concentrations in the groundwater ranging from 4 to 23 mg/l in cultivated farmland. Forty percent of the area exceeded the 10 mg/l drinking water standard.

Nitrate contamination of the soil can be controlled by the rate and timing of the application of fertilizer if the following guidelines are followed:

1. The amount of fertilizer should be matched with the needs of the plants.
2. Fertilizing should occur only at the proper time during the growing season.
3. Fertilizers that slowly release nitrates should be used.

The conversion of nitrates to nitrites in the stomach can result in methemoglobin, particularly in infants. The nitrates can react with chemicals in food to produce nitrosamisis, which are suspected carcinogens. The Environmental Protection Agency (EPA) has begun monitoring the health effects of nitrate concentrations in water.

LIVESTOCK The production of livestock and poultry has become a major source of water pollutants in the United States. Problems are particularly severe from runoff and infiltration from animal feedlots and associated treatment and disposal facilities. In the United States there is a combined total of approximately 170 million cattle, sheep, and hogs, and another 8 billion chickens and turkeys. It is estimated that these animals produce about 160 million metric tons of animal waste annually.

Animal wastes can contaminate groundwater with both nitrates and bacteria. If the animals are concentrated in feedlots, local contamination can be severe. In the United States there are between 1,900 and 2,000 cattle feedlots with a 1,000 head or more capacity containing about 17 million animals. With the manure in a feedlot containing about 5 percent nitrates, about 23 tons of this nutrient are produced annually per acre in a typical feedlot. Although some of the manure may be removed to be spread on other farmland, the nitrate level remains high. Contamination depends on the permeability of the ground and the water handling system.

Feedlots are not the only source of potential groundwater contamination by livestock. It is estimated by the U.S. Department of Agriculture that 25 percent of the 718,000 farms with fewer than 300 head of cattle have the potential to degrade water quality locally. In areas where the soils are porous or in limestone areas where surface water flows easily underground, bacteria, viruses, hormones, salts, and other constituents enter the water system.

The effects of livestock on water quality can be lessened by careful siting and management of animal feedlots. Wastes from feedlots can be collected before runoff contaminates the water supply. The siting of feedlots can be helpful in controlling runoff of wastewater. Other preventative measures include storage and disposal of animal wastes, and the containment of runoff.

PESTICIDES Modern agriculture has become increasingly dependent on pesticides to protect crops from insects, diseases, weeds, and other problems. There are some 50,000 different pesticide products, using 600 active ingredients, registered in the United States. The use of pesticides has increased rapidly in the past 20 years, and it is estimated that over 1 billion pounds are used annually. Of the pesticides, herbicide use has grown most rapidly as chemicals replace mechanical methods of controlling weeds. For example, in 1971, about 71 percent of all U.S. farmland acreage planted in row crops and 38 percent of the acreage planted in small grain crops had herbicide controls. By 1982 these figures had increased to 91 and 44 percent respectively, and they continue to rise.

The Safe Drinking Water Act of 1974 (SDWA) established standards that are being implemented by the Environmental Protection Agency as a result of pesticides being found in drinking water. To illustrate, alachlor, which has proven to be a carcinogen in animals, is the most widely used herbicide in the United States. More than 80 million pounds of alachlor is used annually to kill weeds. An EPA survey found alachlor in drinking water in Maryland, Pennsylvania, Ohio, Iowa, and Minnesota. The SDWA requires regulation of concentrations of alachlor in community water systems, and EPA has restricted its use.

Another type of herbicide is dichlorophenoxyacetic (2,4-D), which was used as a defoliant in the Vietnam War and came to be known as Agent Orange. Today 80 million pounds are produced annually to be sprayed on the land in the United States, although the compound is recognized as playing a role in paralysis and chromosome damage, and recently has been linked to neuropathy and changes in the nervous system. The SDWA requires regulation of concentrations of 2,4-D. Other pesticides controlled by SDWA include endrin, dinoseb, dioxin, epichlorohydrin, ethylene dibromide, hexachlorobenzene, pentachlorophenol, toxaphene, and triazines.

One of the most toxic pesticides is carbonate. It is used on potatoes, peanuts, sugar beets, citrus crops, and cotton and has been found in water in all parts of the nation. Carbonate affects the nervous system of both animals and humans, and there is evidence that the immune system is also affected.

The potential for groundwater contamination from pesticides is particularly high in areas with frequent exchange between surface water and groundwater. There are several ways to control contamination of groundwater by pesticides. The most effective means is to take greater care where, when, and how pesticides are applied to crops. The timing and intensity of pesticide application can be controlled to limit the potential of leaking into groundwater. There is a growing recognition that there must be a shift toward integrated pest management, which promotes the use of nonchemical means to control pests in conjunction with carefully timed chemical sprayings.

Mineral Industries Contamination

The extraction of fuels and minerals creates many opportunities for groundwater contamination. In the mining process potential contamination stems from the actual mining, disposal of mined wastes, and the processing of the ores. Oil and gas production also pose a threat of contamination of groundwater. In 1990, within the United States, there were about 60,800 producing wells, 6,000 nonmetallic mines, 5,000 coal mines, and 1,500 metal mines.

ACID MINE DRAINAGE Acidification of aquatic ecosystems is an old problem that has received renewed attention with the current focus on environmental problems. Acid mine drainage comes from the conversion of elemental sulfur to sulfuric acid when mineral deposits are exposed during mining operations. Acid mine drainage may result from the extraction of a variety of minerals, including lead, zinc, gold, and manganese, but the principal source in the United States is associated with the production of anthracite and bituminous coal.

In coal fields acidification comes from two sources: natural acidity and acid mine drainage. Natural acidity can be defined as that which arises from coal beds or bogs that are undisturbed by humans.

Acid mine drainage originates from these natural deposits that are exposed during mining operations.

The impact of acids generated from naturally exposed mineral strata is significant. In some places, specific soil and climatological conditions give rise to bogs. In such areas of poor drainage, plant production exceeds decomposition and chemical changes favor the invasion of acid-tolerant species. Weak organic acids may be leached from decaying vegetation by carbonate acid and small amounts of sulfuric acid may be produced by sulfur bacteria. Streams flowing from bogs are often coffee colored, with small populations of invertebrates and fish; these streams are also low in biological productivity, lacking plant life and algae. The principal factor limiting these streams is the concentration of the hydrogen ion (H^+), which inhibits the cycling of nutrients and can exceed the tolerance levels of particular species.

Acid mine drainage arises from both underground and surface mines, but is most widespread in surface mine areas. In surface mining, large areas of lands are denuded of vegetation and the coal or other deposits lying close to the surface may contain sulfur. It is estimated that in the United States there are over 13,000 miles of streams degraded by acid mine drainage. Of these, over 78 percent are located east of the Mississippi River in Pennsylvania, West Virginia, Ohio, eastern Kentucky, Tennessee, Maryland, and Alabama.

Acid drainage also originates from abandoned mines and from the earth debris of rocks and stones (also known as culm banks) caused by strip mining. Abandoned mines filled with water constitute a long-term source of acid mine drainage to the surrounding watershed. In strip mining areas pools of water become acidic as sulfur leaches from the mined refuse. Pollution problems are particularly severe during periods of abundant precipitation when large quantities of acid may be flushed into streams. It has been estimated that 40 percent of acid mine drainage comes from active surface and subsurface mines, with the remaining 60 percent flowing from abandoned workings. Of the 60 percent from abandoned workings, 35 percent is due to shaft and drift mines and 25 percent to surface mines. It has been estimated that it will take 800 to 3,000 years to leach all presently exposed residue materials from the Appalachian coal region.

The quantity and chemical composition of mine drainage is the result of an interplay of a variety of geological, climatological, and biological factors. Chemical changes to the water may include the addition of acids, metals, sulfates, and other dissolved solids. Temperature changes and the addition of silt are major physical changes.

PETROLEUM AND NATURAL GAS WELL POLLUTION Since the first oil well drilled in 1859, more than 1 million wells have been completed in the United States. These wells cause groundwater contamination in a number of ways. During their productive years the wells, if not properly installed and operated, can contaminate the groundwater with brine. A more serious form of contamination arises from the disposal of larger amounts of brine that are produced with the oil and/or natural gas. Brine, containing more than 100,000 mg/l of salts, is often found mixed with the crude oil. For decades this brine was released on the ground in the vicinity of the well. The brine seeped into the ground and contaminated the groundwater, then flowed into the streams destroying ecosystems.

In many oil fields, as primary production declines, secondary recovery efforts are initiated. Water and/or gas are pumped into the oil sands to force out more oil. Although the process enhances oil recovery, the increased pressure in the oil sands also forces contaminants into aquifers that contain drinking water. For example, in 1985, gas migrated through underground fractures in a 3,000 square mile area of a Pennsylvania oil field, contaminating the drinking water of 10,000 people ("Federal Cleanup of Oil Spill in Hills of Pennsylvania Puts Residents at Odds," *New York Times*, December 13, 1985).

After production stops the oil or natural gas wells must be properly plugged to prevent brine from entering the ground water. If the well casing corrodes over time, however, the brine can escape the well and enter an aquifer.

States producing oil and natural gas now attempt to control brine. The preferred approach is to reinject the brine into the deep aquifers through injection wells to cause the brine to be naturally filtered and prevent groundwater contamination. Although technology has been developed to control brine, contamination still occurs. Seventeen states have recently documented brine-related groundwater contamination. In Texas there were 23,000 cases of contamination reported. This remains a major water pollution problem, particularly in the older oil-producing regions.

OIL SPILLS Oil has always polluted the environment, but in rather small quantities until recent decades. Only since the large-scale production of oil has the environment been threatened. There are four major sources of oil pollution of waters:

Natural Seepages In many places in the world oil is able to penetrate the surface through fissures from oil sands. This seepage is carried away by running water, resulting in its pollution. This is a natural phenomenon and will continue indefinitely. Although no reliable figures are available, it is estimated that, worldwide, from 1.5 million to over 4 million barrels enter the environment each year from seepages.

Pollution from Land Sources Oil pollution also results from disposal of oil on land and from accidents. Many factories discharge oily wastes directly into rivers, either accidentally or deliberately. Large quantities of oil are poured into drains or on the ground in the course of motor vehicle maintenance activities. In the operation of vehicles, some oil is expelled through the exhaust system and seeps into the roadbed. Much of this oil enters the water system before reaching the oceans.

Spills from Oil Exploitation Oil may also be accidentally spilled. The failure of a storage tank in January 1988 at Ashland, Pennsylvania, on the Monongahela River, illustrates this situation. More than 3 million gallons of oil flowed into the river. The Pennsylvania Emergency Management Administration and local agencies immediately began the process of controlling the spill. This was critical as numerous towns downstream secured their water supplies from the river. For a short period alternative sources of water were used. In the cleanup process over 2 million gallons of oil were recovered from the site and the river. The Pennsylvania Department of Environmental Resources estimated that 11,000 fish and 2,000 waterfowl died on the Monongahela and Ohio rivers. The problem was serious, but within two weeks water quality had been restored due to a massive, coordinated effort to control the spread of the oil and to clean up the oil where required.

Spills from Tankers Due to Operational Activities and Accidents Most of the oil fields of the world are located far from the major consuming countries. The oil must be transported as either crude or refined products. From most of these oil fields, including those in the Middle East, North Africa, Nigeria, and Venezuela, the only means of shipment is by tanker. Oil spills can occur when bilge water is discharged while ballasting or cleaning the oil tanks.

Due to the large amounts of oil carried and the heavy volume of tanker traffic, accidents occur. Oil spills may be caused by structural failure of the vessel, collision of two vessels, fire on the vessel, or the accidental grounding of the vessel. When a tanker grounds the spillage of oil is usually massive. The first supertanker to ground was the *Torrey Canyon* on the Seven Stones Shoal about 21 miles off Cornwall's Land's End in England on March 18, 1967. The tanker lost about 830,000 barrels of Kuwaiti oil in the sea. Since then there have been scores of accidents creating oil spills in the ocean. The most recent major oil spill occurred on March 24, 1989, when the *Exxon Valdez* ran aground on Bligh Reef in Prince William Sound, shortly after leaving Valdez, Alaska.

Oil spills create environmental disaster. For example, the *Exxon Valdez* oil spill occurred in one of the world's richest biological areas. A U.S. Fish and Wildlife Service biologist estimated that within one week of the oil spill, more than 15,000 seabirds had been exposed to oil. There was also concern for the safety of the area's 5,000 sea otters. Of the dozens of otters that were cleaned of oil in the weeks after the spill, only a few survived. Hundreds died in the first days after the spill on remote, oil-fouled shores. By the fall of 1989 an initial evaluation of the damage to the ecosystem could be made. It was conservatively estimated that in the Prince William Sound area 33,000 birds, 980 otters, 30 harbor seals, 17 gray whales, and 14 sea lions had died. An observer of aberrant bird behavior indicated that 75 percent of the bald eagles of the area failed to nest.

The Prince William Sound area is a rich area of fishing for salmon, herring, halibut, and shellfish. In 1988, the year before the oil spill, fishing yielded $131 million to the area. The spill occurred just prior to the herring season, which had to be canceled with a loss to the local fishing industry of about $12 million. The effects of oil pollution on the water will be felt for years by the fishing industry.

Acid Precipitation

In the 1950s and 1960s studies conducted by scientists in the United States and Europe found evidence of acid rain. By the early 1970s sufficient evidence had been gathered to reveal that acid rain was prevalent over the entire northeastern United States and northern Europe. Today the precipitation falling over eastern North America and northern Europe, as well as other industrial areas, ranges from 10 to 100 times more acidic than normal precipitation. According to the Environmental Protection Agency, the acidity of rainfall over eastern North America now averages about 4.5 on the pH scale. In particular precipitation periods, the pH rating varies tremendously. In the White Mountains of New Hampshire, for example, meteorologists have reported that at least once every year samples of rain have had an acidity of pH 2.8, as acidic as vinegar. Although the data are still incomplete, there is increasing evidence that the acidity of precipitation is increasing steadily.

Acid precipitation is the result of sulfur dioxide and nitric oxides combining with water in the atmosphere to form weak sulfuric and nitric acids. According to the U.S. National Research Council, about one-half of the SO_2 in the atmosphere comes from natural sources. It is estimated that 100 million tons of SO_2 are emitted worldwide every year from coal- and oil-fired power stations, industries that consume fossil fuels, and smelters. More than 90 percent of the nitric acid (NO_x) emitted into the atmosphere comes from human sources, of which the single largest is the combustion of fossil fuels. Coal contains about 1 percent nitrogen by weight. During high temperature combustion, the nitrogen combines with oxygen to form NO_x. About 55 percent of the NO_x emissions comes from power plants, and most of the remaining 45 percent comes from mobile sources, such as motor vehicles, aircraft, and trains. Of the amount emitted, about 6.6 million tons are deposited in North America as acid nitrates in precipitation; another 5.6 to 8.8 million tons are dry deposited.

The distance pollutants are transported depends upon a number of factors: wind speed, weather conditions, the chemical state of the pollutants, and the height of the industrial smokestacks. A high wind can transport pollutants hundreds of miles in a few days, but weather conditions are also a significant factor. If the skies are clear, the possibility of long-distance transport of pollutants is enhanced; in contrast, if there is heavy precipitation, the pollutants are quickly removed from the atmosphere. The chemical state of the pollutant also affects the distance it is transported. For example, acid sulfate is absorbed more readily by the ground than sulfur dioxide. As a consequence, sulfur dioxide is normally carried much greater distances by wind than are sulfates.

One of the most important factors in the atmosphere's transport of SO_2 and NO_x is the height at which the substances are emitted into the atmosphere. If the pollutants are emitted into the "mixing layer" of the atmosphere—that is, the layer closest to the ground where there is good vertical mixing—the pollutants will not be transported great distances. The mixing layer is typically below 3,000 feet and is visible as a blanket of polluted air covering an urban area. Pollutants in this lower layer fall to the ground relatively soon. Pollutants that are emitted above the mixing layer are effectively removed from ground contact and can be transported great distances by the strong winds aloft.

Acid precipitation is not only changing the quality of the water in many areas, but entire ecosystems. As acid rain pollutes water, there is an absolute need to determine the changes that have taken place and to develop remedies to control future deterioration. The major adverse effects that have been measured include the acidification of streams and lakes, damage to the forest environment, degradation of the soils, and damage to human-made materials.

The effects of acid precipitation on lakes are widespread. In Ontario, for example, the Ministry of the Environment has reported that 1,200 lakes have lost their fish populations because of acidification, about 3,400 are approaching this state, and an additional 11,400 are at risk. In Massachusetts, a number of streams supporting sport fishing and a number of reservoirs providing drinking water have problems of acidification.

In addition to killing fish, an acidified body of water also loses many other organisms, including certain types of algae, crustaceans, mollusks, and aquatic insects, many of which are important to the fish food chain. Each species is affected by a different pH level; for example, stone flies and mayflies generally expire at a pH 5 acidic level.

An acidified lake is often a beautiful crystalline blue because no biological material is present. Many such lakes are carpeted with mats of algae that can survive in acidified water, and beneath the algae, bacteria that can live without oxygen thrive, producing gases that bubble up to the surface. Leaves that fall into acidified lakes are preserved because the bacteria and fungi that normally break down the leaves are not present. As a result, leaf mulch can build up to the point of choking the body of water.

The acidification of a body of water not only affects its quality, but destroys important commercial and recreational activities. There is increasing interest in restoring acidic lakes, and attempts have been made to neutralize acidic lakes and streams in Sweden, Norway, Canada, and the United States. In Sweden, initial attempts began as early as the 1920s, and in the United States, liming has been practiced on a limited scale since the 1950s. The most commonly used neutralizing agent is some form of limestone ($CaCO_2$ or $CaCOH_2$)—hence the term liming. The most common method is to apply the neutralizing agent directly to the surface of the water. Other methods include injecting the agent into the lake sediments, applying it to the watershed surrounding the lake, or placing it in biologically critical areas.

There are several advantages to liming. Limestone is relatively inexpensive and widely found, the technique is simple, there are reasonably accurate methods for determining the dose required to meet specific water quality targets, and the response of the ecosystem to treatment is rapid.

Although the liming technique is several decades old, the effects are still in question. Conclusive evidence has not been obtained that liming is a permanent solution to the problem of acidified lakes. Some temporary results are encouraging, but local conditions are important. If acid precipitation continues to fall on the water of the lake, acidification may not be reduced in the long run. When toxic metals, such as nickel, copper, or zinc are present, liming programs have not been successful. Each lake must be individually analyzed to determine if a treatment can be successful.

Industrial Chemical Pollutants

The extent and consequences of industrial pollution have never been calculated. There are thousands of potential industrial pollutants. Effluents from meat, milk, butter, and cheese production can release toxic wastes into water systems. Paper mills, steelworks, and chemical plants discharge other toxic materials into water systems.

Many industrial chemicals can pollute water supplies. For example, acrylamide is a synthetic organic chemical used in the production of plastics and textiles. The SDWA requires regulation of concentrations of acrylamide in community water systems because it causes disorders of the central nervous system in animals.

The hydrocarbon benzene, found in gasoline and other petroleum products, is widely used as a solvent for paints, inks, oils, plaster, and rubber cement as well as for production of detergents, explosives, and drugs. It is a suspected carcinogen, mutagen, and teratogen, and the SDWA requires regulation of concentrations of benzene in community water systems.

Carbon tetrachloride is a volatile organic chemical primarily used as a solvent, dry-cleaning fluid, fire retardant, and in the production of pesticides. In animals, carbon tetrachloride is a proven carcinogen and causes damage to the nervous system, liver, and kidneys. The SDWA requires regulations of concentrations of carbon tetrachloride in community water systems.

Other industrial chemicals that are toxic pollutants include bromobenzene, chloratoluene, dichlorobenzene, dichloroethane, ethylbenzene, polynuclear aromatic hydrocarbons, styrene, toluene, vinyl chloride, and xylene. All of these plus many more have contaminated water systems across the United States. Many of these chemicals are suspected carcinogens and negatively affect the health of humans in multiple ways. A major problem is the large quantities of residues that remain in the soil as a result of limited regulation of these chemicals in the past.

Urban Runoff

In the large urban centers, surface water contamination can result when water in vast quantities flows from streets and percolates into the ground. The contaminants of this water come from a great variety of sources, such as automobiles, oils, road salting, pet wastes, industrial pollutants, and erosion from construction sites. Organic and inorganic compounds, bacteria, heavy metals, and pathogens are found in urban runoff water.

Surface water contaminants are carried to underground aquifers, thereby contaminating the groundwater. Studies have revealed that groundwater contamination frequently originates from surface waters. Efforts to reduce the pollutants in surface waters will help prevent groundwater contamination. If the surface water can be directed to vegetated areas or wetlands, the soil will absorb and filter out part of the pollutants. Large catchment basins in urban areas where the waste materials can settle out of the water by slow release will aid in solving this problem.

The use of salt on highways to melt ice and snow presents a particular type of pollution. The "salt" is primarily sodium chloride, but ferric ferrocyanide and sodium ferrocyanide are sometimes added to minimize caking during storage. Chromate and phosphate additions reduce the salt's corrosiveness. The amount of salt used in a winter season depends somewhat on the severity of the weather, but it is usually 9–10 million tons of dry salt and 2 million tons of liquid salt within the United States alone. Many highways receive as much as 100 tons of salt per road mile during a winter season.

This heavy use of salt has contributed to groundwater pollution. The salts are carried by runoff during melting periods considerable distances off the highway. Also stockpiled road salt can dissolve and percolate into the ground. Chloride levels on roadsides have ranged from 1,000 to 25,000 mg/l.

To reduce the effects of salt de-icing, alternative methods are now being used, such as sand and/or cinders. A number of states now require that the salt be stored in sheds, reducing the dissolution of the salt. It has also been shown that salt applications to the roads could be lighter with the same de-icing results.

Metal Contaminants

There are a large number of metal pollutants possible in groundwater. All metal concentrations in drinking water are regulated by the Safe Drinking Water Act because they have potential deleterious effects on human health. The following examples of some common metal pollutants illustrate the health effects of such minerals in water.

Aluminum is the fourth most common element in the Earth's crust, and is found in water everywhere. There is evidence that aluminum causes damage to the nervous system in humans and possible brain disorders, such as Alzheimer's disease. Aluminum is known to cause epilepsy in animals.

Arsenic compounds are widely distributed in the Earth's crust and the biosphere. Zinc, copper, and lead smelters emit arsenic, which often contaminates water supplies. Runoff from mines and mine tailings will contaminate nearby streams. Plants watered with arsenic-laden water can be contaminated. Chronic exposure to arsenic can result in fatigue, inflammation of the stomach, kidney degeneration, bone marrow degeneration, nervous system damage, and severe dermatitis. It is also suspected that arsenic may cause skin cancer.

Barium metal is found naturally in surface water throughout the nation, with the highest concentrations occurring in the lower Mississippi Basin. It has a number of productive uses, including the making of paints, paper, and pesticides and the purification of beet sugar. Barium is also used to kill rats. Barium affects the nervous system and increases blood pressure. Several toxic barium compounds can be found in water; epidemiological studies of a number of communities with higher concentrations of barium in drinking water have indicated a higher death rate due to cardiovascular disease.

Beryllium is sometimes found in acid drainage from mines. It is a proven carcinogen in animals and a suspected cause of bone cancer in humans. It also affects the respiratory system, heart, liver, and spleen.

Although pure boron is not found in nature, boron compounds, such as borax, boric acid, and sodium borate, are widely distributed in dry regions. Boron compounds in drinking water irritate the stomach and cause loss of appetite, rashes, and kidney disorders.

Cadmium is associated with copper, lead, and zinc ores. It is used in pipe production and is found as an impurity with zinc in galvanized pipe. Corrosion of such pipe is a primary source of this metal in drinking water. Cadmium is a suspected cause of cancer, testicular tumors, high blood pressure, and inhibition of growth. Its irritating effect on the stomach causes nausea and diarrhea.

Chromium has both positive and negative health effects, depending upon its concentration. Trivalent chromium is an essential trace mineral and helps protect the body from effects of vanadium. In contrast, high levels of hexavalent chromium causes ulcers, respiratory disorders, and skin irritation.

Copper from natural and industrial sources is found in drinking water throughout the country. Most of the copper in drinking water systems comes from copper plumbing. Copper is essential for health. Five to ten percent of the minimum daily requirement of 2 to 3 mg typically comes from drinking water. At higher levels copper is an irritant to the stomach and can prove fatal.

Lead is also found in drinking water in all parts of the country. Sources of lead in water supplies include mines and mine tailings, junked batteries, and runoff from streets. Lead pipes in home water systems are the most likely major source of lead in drinking water. Recent studies show that even low concentrations of lead in drinking water can cause loss of hearing, and learning disabilities in children. The EPA has estimated that at least 143,500 children suffer reduced intelligence as a result of lead poisoning, and it is the cause of over 118,000 cases of hypertension among middle-aged whites. Symptoms of lead poisoning in children include hyperactivity, chronic fatigue, apathy, dullness, and insomnia. In children and adults lead poisoning can manifest in stomach pains, nausea, vomiting, and constipation.

Mercury is a rare mineral and the compound methyl mercury is the form of mercury which is of greatest concern in drinking water contamination. Consuming fish from contaminated water is a more common source of mercury poisoning than drinking polluted water because fish concentrate the mercury in their tissues. Methyl mercury causes a deterioration of the human nervous system resulting in impairment of speech, hearing, and thought. Breathing mercury fumes can also affect the brain and nervous system. At one time leather was softened by using mercury, and hatmakers became demented, giving rise to the expression "mad as a hatter."

Molybdenum is also an essential mineral for both humans and animals. An overdose, however, causes gout and bone disorders. Symptoms of molybdenum poisoning include loss of appetite, anemia, and bone defects. The principal source of molybdenum in drinking water is from molybdenum and chromium mines and mills.

Nickel is an essential trace element, but high concentrations in drinking water affect the nervous system and cause stomach disorders and excess blood sugar in humans. Nickel has been found to cause disorders of the heart, brain, liver, and kidneys in animals.

About one-quarter of the nation's surface water has measurable traces of vanadium, with the highest concentrations in the Southwest. The leading sources of vanadium are mine tailings and auto exhaust. Vanadium accumulates in humans in the liver and bones, but is not thought to be harmful in the concentrations presently found in drinking water.

Zinc is an essential trace element needed by humans, but poisoning can result from excessive amounts. Zinc is found in drinking water, usually due to long-term use of galvanized pipes. Poisoning manifests itself in loss of appetite, nausea, stiffness of muscles, and irritability.

Waste Disposal

The disposal of waste is the greatest threat to surface and groundwater quality relating to human health. The United States generates about 4.5 billion tons of solid waste annually, or roughly 100 pounds per person per day. Hazardous industrial wastes and radioactive wastes together make up only about one percent of the total. Among the waste disposal technologies that pose water contamination problems are landfills, on-site sewage disposal, injection wells, and surface impoundments.

LANDFILLS The traditional way of disposing of wastes is in dumps and sanitary landfills. The number of these disposal sites in the United States is unknown but could total 30,000 to 40,000. Most of these sites are small, but a few receive as much as 1,000 tons of waste per day. Wastes discarded in the environment range from common household trash to complex materials such as industrial wastes, hazardous materials, and pathogens.

In the past one of the major ways to remove waste from a dump site was to burn it. This practice has nearly stopped due to the Clean Air Act, which regulates air pollution. Open dumps were also commonly used to dispose of tires, automobiles, vegetation, and other materials. This type of disposal has been largely illegal since 1976 and the passage of the Resource Conservation and Recovery Act, although the practice persists in remote rural areas.

Older sanitary landfill sites and open dumps were often unlined, allowing wastes to seep into the soil and ultimately the groundwater. In addition, many early landfills were located in marshlands, gravel pits, or abandoned strip mines. Water infiltration through these sites is a source of groundwater contamination.

Modern landfills are lined with an impermeable material, such as plastic or cement, and sealed so that no moisture escapes. Due to weathering, however, percolation of leachates from sanitary landfills is common. Nevertheless, groundwater contamination can be minimized by improved landfill design, construction, operation, and maintenance.

The millions of tons of waste generated in the United States has created the problem of locating landfills that can contain it. Waste is now transported hundreds of miles to landfills. Recycling of wastes is important to reduce the need for new landfills.

ON-SITE SEWAGE DISPOSAL It is estimated that 20 million homes in the United States have on-site disposal systems for residential sewage. These include septic systems, cesspools, and the traditional pit privy; all of them have a high potential for contaminating groundwater.

The septic system consists of a subsurface tank and drainage field. In the tank the liquids separate from the solids by gravity. The liquids drain into the rock layers of the drainage field and the solids remain in the septic tank and are decomposed by bacteria. A cesspool is an older type of installation, in which the liquids drain into the surrounding soil from the bottom of the tank.

On-site sewage systems are a major source of pollution. The contaminants include nitrates, phosphates, heavy metals, inorganic materials, and other toxic organic substances. Septic tanks have been reported as sources of contaminants in most parts of the United States.

Groundwater contamination from septic tanks results from inappropriate siting, inadequate design and construction, improper cleaning, and poor maintenance. Locations that have slow percolation rates, shallow soils, soils over permeable aquifers, and high water tables are most likely to experience problems.

INJECTION WELLS In the 1930s the practice began of disposing of brine from oil wells by injecting these fluids back into the producing horizon. This practice began to be used to dispose of industrial wastes in the 1950s. Today there are about 200,000 injection wells in the United States. Injection wells are used to dispose of wastes into geological strata below the depth of usable groundwater supplies.

The Environmental Protection Agency has identified five classes of injection wells. Class I wells handle industrial and municipal wastes.

They inject the wastes below the lowest aquifer containing potable water, typically at depths of 3,000 to 12,000 feet. The chemical industry accounts for one-half of the injected volume. The hazardous materials most commonly injected include corrosives, organic compounds, metals, and reactive wastes.

Class II wells handle waste fluids associated with oil and gas production, such as brine. The injection wells are used to enhance oil recovery, with the brine injected into the oil sands increasing the pressure and forcing oil into producing wells. There are about 140,000 of these wells in the United States with an annual disposal rate of 500 billion gallons.

Class III wells are used for injecting leaching solutions into the earth for on-site mining of oil shales, uranium, sulfur, salt, and potash. The leaching solution dissolves or otherwise mobilizes the resource, which is then pumped to the surface and removed. There are about 12,000 solution wells in operation today.

Class IV wells are used for the injection of hazardous and radioactive wastes into shallow formations above the potable water supply. All remaining wells are Class V. These include municipal, industrial, and agricultural wastewater wells, and injection wells for subsidence control, disposal of nuclear wastes, and the prevention of saltwater intrusion.

When properly installed there is little danger of contamination of groundwater from these wells. Contamination can result from failure due to poor design, construction, placement, and operation of wells and to their improper abandonment.

SURFACE IMPOUNDMENTS A surface impoundment is a pit or basin that is used to store, treat, or dispose of wastewater. In many instances, these are used to discharge clarified wastewater into surface streams after the solid materials have settled out. It is estimated that there are about 190,000 of these impoundments in the United States. They are used mostly by the paper, chemical, primary metal, petroleum, and coal product industries. They are also used for storage or disposal of municipal sewage plant sludge, as well as for animal and other farm wastes.

Contamination of the surface water can occur when there is a sudden release of liquids as well as through general leakage. Well over 90 percent of these impoundments have a high potential for contamination of water. In recent years liners have been used to contain the liquids, but these cannot guarantee freedom from pollution. Liners are easily torn and deteriorate over time, releasing contaminants.

Ocean Pollution

It was once thought that the oceans were so vast that they could not be polluted. This assumption is not correct. When water from rivers and streams flowing into the ocean contains waste material, the shorelines can become highly polluted. The pollution can become critical along shorelines and in enclosed bays. For example, the Chesapeake Bay has become highly polluted, and the oyster and fish yield from the bay has been greatly reduced. A major program supported by the affected states to remove and control the pollution is now under way.

The oceans have also been used as places to dump the waste of metropolitan centers. Wastes from the New York metropolitan area have been transported by barge approximately 40 to 50 miles offshore and dumped. The volume of the wastes has been so great that the currents have not been able to remove this material. This offshore area, the New Jersey Bight, is now a dead sea and will require years to recover.

The problem of ocean pollution is an international one. In a few areas of the world its seriousness is recognized. In 1980, 17 nations signed an agreement to begin cleaning up the Mediterranean Sea, a body of water with a very sluggish self-flushing action. In the same year a group of ecologists met to evaluate the pollution problem of the shoreline waters of the Persian Gulf. Nevertheless, little is being done on a global basis to deal with the environmental problems of the ocean. As the American Chemical Society has noted, the sheer size of the marine environment makes it potentially the biggest water pollution problem of all.

Control of Water Pollution

The removal of contaminants from water has received much attention in the past 20 years, and the public effort to protect water resources has involved great expense. New sewage plants and improved municipal and industrial waste dumps have been constructed to reduce the volume of contaminants in water. For example, in 1984 some $12 billion was spent on the construction and operation of municipal sewage treatment plants. An additional $7.5 billion was spent on industrial wastewater treatment plants.

In spite of these efforts the overall volume of pollutants from sewage plants has remained relatively constant, and the number of artificial chemicals is increasing. Although the quality of the water from these facilities has improved, the total volume of waste and variety of contaminants has increased. It is now estimated that it will cost well over $100 billion to improve the nation's sewage treatment plants sufficiently to end significant pollution of the nation's waters from that source.

An awareness of the threat to health posed by polluted wastes has grown in recent years. As a result, legislation has been enacted at the state and national levels to protect the purity of water. Of the federal laws, only the Safe Drinking Water Act of 1974 directly protects the quality of our drinking water. The Clean Water Act, the Comprehensive Environmental Response, Compensation and Liability Act (CERCLA; commonly known as the Superfund), and the Resource Conservation and Recovery Act (RCRA) have an impact on water, but their primary focus is more broad.

The 1974 Safe Drinking Water Act was passed after a study linked the consumption of New Orleans's water to increased incidence of cancer of the digestive tract. Cleaning up the nation's drinking water was considered at that time to be a task that would take only a few years. This has proven to be entirely incorrect.

In 1986 the Safe Drinking Water Act was strengthened due to growing public pressure to improve the quality of the nation's waters. The implementation of the act is the responsibility of the Environmental Protection Agency (EPA), which has established two sets of standards governing toxic elements that are regulated because they pose a potential health threat; secondary standards apply to materials that may give an undesirable taste or odor to the water, stain laundry and fixtures, or have some other undesirable effect, not necessarily affecting health.

In order to implement the act the EPA is required to:

1. Set a Maximum Contaminant Level (MCL) for safe drinking water and a Maximum Contaminant Level Goal (MCLG) for each contaminant. The MCL is the maximum level allowed in water supplied by community water systems.
2. Require each state to develop a program to safeguard the recharge areas of aquifers used by community water systems.
3. Develop more comprehensive management of watersheds that supply drinking water.
4. Require that all factories, schools, and restaurants with self-contained water systems supplying 25 or more people meet the act's requirements.
5. Ban the use of lead pipe and lead solder in water supply and plumbing systems.
6. Make it a federal crime to introduce a contaminant intentionally into, or otherwise tamper with, a public water system with the intent of harming its users.

7. Tighten the regulation of waste pumped into the ground through injection wells.
8. Establish new standards for the treatment of drinking water.
9. Establish granular activated carbon filtration as the standard to which other waste treatment methods will be compared.
10. Allow the use of "point-of-entry" water purification devices to purify all the water entering a house as a way to meet water quality standards.
11. Require monitoring for contaminants that are not now regulated. The results of this monitoring will help determine which toxics should be limited in public water supplies in the future.

The act has received little support in many communities. There is particular concern over the filtration and testing requirements because of their high cost. It is estimated that an activated carbon system sufficiently large to filter the water used by a city of 50,000 will cost about $3 million. A larger system for a major city will cost billions.

Although the total cost may be high for an urban area, treatment costs are much higher per capita for small water systems. As a result, utilities serving fewer than 500 customers are permitted to apply for an indefinite number of exemptions from the act's requirements, as long as they can show that they are improving the quality of their water system.

The cost of complying with the Safe Drinking Water Act for a typical customer is estimated to range from $4 per month in a large urban area to $32 in a system serving 500 customers. In a survey by the American Water Works Association in 1985, most people recognized that clean water was no longer free, and they were willing to pay rates high enough to provide clean water.

TABLE 1.2

Potential Sources of Water Pollutants

There are 187,000 public drinking water well systems in the United States. Potential contamination sources include the following:

23 million septic tank systems

9,000 municipal landfills

190,000 surface impoundments

280 million acres of cropland treated with pesticides annually

50 million tons of fertilizer applied to crops and lawns annually

10 million tons of dry salt applied to highways every winter

2 million gallons of liquid salt applied to highways every winter

Source: *EPA Journal* (May 1987). Volume 13, p. 148.

Water in the Western United States

The region extending from the grasslands of the Great Plains to the Pacific Coast is the most arid in the country. The Great Plains states, at the center of the continent with their vast irrigated and dryland farms, are subhumid to semiarid in nature. To the west the Rocky Mountains, extending from New Mexico to Idaho and Montana, form a massive mountain barrier that contains the headwaters of entire river systems. The Rocky Mountains supply water to drylands both to the east and to the west, where the mountains give way to the arid lands of Arizona, Nevada, Utah, California, Oregon, and Washington.

Most of the West receives less than 20 inches of precipitation per year, and this precipitation is unevenly distributed. For example, the Cascade Mountains on the coast of Washington receive about 80 inches of precipitation annually, whereas the interior of Washington state averages less than 10 inches of precipitation annually. Within the western states variations in precipitation from season to season, year to year, and even decade to decade, are equally as important as the quantity of rainfall in any single year. For example, southern Arizona, where most of the state's people live and where most of the agriculture is located, receives most of its precipitation as heavy thunderstorms between July and September. This is the time of the year when daily temperatures exceed 100° Fahrenheit, and much of the precipitation is lost through evaporation and transpiration. The surface runoff is substantial, but groundwater recovery is limited and unpredictable (see precipitation map, page 403).

In the semiarid and arid West, water is one of the most precious resources. In most areas there is too little water, and water is phenomenally expensive to move. John Wesley Powell, evaluating the water availability, reckoned that if the surface water was evenly distributed between the Colorado River and the Gulf of Mexico, you would still have a desert, almost indistinguishable from the one that is there today. But the early settlers coming from the wet East believed that rain would follow the plow. In the 1880s such beliefs amounted to biblical dogma. When precipitation did not come, the settlers turned to irrigation agriculture which required the building of gigantic dams. The "greening" of the West took the form of a religious ideal, but in spite of the expenditure of billions of dollars, only an area equal to about the size of Missouri has been converted to irrigated farming, and most of the water for the endeavor comes from nonrenewable groundwater.

Water Demand

The demand for water varies greatly in the West. Irrigated agriculture consumes about 90 percent of the total while all other users, including urban populations, consume only about 10 percent. Agricultural water use also has the highest consumption-to-withdrawal rates, which means that relatively more of the water taken from streams, lakes, or underground aquifers, evaporates instead of returning to the source for reuse. This ratio averages about 60 percent, compared to 25 percent for urban use and 5 to 25 percent for individual households.

Irrigated Agriculture

Irrigated agriculture has dominated farming in the West (see Table 1.3, page 44). The economic value of irrigation water depends upon such factors as type of crop grown and efficiency of water use. Low-value crops, such as alfalfa, corn, sorghum, wheat, and other grains, account for about 74 percent of the irrigated acreage in the West. High-value crops like vegetables and fruits have above average irrigation water requirements, but comprise only 9 percent of total irrigated acreage. In the middle range of values are such crops as cotton and soybeans with about 17 percent of the acreage.

The semiarid to desert climate provides an environment that is highly favorable to a great variety of crops—if water is available.

Evolution of Western Irrigation

The Mormons instituted irrigation with their initial settlements in 1847. By the 1870s and 1880s, hundreds of irrigation companies, founded mostly with eastern capital, were formed to reclaim the arid lands. Most of these companies operated in the arid regions of the Central Valley of California, and in Nevada, Arizona, southwestern Colorado, and southern Arizona. Almost none of them survived more than ten years. Lack of water, inept design and dam construction, and limited capital were major factors in their failure.

To rescue the farmers, California was the first state to enact regulations within the nineteenth century with the Wright Act. This act, which took its inspiration from the township governments of New England, established self-governing ministates, called irrigation districts, whose sole function was to deliver water into arid lands. It was a good idea that floundered in practice. The districts issued bonds that would not sell, built reservoirs that would not fill, and allocated water unfairly. Other states, including Colorado and Wyoming, were equally unsuccessful.

After long and bitter debate the federal government passed the Reclamation Act of 1902. Under this legislation, projects were to be financed by a reclamation fund, which was to be created entirely by revenue from the sale of federal land in the western states, and then paid back through the sale of water to farmers. The program attracted some of the best engineering graduates in the nation, who built dams that were engineering marvels. Developing irrigation agriculture, however, presented additional problems. There was a lack of scientific information on the effects of irrigation on soils, drainage, and salinity. There were also economic problems.

The productivity of the land was largely ignored in land distribution under the Reclamation Act, which allocated 160 acres to each farmer without recognizing the great variation in land productivity. For example, on a vegetable farm in California a farmer could grow wealthy on 160 acres, whereas 160 acres of irrigated pasture in Montana was paramount to economic disaster. Further, most of the early farmers had little experience in irrigated farming.

To rectify the early problems, the Reclamation Act was amended a number of times over the next 20 years. Scientific information was gradually accumulated. The economic system was reformed so that farmers could operate. In the 1930s and 1940s dozens of new dams were built that increased the area of irrigated land. This was the period when the concept of river basin development evolved and dams, canals, and irrigation projects grew as total systems. As a consequence, the natural landscape of the West underwent a transformation, the like of which no desert civilization had ever seen.

Sources of Irrigation Water

Water for irrigation comes from a variety of sources pursuant to a number of institutional arrangements. Much of the water originates from state or federal water projects developed by the U.S. Bureau of Reclamation. This water is subsidized by taxation. With the traditional philosophy of use-it-or-lose-it provisions in state water laws, combined with restrictions on the amount and timing of return flow, governmental subsidies discouraged conserving water. As long as water is relatively abundant it will be used inefficiently.

The users of surface irrigation water may own certain appropriative rights, but are more likely to obtain water through an irrigation district or water company that holds rights of use, or contracts for water, from federal or state projects. The farmer purchases a given amount of water for irrigation purposes from either a public or private source.

When irrigation is practiced from groundwater sources, the farmer merely pumps the water as required, subject to state law on pumping rates and well spacing. Energy costs comprise the major share of water procurement expenses. As energy costs rise, projects decline, and farmers are forced to consume less water. In a number of areas higher energy costs combine with ever-greater pumping depths as the water is removed from aquifers. To combat rising costs, more efficient pumps and irrigation systems have been developed, with management practices adapting to increased water scarcity. In some places, there has been a shift to crops that are more profitable. In other places where water costs exceed profits, there has either been a return to dryland farming or the land has been abandoned. For example, in the High Plains region of Texas that overlies the Ogallala Aquifer, as well as in other isolated areas of the West, the shift has been to dryland farming.

TABLE 1.3

Irrigated Acreage and Water Use

State	Irrigated Land (Acres) 1982	Irrigated Land as Percent of Total Cropland 1982	Consumption Water Use (million gallons/day) 1980	Percent of Total Use 1980
Arizona	1,153,478	74	4,000	89
California	8,460,508	75	23,000	92
Colorado	3,200,942	30	3,600	90
Nevada	829,761	96	1,500	88
New Mexico	807,206	36	1,700	89
Utah	1,082,328	56	2,400	83
Texas	5,575,553	14	8,000	80

Source: U.S. Department of Commerce (1982) for acreage and land use data; U.S. Geological Survey (1984) for water use data.

The economic value of irrigation water depends upon such factors as type of crop grown and efficiency of water use. Low-value crops, such as alfalfa, corn, sorghum, wheat, and other grains, account for about 74 percent of the irrigated acreage in the West. High-value crops like vegetables and fruits have above average irrigation water requirements, but comprise only 9 percent of total irrigated acreage.

In the middle range of values are such crops as cotton and soybeans with about 17 percent of the acreage.

Declining Groundwater Supplies

In a number of areas in the West, notably the Ogallala Aquifer of the High Plains states, central Arizona, and the San Joaquin Valley of California, water is being drawn from aquifers at rates faster than nature replaces it. The overdraft of the groundwater is of concern both for municipal water suppliers, and particularly, for the future of irrigated agriculture (see groundwater overdraft map, page 404).

OGALLALA AQUIFER The Ogallala Aquifer is a formation of sand, gravel, and silt underlying 115 million acres of the High Plains from Texas to Nebraska. Irrigation in the region began in the early 1900s, but was of little importance until the 1950s. As demand for agricultural products rose, the seemingly unlimited supply of water from the Ogallala formation, combined with fertile soils, newly developed hybrid grain, sorghum, and other crops, a favorable climate, and a progressive agribusiness encouraged the expansion of irrigation agriculture. By the 1950s, acreage irrigated from the Ogallala Aquifer accounted for more than 25 percent of all irrigated land in the United States.

As the water table dropped, concern grew over the agricultural productivity of the region, and particularly the livelihood of the farmers in the area. In 1976, Congress initiated a study of the present and future status of the aquifer. Projections were to be made for the time period 1977 to 2020 for dryland and irrigated acres of crops, value of agriculture output, input costs, employment, and income for each state under several employment categories. Researchers considered the interaction between crop yields, water use, improved technology, declining well yields, rising pumping costs, competing crops, and agricultural practices.

It was recognized that water use would vary among the states depending upon availability. Significant differences in water supply exist among the High Plains states. The aquifer is estimated to contain about 21.8 billion acre-feet of water which, if evenly distributed, would amount to about 190 acre-feet of water under each acre of land. Texas, Oklahoma, and New Mexico, however, contain about 30 percent of the land area of the aquifer but only about 15 percent of the water, whereas Nebraska contains 36 percent of the aquifer and 64 percent of the water. Consequently, the irrigated area of Nebraska could continue to increase rapidly, although in other states the area of irrigated land would decrease. Natural recharge of the aquifer in Nebraska is higher than in any other area of the region. The large expansion of irrigated land associated with intensive crop production using fertilizers and pesticides creates a considerable potential for pollution of the groundwater supply. The likelihood of nutrients and salts leaching into the aquifer will intensify with continuing development, and good management systems addressing such problems will be needed.

Projections for Kansas and Texas show substantial decreases in irrigated acreage and corresponding increases in dryland acreage. The increases in agricultural land will come from marginal areas where productivity will be low and with perhaps a higher than average erosion potential. It is predicted that much of the irrigated land associated with the Ogallala formation will go out of cultivation. Most of the land is not suitable for dryland farming, and if agriculture does not develop, serious environmental consequences could occur. Without revegetation, serious soil erosion is likely on abandoned irrigated farmland.

In areas where dryland farming existed before irrigation, there will be few problems reverting back to the earlier form of farming. Production from dryland farming is highly variable, but in general it is much lower than from irrigated farming. Dryland farm sizes are much larger, and the population density is much lower. It is not likely that a return to dryland farming will re-create the Dust Bowl conditions of the 1930s. The development of large farming equipment allows more timely and effective cultivation. Crop varieties have been developed that are better able to withstand dry conditions, and other technologies are emerging that are adapted to dryland farming systems.

The slow recognition that water from an aquifer is a depletable resource has created an unstable economy in the High Plains. The conversion of irrigated farming to dryland farming will vary depending upon the available water supply. If water availability is the primary restraint, farming techniques will gradually move from full irrigation to limited irrigation to dryland farming. The future of this region lies in the development of a sound management plan for utilization of the remaining water of the Ogallala formation.

Irrigation Methods

There are three primary methods of irrigation in common use: surface irrigation, subirrigation, and sprinkler irrigation.

SURFACE IRRIGATION Surface irrigation is the oldest form of irrigation in the West. It was practiced by the Hohokam people as early as A.D. 500 and by the Mormons beginning in 1847. In surface irrigation the water is brought to the crops by a series of increasingly smaller ditches, the largest of which, for primary distribution, is usually referred to as a *canal*. Taking off from the canal are *laterals* that carry the water to individual fields. Laterals generally follow property lines or edges of fields. Secondary laterals, if present, cross fields to provide water where it is most needed. The canals and laterals may be open ditches or they may be lined with concrete or other material to prevent seepage. Siphon tubes are an innovation that transfers the water over the banks by siphonic action to furrows. This provides excellent control of the water.

Surface irrigation can be grouped into two broad methods: flood and furrow. In flood irrigation, the entire surface of a field is covered with a continuous sheet of water retained in place by soil strips lying between dikes. This method is effective for fairly level ground in plots from 400 to 600 feet in length. It is well adapted for close-growing crops, such as hay or grains, and for pasture.

In furrow irrigation, closely spaced ditches distribute the water laterally across the field and downward to moisten the plant root system. If the slope of the land is too great, erosion can occur. This is the common method to irrigate truck crops, orchards, and row crops, such as sugar beets, corn, potatoes, and tomatoes.

By either method of surface irrigation, great quantities of water are lost through evaporation in the hot climates of southwestern United States. When water was plentiful there was little consideration given to these losses, but as water becomes more scarce, irrigators must utilize the water as effectively as possible.

SUBIRRIGATION In subirrigation, water is applied beneath the ground rather than at the surface. It is a method of artificially regulating the elevation of the groundwater table to support plant growth. This method is generally associated with existing drainage systems that can be controlled during periods of dryness to hold the water table sufficiently high to support water for the crops.

Subirrigation is practiced where the water table is near the surface. In the West subirrigation is found in California, Idaho, Colorado, Utah, and Wyoming. For example, at the convergence of the Sacramento and San Joaquin rivers in California, about 160,000 acres of low-lying delta lands are subirrigated. Before diking, this area periodically flooded. With diking and draining, however, the water table can be maintained either by pumping the excess water into the rivers during the wet season, or by directing the river water by gravity to the fields during the dry summers. The San Luis Valley of Colorado is one of the most extensive subirrigated areas in the West. Along the Platte River in Nebraska a high natural water table favors the production of subirrigated alfalfa. Along the upper Snake River in Idaho, there are 28,000 acres of subirrigated land.

SPRINKLER IRRIGATION The sprinkler irrigation system became an important means of irrigation in the 1950s in both humid and arid regions of the United States. The system takes a number of forms, from pipes that spray water on row crops to large rotating sprinklers that are commonly used on grasslands and grain crops. Sprinkler systems can be portable, semiportable, or stationary.

In a fully portable system, the main lines, laterals, sprinklers, and sometimes even the pumping mechanism is portable. The semiportable system normally has a stationary pump and stationary main lines, but portable laterals and sprinklers. In a stationary system, the main lines and laterals are generally buried and only sprinklers appear above ground. Such a system is expensive and normally used only for high-value, water-sensitive crops such as berries, vegetables, and orchards. Its great advantage is that water can be placed wherever and whenever needed.

Sprinkler systems can also have problems. For example, they may not be able to apply water at rates necessary for best crop growth, or it may be difficult to distribute the water uniformly. The shape of the field may limit sprinkler irrigation, the costs may be too great for a farmer to use sprinklers, and the labor requirements may also be too great. It is also possible to overwater with sprinklers.

Problems of Irrigation

There are a number of problems associated with the use of water in irrigation. These include water utilization and salinity.

WATER UTILIZATION Under present irrigation practices, crops use only about one-half of the water applied to a field. Declining water resources provide a major incentive for more efficient use of irrigation water. There are three major ways to increase irrigation efficiency—the measure of how much of the water applied to a field is actually used by the plants—with available water supplies: (1) reduce evaporation from soil and surface reservoirs, (2) reduce evapotranspiration by growing low-vegetation that covers the earth on farms, and (3) reduce system losses, including drain water and operational spills. Greater production per unit of water consumed can be gained by increasing water use efficiency.

Major changes to conserve irrigation water will require huge capital investments and possibly increased annual operating costs. If farmers cannot increase the area of their irrigated land as a result of greater efficiency, benefits will be limited to the indirect benefits of better water control. In addition, improvement may result in a loss of wet wildlife habitat in poorly drained areas, and benefits, if any, will be attributable to excess water diversion.

As urban populations increase, there will be more pressure to divert irrigation water for municipal use. The immediate requirements of the human population will be more critical than for the use of water in agriculture.

SALINITY The major water quality problem in the semiarid and arid regions of the West is salinity. This is a universal problem that originates partly from natural causes, such as saline springs and salt-containing rock formations, but it has intensified due to irrigation practices. The salinity of most streams increases with return flows from irrigated land. In the Rio Grande River in Texas, for example, salinity concentration increases from 890 to 4,000 mg/l within a stretch of 75 miles (see salinity map, page 405).

In a number of areas, groundwater is contaminated by deep percolation of irrigation water and seepage from irrigation systems. The interactions among soil properties and salts in liquids are numerous.

To avoid a continual increase in salinity in the groundwater, there must be adequate soil drainage to remove excess salts accumulated from irrigation water. Drainage is also required to avoid water logging. To solve these problems, the concept of salt balance has evolved. Salt balance means that the amount of salt brought into a region by irrigation must be removed by the water system.

Salinity increases accompanying continual irrigation can affect crop productivity directly. It is estimated that about 10 million acres, or 20 percent of the irrigated area of the West now suffers from salt-caused yield reduction. The result is reduced productivity and a change to more salt-tolerant but less profitable crops. Crop yield losses begin when the salt concentration in irrigation water reaches 700 to 850 mg/l, depending on soil conditions and crop type (The Environmental Protection Agency sets the upper limit of salt in drinking water at 500 mg).

The cost of salinization in the West is high. The corrosive effect of salinity on plumbing, boilers, and household appliances plagues commercial, residential, and industrial users. To illustrate, in the Colorado River Basin, which carries a salt load of 9–10 million tons annually, the yearly cost is estimated at more than $125 million.

Salinization is not only a persistent problem, but the salinity of rivers has grown worse with the increased development of water resources. Little progress will be made in solving this problem until water quality management schemes are introduced.

Urban Water Management

Although the demand for urban water in the West is far less than for the agricultural sector, the procuring of water supplies for cities has become a major political priority. Vast stretches of the West have low population density, with the people concentrated in large metropolitan centers such as Denver/Colorado Springs, Salt Lake City/Provo, Phoenix, Los Angeles/San Diego, and the San Francisco Bay area. The population living in metropolitan areas ranges from a high of 95 percent in California to about 42 percent in New Mexico, with a regional average of 76 percent. The West is the most urbanized portion of the United States. Municipal water use varies from about 4 percent of consumption in Colorado and Nevada to 7 percent in California and 10 percent in Utah.

The demand for municipal water in the West is different from that in the East. In the West, where it is hot and dry, evaporative air conditioning units and lawns consume large amounts of water. The amount of water consumed by these activities depends on local conditions. For example, in Arizona 61 percent, and in California 41 percent of the water goes for these two uses. In Denver, 51 percent of the municipal water is used to water lawns. In general, the per capita municipal use is higher in the arid West than in the humid East. To illustrate, in Raleigh, North Carolina, the monthly summer water use per household is 8.8 hundred cubic feet whereas the average in Tucson, Arizona, is 16.4 hundred cubic feet.

As the cities expand their population and associated supportive industries, the problem of securing water increases. Traditionally, western cities secured their water by appropriating local instream flows and drilling wells for groundwater, but these methods no longer suffice. By and large, water has to be brought from great distances, as the need for water has long exceeded local surface supplies. In California, with the balance of the population and political power in the south and the water in the north, a great canal system has been built to serve both the urban population and the agricultural regions.

To distribute the water, over 2,000 water districts have been created (see Califonia water supply map, page 406).

As the demand for water increases, urban areas have taken additional measures to assure an adequate water supply. A number of cities, such as Tucson, Arizona, have enacted legislation to condemn or annex surrounding agricultural lands and take over the water previously allocated to irrigated farming. Leases and exchange agreements have also increased the effective yield of water systems. Cities have constructed new water storage and impoundment pools, frequently supported by state and federal funds. The options available to a specific urban center depend to a considerable degree on local conditions and the water laws and institutions of the state.

Water for municipal use is often supplied by an urban water department, public water utility, or water conservation district. The water is then delivered to individual areas through a pressurized system. Most towns have residential water meters and charge a fee for the water used; in general, large-volume users pay a lower rate per gallon for water than small-volume users. Most states have enacted laws preventing urban water systems from making a profit and, as a consequence, funds for development of additional water supplies cannot be accumulated. In a number of cities, suburbs have grown so rapidly that water meters have not been installed. Denver, for example, has 88,000 residences without meters (Office of Technology Assessment, 1983). Such residences normally pay a flat fee for water and sewer service.

As the demand for water increases and the supply decreases, the price of water has risen. Most studies confirm the impression that urban water use responds to change in price although, for example, a 10 percent rise in price does not decrease consumption by 10 percent. In general, outdoor water uses, such as watering lawns, are much more sensitive to price increases than indoor uses. Price elasticity of water demand may be greater at higher water prices and is certainly greater over the long term as capital is invested in water conservation. In other words, people are more likely to respond to price increases if they are already spending a significant portion of their budget on water bills. Price changes may alter habits in the short term, but if conservation is to continue, investment in such water-saving devices as low-flow showerheads and water-saving toilets will be more effective.

In the past the concept prevailed in the West that water for urban use had to be developed at any cost. Few places on the Earth offer a more improbable setting for vast industrialized cities than the western

United States, an incongruity exaggerated when a drought prevails, as in the late 1980s in California. The obvious key to conservation is to change the outdated system of water rights and impose new incentives for more efficient agriculture. Although this is potentially possible, it is by no means certain. William M. Kahri, author of *Water and Power*, a major chronicle of the California water wars, states, "It is an empty exercise to put cities through elaborate conservation exercises that picks the course that causes the maximum inconvenience to the largest number of the deliveries." It is a sobering thought to realize that if every household in California cut its annual water use by 35 percent, overall water use would be reduced by only 2 percent. The viewpoint of most residents was expressed by Carl Boronkay, the general manager of the Metropolitan Water District of Southern California (MWD), when he stated, "I do not have any thought of giving up water—it's a matter of how we get it."

The drought and water crisis of the late 1980s and early 1990s has given California new reason to contemplate the state's future to attempt to prepare for the uncertainty and the growth that seems as inevitable as it is unwanted.

Sources of Urban Water Supplies

The sources of water for the South Coast Basin of California (Los Angeles–San Diego) and Tucson illustrate how two urban areas obtain supplies. The South Coast Basin of California secures its water predominantly from a vast water system that collects surface water from areas throughout the state. In contrast, Tucson is supplied primarily from groundwater sources.

SOUTH COAST BASIN OF CALIFORNIA The metropolitan area of southern California is the world's largest urban region dependent upon imported water. Mean annual rainfall in the basin averages 14 inches, with annual extremes of 5 and 38 inches. In this Mediterranean climate, the rainy season falls between November and April.

The development of remote water supplies has made it possible for southern California to grow in population from 2.9 million in 1940 to over 12 million in 1990. The MWD of southern California predicts a regional population growth of 34 percent between 1980 and 2000. The water demand between 1980 and 2000 is projected to rise from 3.06 million acre-feet (maf) to 3.61 maf in the year 2000.

Three major aqueducts and associated storage reservoirs supply about 75 percent of the region's water. These are the Colorado River Aqueduct, the Los Angeles Aqueduct, and the California Aqueduct of the State Water Project. These aqueducts allow southern California to import water from the Colorado River, the Central Valley of California, and the Owens Valley (see California water supply map, page 406).

These systems were originally capable of delivering 4.79 maf of water to southern California, but because the demand for water has been so great in other areas, the supply for MWD has diminished over the past several decades. To illustrate, water taken from the interstate Colorado River is no longer a certainty as Arizona claims its previously unused share for the Central Arizona Project. Water supplies to the Los Angeles area from the Owens Valley and the Mono Lake Basin in eastern California are also in jeopardy. Owens Valley interests have challenged Los Angeles over the pumping of groundwater at a rate faster than it is naturally restored. To conserve water, the public trust doctrine was invoked on behalf of the Mono Lake Basin. Water transfers to Los Angeles have reduced the level of Mono Lake, increasing its salinity to the point that the bird population is threatened. The water supply to Los Angeles from Mono Lake may be greatly reduced in the future.

Besides lawsuits, economic and political factors have an influence in developing the future supply of water that the southern California basin requires. The water supply could be greatly increased with the completion of the California State Water Project (SWP). Currently only 1.13 maf of water can be assured from the SWP. If it is enlarged by additional storage projects and aqueducts, the supply to the Los Angeles area could increase to 2.01 maf. MWD would then receive in a normal year about 3.47 maf of water by the year 2000. Supplies would not be adequate, however, to meet the projected growth in water demand. If the demand forecasts are correct, MWD will require a continual budgeting of its water supply. There is some possibility that raising the price of water could affect the quantities demanded, but this is by no means assured.

Recognizing the potential problem, California's South Coast Basin is attempting to develop new supplies, but the alternatives are limited. The development of new water systems are expensive, and it is uncertain whether the people of the region will be willing to pay the cost. At present, the region's growing demand for water will certainly increase, and there are only limited prospects of securing new supplies.

SANTA CRUZ RIVER BASIN (TUCSON, ARIZONA) Arizona is a desert state. Precipitation varies from 10 to 35 inches, but is concentrated in the central region. In the arid desert environment there is little surface water available. Consequently, Tucson has turned to the large stores of underground water that have accumulated over the centuries in the sand and gravel of the Santa Cruz River Basin. About 80 percent of Tucson's water supply comes from groundwater pumped in well fields south of the city. The remainder of the water supply is retrieved from agricultural irrigation areas in neighboring valleys and pumped through the Tucson Mountains to the city.

In the years 1975–1980, the average yearly pumpage was 179,000 acre-feet for municipal, industrial, and mining needs and 230,000 acre-feet for irrigation. The groundwater overdraft is 70 percent of total consumption use. It is recognized that this is a critical problem which cannot continue. The Arizona Groundwater Management Act requires that Tucson's Active Management Area (AMA) achieve zero groundwater overdraft by 2025 (see groundwater overdraft map, page 404).

Some steps to address the problem have already been initiated. Additional water will be secured from the Central Arizona Project (CAP). This additional supply will be inadequate and other measures must be taken. Demand management through higher prices has not been very successful in the area. At present, Tucson uses average cost pricing with an inverted block structure, based on a water cost of about $45 per acre-foot, and plans are to increase prices due to the high cost of the Central Arizona Project water. The actual cost of delivery of CAP water to Tucson is about $100 per acre-foot. The price to Tucson users, however, will be only about $58 per acre-foot since the water will be subsidized by federal and state grants.

To balance water supply with demand, Tucson is buying up irrigated agricultural land. This land normally does not revert to dryland farming, but is abandoned and is allowed to return to desert, which is highly subject to erosion. Nevertheless, this practice will continue, and with expected growth in population, could be greatly expanded.

In Tucson there has been a start at recycling the urban water. At present, municipal effluent is used to irrigate public parks and golf courses. It would appear that this practice must be expanded if Tucson is to have sufficient water in the twenty-first century.

Predicted shortages of water may not occur, but only if a number of things change, including greater water conservation, a decrease in population growth, demand management reducing per capita consumption, and a decrease in irrigated farming. These problems must receive immediate attention.

Water Management

Federal-State Relationships

Traditionally, water resource management has been a state responsibility, but the federal government has gradually come to play a significant role in water management. This change in management is due to such factors as the high percentage of federal land ownership, national pressures to preserve scenic and recreational areas, and the fact that most western water projects are built with federal funds. The role of the federal government in electric power development through water power is certain to grow in the future. Given the certainty that electric power will continue to be a high priority, the West will continue to develop its water power resources with the federal government continuing to play an important role in water resource management.

Research and Development

At the present time research on water availability and quality is supported by a wide range of state, federal, and private-sector programs. Over 20 federal agencies conduct water management and R&D activities. Much of the research by the Environmental Protection Agency and the Department of the Interior—each conducting about 30 percent of the water-related federal research—is to support specific responsibilities. Nearly all EPA research has been devoted to water quality, generally focusing on technical research. The Department of the Interior has focused on increasing the supply of water and conservation practices. Coordination of water research related to energy development has been overseen largely by the Water Resource Council. The U.S. Geological Survey has conducted studies to advance understanding of hydrologic systems and processes and the National Science Foundation has been concerned with basic and applied research.

State water resources research is frequently conducted through land grant universities in direct support and evaluation of state water plans. To illustrate, the Water Resource Laboratory at Colorado State University assesses water development projects to determine the economic effects on alternative water development projects, the effects of water utilization on local and regional water availability, and potential changes in water quality. At the Water Research Laboratory of Utah State University, watershed models assess the combined effects of energy and industrial development on the availability and quality of water in Utah's streams. These agencies provide information to other water planning agencies.

There are a number of federal R&D assistance programs in the states. These include the Office of Water Research and Technology (OWRT) of the Department of the Interior. This agency has 53 centers to develop an R&D program to improve management of water resources so that adequate supplies of water are available to meet the needs of the nation. The studies concentrate on basic research and technological development. Most of the research does not currently emphasize information directly useful for managing water resources through the existing allocation systems, nor is the research generally coordinated with specific state water development plans.

Future Research Needs

To develop effective water management strategies, there is a need for additional information in a number of areas. More accurate data on water consumption are required. Industry, government, and independent research provides conflicting data on water consumption as well as on suggestions to reduce consumption. Improvements in these data are critical to resolve issues of how much of the limited water supply is available for each user—energy, agriculture, industry, and direct human consumption. The factors that influence data variability need to be systematically identified and defined, with an emphasis on economics, process design, and operating conditions.

As water becomes more limited there will be greater need for further study of water consumption. Because irrigation is currently the largest consumer of water, there is a significant savings to be gained from reducing agricultural consumption. Studies by the Department of Agriculture need to be closely linked to information about the location of energy facilities. For example, in the intensive-energy-use agricultural area of the San Juan River Basin, the potential for a reduction in agricultural water use is significant.

There is limited information presently available concerning groundwater quality. There is a need to build programs about hydrological conditions as a basis for determining standards appropriate for control technology.

It has long been a common practice for energy conversion facilities to develop holding ponds to reduce or eliminate discharges of pollutants. As water development expands there is a need to develop a greater understanding as to the effectiveness of holding ponds. Research needs to be conducted on such aspects as the failure of holding ponds and the resultant pollution of water, the movement of pollutants through pond liners during normal operation, the long-term stability of holding ponds, and the water consumption of holding ponds.

As management of water grows more complex, there is a need to develop procedures for making and implementing decisions. Little documentation now exists as to how policies and procedures are carried out at local, state, or federal levels. Most decision makers have only limited understanding of the details of water distribution. Research has focused on technical problems, design considerations, or cost/benefit considerations. Too little is known about water consumption patterns among competing users.

Implementation of Water Management

Better research and information will not necessarily mean that the data will be used to address specific problems. A successful management program must integrate information directly into the activities of the water resource agencies, health departments, and other organizations dealing with water resources. This requires that environmental, social, economic, legal, and technical perspectives be considered. Programs must be able to adapt to the water problems of different regions.

Water availability and quality are at the core of any water management program. The rapidly growing demands of agriculture, energy producers, municipalities, Indian tribes, and environmental interests requires sound management procedures if water is to be available for all users. Disputes among the users create an increasingly complex context for creating water policy. In some areas, not all of the demands can be met. In others, competition adds to the uncertainty over what the best use of available water will be. Many water resource problems become increasingly difficult due to the complex combination of state water laws, federal water policies, court cases, interstate agreements, and international treaties. Although this system has dealt successfully with most water problems so far, it discourages a diversity of water uses and creates barriers to change.

Issues and Strategies

It is now recognized that a holistic approach to management of water resources is necessary to guarantee not only an adequate supply of water for the future but also a satisfactory living environment. In this approach water resources are treated as only one element in the total ecosystem. This ecocentric perspective is gradually gaining recognition. To be effective, however, changes must occur as, in the past, management agencies have regarded water development in a technocentric manner.

The days of regarding water as free in the United States are at an end. In essentially all areas metering and pricing are now used to reduce the demand that is derived from unrestricted water use. To implement these new practices, information must be developed that can be readily transmitted to decision makers; scientists in both the natural and social sciences must provide the data needed for policymakers who implement the changes. Problems with information transfer and policy implementation can stymie development.

Water resource systems are increasingly complex. This complexity is a response to new linkages in the physical system as well as the need of management to interface with the engineering systems. To illustrate, in the past several decades the main elements of the water resource system have evolved from linear systems, in which a single source supplied a single demand, to a network system of multiple sources supplying multiple demands. In addition, the complexity is increased by a continually dynamic situation, with short-term changes in demand associated with such aspects as change in weather and economic activity, and long-term changes associated with such factors as population growth.

At the present time there are few examples of integrated resource management, although the management of land can clearly have a significant effect on water resources. For example, management of the vegetation of an area or the introduction of drainage ditches can have a major effect on the water yield of a catchment basin. On a larger scale, the management of the Great Lakes Basin can create long-term benefits in the water quality of the Great Lakes.

Traditionally, emphasis was placed on the quality of water. Although this remains a fundamental objective, the issue of water quality has been broadened to emphasize such aspects as carcinogens and genetic impact. The research priority is now to define the epidemiological effects of water contaminants on the general population. These effects are based on the synergistic mixtures of toxins in the environment, of which some are ingested with water. The exact degree to which health problems can be directly related to water is very difficult to ascertain. Resource managers and policymakers have shown a reluctance to implement water quality controls before quantified health damage has been proven to already be occurring.

Water quality is the most obvious area requiring an ecological perspective. For example, many emerging water quality problems are based on the increased application of fertilizers and pesticides to agricultural lands, causing water pollution over large areas. The epidemiological effects of these pollutants are different from the dramatic and visible effects of the enteric diseases that swept urban areas in the nineteenth century. Often the effects are subclinical and do not become evident until many years of ingestion have taken place. The lack of a simple cause and effect makes integrated pollution control measures more difficult to develop and implement—but also more critical.

2

Chronology

WATER AVAILABILITY IS ONE OF THE NATION'S most important problems. The following listing of important topics and dates may further an understanding of some of the major issues in water quality and utilization.

Development of Irrigation

A.D. 500–600 Elaborate irrigation systems developed by the Hohokam people who inhabited the Gila and Salt River basins of southern Arizona.

1500s Spanish missionaries in the Southwest endeavored to establish among the Indians an agricultural way of life based upon irrigation.

1847 July 23. Mormon pioneers directed the waters of City Creek onto the parched lands of Salt Lake Valley, initiating modern irrigation in the West.

1847 Irrigation of fields by Protestant missionaries near Lewiston, Idaho.

1863 Missionaries irrigated fields near Walla Walla, Washington.

1866 July 22. Congress passed legislation granting rights-of-way for ditches and canals on public lands to acknowledged holders of valid water rights.

1877 Desert Land Act provided for the reclamation of arid land in the West. This act authorized the sale of 640 acres of land at $1.25 per acre to any person who would irrigate it within 3 years.

1891, 1893 The first National Irrigation Congress in Salt Lake City and the second in Los Angeles went on record as endorsing the cession of public land to the states.

1894 Carey Act permitted the federal government to give western states an amount of land not to exceed 1 million acres. The states in turn were to assume the responsibility for settlement, irrigation, and cultivation of the land.

1900 A total of 7.5 million acres were under some form of irrigation in the western states.

1902 June 17. The Reclamation, or Newlands, Act brought federal control of irrigation and charged federal officials with the responsibility of solving the problems of water rights, distribution of water, construction of water systems, and settlement of the reclaimed areas.

1924 Between 1902 and 1923 there were no less than 500 hearings on the Reclamation Act. Finally, in 1924, a committee of special advisers on reclamation, known as Fact Finders, was appointed to review the mistakes and success of reclamation. This committee found that the project costs had far overrun estimates, that the costs to the settlers of farm developments far exceeded expectations, and that inadequate consideration had been given to such matters as soils, project economics, and drainage. The committee also found that there had been a considerable amount of speculation in connection with the sale of private lands within the project area.

1925–1939 On the basis of the report of the Fact Finders committee, legislation was enacted to alleviate some of the problems. Over $14 million in construction charges were written off, repayment contracts were extended to a maximum of 40 years, and the antispeculation provision of the reclamation laws was re-enforced.

1925–1939 *cont.*

The legislation included numerous other provisions to protect both the government and the settlers, and provided a basis for continuation of the reclamation program until the evolution of multiple-purpose concepts necessitated a revision of reclamation laws in 1939.

1928

The single-purpose irrigation projects were obviously the first to be developed. As the population of the West grew, however, the demand to use water to serve other purposes increased. The first of the major multiple-purpose projects was the Boulder Canyon Project. The intensification of agriculture in southern California posed several related hydrological problems: the need for flood control and irrigation, the need for hydrological power and municipal water supplies throughout the area, and the need for better protection of the fish and wildlife resources.

1939

Reclamation Project Act. This act brought up to date reclamation laws pertaining to multiple-purpose projects, variable payment plans, classification of land as to irrigability and productivity, contracts for the sale of water for irrigation and municipal water supply purposes, and the sale of electric power. Several multiple-purpose projects were started under the Reclamation Act. World War II forced a discontinuance of the program.

1940s

Congress became concerned with the possibility of postwar unemployment and authorized multiple-purpose river basin developments so that a backlog of construction projects would be available. Among these was the Pick-Sloan Plan for the Missouri River Basin. This was a comprehensive multiple-purpose river basin development program participated in by two governmental agencies: the Bureau of Reclamation had responsibility for irrigation development and the Corps of Engineers was responsible for flood control and power generation.

1950s

Development of sprinkler irrigation throughout the United States.

Major Droughts

1931–1938 Extended drought in the Great Plains of the United States created a great Dust Bowl in such states as Oklahoma and Kansas.

1950–1954 A major dry period occurred in the southern Great Plains and the Southwest.

1960–1963 Drought affected most of northeastern United States for several years.

1975–1977 A lack of winter snowfall created extensive dry conditions for areas in western United States.

Major Floods

Date	Location	Deaths
May 31, 1889	Johnstown, PA	2,200
September 8, 1900	Galveston, TX	5,000
June 15, 1903	Heppner, OR	325
March 25–27, 1913	Ohio–Indiana	732
August 17, 1915	Galveston, TX	275
March 13, 1928	Saugus, CA	450
September 13, 1928	Okeechobee, FL	2,000
January 22, 1937	Ohio–Mississippi valleys	250
April 1, 1946	Hawaii–Alaska	159
January 18–26, 1969	Southern California	100
August 20–22, 1969	Western Virginia	189
February 26, 1972	Buffalo Creek, WV	118
June 9, 1972	Rapid City, SD	236
June 19–29, 1972	Pennsylvania–New Jersey	118
June 5, 1976	Teton Dam collapse, ID	11
July 31, 1976	Big Thompson Canyon, CO	139
July 19–20, 1977	Johnstown, PA	68
February 13–22, 1980	Southern California–Arizona	26
June 6, 1982	Southern Connecticut	12
December 2–9, 1982	Illinois, Missouri, Arkansas	22
February–March, 1983	California coast	13
April 6–12, 1983	Alabama, Louisiana, Mississippi, Tennessee	15
May 27, 1984	Tulsa, OK	13
June 14, 1990	Shadyside, OH	22

Largest Embankment Dams

Year Completed	Name	River	State	Volume (cubic yards in thousands)
1937	Fort Peck	Missouri	MT	25,624
1941	Kingsley	North Platte	NE	31,999
1952	Fort Randall	Missouri	SD	49,962
1953	Garrison	Missouri	ND	6,498
1958	Oahe	Missouri	SD	91,996
1959	Earthquake Lake	Madison	MT	49,998
1967	San Luis	San Luis Creek	CA	77,897
1968	Oroville	Feather	CA	77,997
1973	Castaic	Castaic Creek	CA	3,998
1973	Ludington	Lake Michigan	MI	37,699
1975	Cochiti	Rio Grande	NM	65,693
1982	Warm Springs	Dry Creek	CA	9,977

Highest Dams

Year Completed	Name	River	State	Type*	Height (feet)
1936	Hoover	Colorado	NV	A	725
1942	Grand Coulee	Columbia	WA	G	551
1945	Shasta	Sacramento	CA	G	600
1949	Ross	Skagit	WA	A	541
1953	Hungry Horse	South Fork Flathead	MT	A	564
1958	Swift	Lewis	WA	E	610
1966	Glen Canyon	Colorado	AZ	A	708
1968	Mossyrock	Cowlitz	WA	A	607
1968	Oroville	Feather	CA	E	754
1970	New Bullards Bar	North Yuba	CA	A	636
1973	Dworshak	North Fork Clearwater	ID	G	718
1979	New Melones	Stanislaus	CA	R	626

* Legend
A = Arch
E = Embankment, Earth fill
G = Gravity
R = Embankment, Rock fill

Source: U.S. Army Corps of Engineers, Committee on Register of Dams.

Largest Artificial Reservoirs

Year Completed	Dam	Reservoir	Location	Reservoir Capacity (acre-feet)
1936	Hoover	Lake Mead	NV	28,253,000
1937	Fort Peck	Fort Peck Lake	MT	17,933,000
1942	Grand Coulee	F. D. Roosevelt Lake	WA	9,558,000
1945	Shasta	Lake Shasta	CA	4,548,000
1951	Wolf Creek	Cumberland Lake	KY	3,997,000
1952	Fort Randall	Lake Francis Case	SD	4,621,000
1953	Garrison	Lake Sakakawea	ND	22,635,000
1958	Oahe	Lake Oahe	SD	22,238,000
1964	Flaming Gorge	Flaming Gorge Reservoir	UT	3,786,000
1966	Glen Canyon	Lake Powell	AZ	26,997,000
1968	Toledo Bend	Toledo Bend Lake	LA	4,475,000
1973	Libby	Lake Koocanusa	MT	5,813,000

Major Oil and Chemical Spills

Besides the major oil spills, there have been hundreds of small accidental spills on land and in the oceans. In addition, there has been deliberate dumping of waste oil in the oceans and on land.

1967 March 18. The tanker *Torrey Canyon* grounds on the Seven Stones Shoal off the coast of Cornwall, England, spilling 830,000 barrels of Kuwaiti oil into the sea. This was the first major tanker accident.

October 15. A dragging ship anchor punctures a pipeline in the Mississippi River at West Delta, Louisiana, spilling 160,000 barrels of oil.

1968 March 3. The tanker *Ocean Eagle* grounds at the entrance to San Juan, Puerto Rico, harbor spilling its cargo into the harbor.

June 13. The tanker *World Glory* experiences hull failure off South Africa and spills nearly 320,000 barrels of oil in the South Atlantic Ocean.

1969 January 28. An oil spill occurs in the coastal waters of Santa Barbara, California, because of lack of control of underground oil pressure during drilling operations.

November 4. A storage tank ruptures at Sewaren, New Jersey, spilling 200,000 barrels of oil on land surfaces.

November 5. The tanker *Keo* experiences hull failure off the Massachusetts coast and spills 210,000 barrels of oil into the ocean.

1970 March 20. The tanker *Othello* experiences a collision in Tralhavet Bay, Sweden, spilling more than 420,000 barrels of oil.

1971 November 30. An oil tanker off the coast of Japan breaks in half, dumping 150,000 barrels of oil in the Pacific Ocean.

1972 December 19. The tanker *Sea Star* experiences a collision in the Gulf of Oman and spills 800,000 barrels of oil.

1976 May 12. The tanker *Urquiola* strikes an underwater obstruction entering the harbor of La Coruna, Spain, and spills 700,000 barrels of oil.

December 15. The tanker *Argo Merchant* grounds off Nantucket, Massachusetts, spilling over 180,000 barrels of oil.

1977 February 28. The tanker *Hawaiian Patriot* catches fire and has to dump 690,000 barrels of oil in the North Pacific Ocean.

April 22. The Ekofisk oil field in the North Sea experiences a well blowout; 195,000 barrels of oil flow into the sea.

December 16. The oil tankers *VenOil* and *VenPet* collide off Port Elizabeth, South Africa, spilling some 210,000 barrels of oil.

1978 March 16. *Amoco Cadiz*, carrying a cargo of 1,540,000 barrels of light Arabian crude oil, experiences rudder failure near the English Channel and grounds on the Brittany coast, spilling its entire cargo.

1979 June 3. Ixtoc fields Number 1 well in the Gulf of Mexico experiences a blowout, spilling 86,000 barrels of oil.

1979 *cont.*

July 19. The tankers *Atlantic Empress* and *Aegean Captain* collide off Trinidad and Tobago, depositing 2.1 million barrels of oil in the sea.

November 1. The tanker *Burmah Agate* collides with another vessel in Galveston Bay, Texas, spilling about 280,000 barrels of oil.

1983

February. A blowout of a well in the Newraz oil field in the North Sea deposits 4.2 million barrels of oil in its vicinity.

August 6. The Spanish supertanker *Castillo de Bellver,* laden with 1.75 million barrels of Persian Gulf oil, bursts into flame about 80 miles northeast of Capetown, South Africa. The tanker breaks apart, spilling its entire cargo of oil into the ocean.

1984

January 12. The Danish ship *Dana Optimor* had its cargo of 80 drums of the chemical dinoseb swept overboard in the North Sea during a storm.

1987

March 6. British car ferry *Herald of Free Enterprise* sinks in the Belgian harbor of Zeebrugge with five vans containing hundreds of different chemicals.

1988

January 2. At Ashland, Pennsylvania, on the Monongahela River, an oil bulk storage facility fails, spilling about 90,000 barrels of Number 2 fuel oil onto the river's edge. The fuel oil contaminates the Monongahela and Ohio rivers to the Ohio–Pennsylvania border.

1989

March 24. The *Exxon Valdez* tanker grounds on Bligh Reef in Prince William Sound near Valdez, Alaska, spilling nearly 250,000 barrels of crude oil and creating the worst oil spill in waters of the United States.

June 23. Greek tanker *World Prodigy* spills oil off the coast of Rhode Island at Brenton Reef.

1991

Persian Gulf War. The Iraq army opened the flowing oil wells of Kuwait. Although most of the oil burned, thousands of barrels were spilled on the land and into the ocean.

Laws and Regulations

Environmental Protection Laws

1937	Water Facilities Act
1948	Water Pollution Control Act
1958	Water Supply Act
1964	Water Resources Research Act
1965	Water Resources Planning Act
1969	National Environmental Policy Act
1970	Water Bank Act
1974	Safe Drinking Water Act
1977	Federal Water Pollution Control Act (Clean Water Act)
	Surface Mining Control and Reclamation Act
1979	Toxic Substance Control Act
1980	Comprehensive Environmental Response, Compensation, and Liability Act (Superfund)
1984	Water Resources Research Act
1986	Lower Colorado Water Supply Act
	Safe Drinking Water Act Amendment
1987	Water Quality Act
1988	National Drinking Water Week Act
	Disaster Assistance Act

Irrigation

1877	Desert Land Act

1894	An Act to Provide for the Sale of Desert Land in Certain States and Territories (Carey Act)
1902	Reclamation Act (Irrigation of Arid Land)
1911	Reclamation Act
1939	Reclamation Project Act
1972	Reclamation Project Authorization Act
1982	Reclamation Reform Act

Flood Control

1917	Flood Control Act
1954	Watershed Protection and Flood Protection Act
1968	National Flood Insurance Act
1973	Flood Disaster Protection Act
1974	Water Resources Development Act

Water Power

1920	Federal Water Power Act

Recreation

1965	Federal Water Projects Recreation Act

Ocean Pollution

1972	Marine Protection, Research and Sanctuaries Act
1978	National Ocean Pollution Research and Development and Monitoring Planning

Oil Pollution

1990	Oil Pollution Act

3

Laws and Regulations

A VAST BODY OF LAWS AND REGULATIONS HAVE evolved in the United States because of the importance of water to everyday life. In the water-deficient western states, laws controlling water use were among the first passed by state legislatures. The earliest federal laws were concerned with the control of floods and the use of water to produce power. In more recent years, as demand for water has grown even as our industrial society increasingly pollutes the water supply, legislation has been directed to assuring water supplies not only in adequate amounts but also of acceptable quality. The following list is selective because of the sheer number of laws and regulations relating to water. Most have been amended over time; amendments follow the explanations of the core laws.

Environmental Protection Laws

A number of environmental laws include provisions affecting water resources.

National Environmental Policy Act of 1969 (Public Law 91-190, January 1, 1970)

In the National Environmental Policy Act of 1969 (NEPA), Congress established a national policy to encourage a productive and enjoyable harmony between humans and the environment, to promote efforts to prevent or eliminate damage to the environment and biosphere and stimulate the health and welfare of humans, to enrich the understanding of the ecological systems and natural resources important to the nation, and to establish a Council of Environmental Quality. This act was fundamental to the establishment of the Environmental Protection Agency (EPA) by the President in 1970.

In adopting this act, Congress recognized the profound impact of human activity on the interrelations of all components of the natural environment—particularly population growth, high-density urbanization, industrial expansion, resource exploitation, and new and expanding technologies. Congress recognized further the critical importance of restoring and monitoring environmental quality and called for the cooperation of state and local governments and other concerned public and private organizations to use all practical means, including financial and technical assistance, establish conditions under which humans and nature can exist in productive harmony.

Subsequent amendments to this legislation include:

Public Law 94-52, July 3, 1975

Public Law 94-83, August 9, 1975

Comprehensive Environmental Response, Compensation, and Liability Act of 1980 (Public Law 96-510, December 11, 1980)

The Comprehensive Environmental Response, Compensation, and Liability Act of 1980 (CERCLA) authorized the federal government to develop programs to protect the environment from the release or threatened release of hazardous materials into the environment. Given that many incidents represent a threat to water resources, CERCLA is an important statute for water protection and remediation.

Under CERCLA the federal government can take direct action to remove hazardous substances released into the environment—including contaminants in groundwater. Remedial measures are to be carried out under regulations established by EPA in the National Contingency Plan. The government may seek reimbursement for response costs, and where the responsibility for water-causing contamination can be traced to companies with financial resources, CERCLA requires that the financial responsibility for the cleanup be placed on those companies.

The National Contingency Plan is of particular importance to water protection because it outlines the manner in which the federal government is to investigate releases and threatened releases to determine if remedial action is necessary. This assessment procedure consists of:

Phase 1 Discovery and Notification

Phase 2 Preliminary Assessment

Phase 3 Immediate Removal of Most Hazardous Materials

Phase 4 Evaluation and Determination of Appropriate Action for Federal, State and Local Governments and for Interstate and Nongovernmental Entities in Affecting the Plan

Phase 5 Planned Removal

Phase 6 Remedial Action

Phase 7 Documentation of Removal of the Hazardous Materials

Based on these criteria EPA has developed a Hazard Ranking System (HRS). The HRS "score" establishes the foundation for determining whether further response actions are warranted at a site. The HRS takes into account the population at risk, the hazard potential of hazardous substances at a site, the potential for contamination of drinking water supplies, the potential for direct human contact, and the potential for destruction of sensitive ecosystems. From these determinations the National Priority List (NPL) is prepared.

A critical question in deciding on the appropriate remedial response to release of hazardous substances into water resources is the standard that cleanup should attain. In 1985, EPA revised the NPL to include consideration of other relevant and appropriate state laws in determining the appropriate extent of cleanup. This revision was incorporated by Congress into the 1986 amendment to CERCLA, and now both federal and state standards are considered in CERCLA determinations. For remedial actions for groundwater that is to be used in the future for drinking water, the new regulations require the application of the Maximum Containment Level Goals (MCLGs) published by EPA under the Safe Drinking Water Act. Because the MCLG for many contaminants is zero, EPA cannot determine a threshold health effect and such a level may not be feasible to measure for technological or economic reasons, it will be important to see how

Applicable or Relevant and Appropriate Requirements (ARARs) are applied to groundwater remedial actions in the future.

Amendments to CERCLA include:

Public Law 98-80, August 23, 1983

Public Law 99-499, October 17, 1986

Public Law 100-202, December 22, 1987

Public Law 100-707, November 23, 1988

Public Law 101-508, November 5, 1990

Public Law 101-504, November 15, 1990

Surface Mining Control and Reclamation Act of 1977 (Public Law 95-87, August 3, 1977)

In 1977 Congress enacted the Surface Mining Control and Reclamation Act (SMCRA) to protect the environment from the adverse effects of surface mining. In passing this law Congress found that:

1. Extraction of coal and other minerals from the Earth can be accomplished by various mining methods including surface mining.
2. Many surface mining operations result in disturbances of surface areas that burden and adversely affect commerce and the public welfare by destroying or diminishing the utility of land for commercial, industrial, residential, recreational, agricultural, and forestry purposes, by causing erosion and landslides, by contributing to floods, by polluting the water, by destroying fish and wildlife habitats, by impairing natural beauty, by damaging the property of citizens, by creating hazards dangerous to life and property by degrading the quality of life in local communities, and by counteracting governmental programs and efforts to conserve soil, water, and other natural resources.
3. Surface mining and reclamation technology is now developed so that effective and desirable regulation of surface coal-mining operations by the state and the federal governments is an appropriate and necessary means to minimize the adverse social, economic, and environmental effects of such mining operations.
4. Because of the diversity in terrain, climate, biological, chemical, and other physical conditions in areas subject to mining operations, the primary governmental responsibility for

developing, authorizing, issuing, and enforcing regulations for surface mining and reclamation operations should rest with the states.

5. There are a substantial number of areas of land throughout major regions of the United States disturbed by surface and underground mining on which little or no reclamation was conducted, and the impact from these unreclaimed lands imposes social and economic costs on residents in nearby and adjoining areas as well as continuing to impair environmental quality.

To administer the act an Office of Surface Mining Reclamation and Enforcement was established in the Department of the Interior. Within the states, Mining and Mineral Resources and Research Institutes were established to provide reclamation information. A trust fund was created to pay for reclaiming abandoned strip-mined land.

Amendments to this legislation include:

Public Law 95-240, March 7, 1978

Public Law 95-343, August 11, 1978

Public Law 95-617, November 9, 1978

Public Law 96-581, December 11, 1980

Public Law 97-98, December 22, 1981

Public Law 98-473, October 12, 1984

Public Law 99-500, October 18, 1986

Public Law 99-591, October 30, 1986

Public Law 100-34, May 7, 1987

Public Law 100-71, July 11, 1987

Public Law 101-58, November 5, 1990

Toxic Substance Control Act of 1976 (Public Law 94-469, October 11, 1976)

The Toxic Substance Control Act of 1976 is a comprehensive law designed to protect humans and the environment from toxic substances. As water is often the means of transporting these substances, this legislation affects water issues.

Congress found these measures necessary because:

1. Human beings and the environment are exposed each year to a large number of chemical substances and mixtures.
2. Among these chemical substances are some whose manufacture, processing, and distribution in commerce may present an unreasonable risk of injury to human health or the environment.
3. Effective regulation of such chemical substances and mixtures necessitates their regulation in intrastate commerce.

To implement this act, Congress called for:

1. Data to be developed with respect to the effect of chemical substances and mixtures on health and the environment.
2. Regulation of chemical substances and mixtures that present an unreasonable risk of injury to health or the environment.
3. Authority over chemical substances and mixtures to be exercised in such a manner as not to impede unduly or create unnecessary economic barriers to technological innovations.

Subsequent amendments to this legislation include:

Public Law 97-129, December 29, 1981

Public Law 98-80, August 23, 1983

Public Law 98-620, November 8, 1984

Public Law 99-419, October 22, 1986

Public Law 100-368, July 18, 1988

Public Law 100-418, August 23, 1988

Public Law 100-551, October 28, 1988

Public Law 101-508, November 5, 1990

Public Law 101-637, November 25, 1990

Water Supply

Water Supply Act of 1958 (Public Law 85-500, July 3, 1958)

Public Law 85-500 enacted comprehensive legislation for the construction, repair, and preservation of public works in rivers and harbors for navigation, flood control, and other purposes; Title III of the act is the Water Supply Act of 1958.

This legislation recognized the primary responsibilities of the states and local interests in developing water supplies for domestic, municipal, industrial, and other purposes, but provided for federal participation and cooperation in developing water supplies in connection with the construction, maintenance, and operation of federal navigation, flood control, irrigation, and multiple-purpose projects.

The act dealt largely with questions of financing, providing:

1. That before construction or modification of any project including water supply provisions is initiated, state or local interests shall agree to pay for the cost of such provisions.
2. That an amount not to exceed 30 percent of the total estimated cost of any project may be allocated to anticipated future demands where states or local interest give reasonable assurances that they will contract for the use of storage for anticipated future demands within a period of time that will permit paying out the costs allocated to water supply within the life of the project.
3. That the entire amount of the construction costs allocated to water supply shall be repaid within the life of the project.

Amendments to this legislation include:

Public Law 87-88, July 20, 1961

Public Law 99-662, November 17, 1986

Water Resources Planning Act of 1965 (Public Law 89-80, July 22, 1965)

The Water Resources Planning Act of 1965 was intended to provide for the optimum development of the nation's natural resources through the coordinated planning of water and related land resources by establishment of a water resources council and river basin commission and by providing financial assistance to the states to increase state participation in planning. Congress called for the conservation, development, and utilization of water and related land resources of the country on a comprehensive and coordinated basis by the federal government, states, localities, and private enterprise, with the cooperation of all affected federal agencies, states, local governments, individuals, corporations, business enterprises, and others concerned.

The responsibilities of the Water Resources Council were to:

1. Maintain a continuing study and prepare an assessment at least biennially of the adequacy of supplies of water necessary to meet the water requirements in each water resource region.
2. Maintain a continuing study of the relation of regional or river basin plans and programs and relates these requirements to larger regions of the nation. The act also considers the adequacy of administrative and statutory means for the coordination of the water and related land resource policies and programs of federal agencies.
3. Appraise the adequacy of existing and proposed policies and programs to meet water requirements.
4. Make recommendations to the president with respect to federal policies and programs.

It is also the responsibility of the council to establish federal project procedures. When a plan is received from a river basin commission the council reviews the plan as to:

1. The efficiency of the plan or revision in achieving optimum use of the water and related land resources in the area involved.
2. The effect of the plan on the achievement of other programs for the development of agricultural, urban, energy, industrial, recreational, fish and wildlife, and other resources.
3. The contributions the plan or revision will make in obtaining the nation's economic and social goals.

It is also the responsibility of the council to recommend to the president the establishment of a river basin water and land resources commission.

Requests must:

1. Define the area, river basin, or group of related river basins for which a commission is requested.

2. Be made in writing by the governor or other state official to the council.
3. Be concurred in by the council and by not less than half the states within which the basin(s) are located.

The purposes of the commission are:

1. To serve as the principal agency for the coordination of federal, state, interstate, local, and nongovernmental plans for the development of water and related land resources in its area, river basin, or groups of river basins.
2. To prepare and keep up to date a comprehensive, coordinated, joint plan for federal, state, interstate, local, and nongovernmental development of water and related resources, including an evaluation of all reasonable alternative means of achieving optimum development of water and related land resources of the basin.
3. To recommend long-range schedules of priorities for the collection and analysis of basic data and for the investigation, planning, and construction of projects.
4. To foster and undertake such studies of water and related land resources problems in the area, river basin, or group of river basins as are necessary in the preparation of the plan.

Amendments to this legislation include:

Public Law 90-547, October 2, 1968

Public Law 92-27, June 17, 1971

Public Law 92-396, October 20, 1972

Public Law 93-55, July 1, 1973

Public Law 94-112, October 16, 1975

Public Law 94-285, May 12, 1976

Public Law 95-41, June 6, 1977

Public Law 95-404, September 30, 1978

Public Law 97-449, January 12, 1983

Water Resources Research Act of 1964 (Public Law 88-379, July 17, 1964)

The basic purpose of the Water Resources Research Act of 1964 was to assure the nation of a permanent supply of water in quantity and quality to meet the requirements of an expanding population. Goals included the stimulation, sponsoring, providing for, and supplementing of existing programs, the conducting of research, investigations, and experiments, and the training of scientists in the fields of water and resources that affect water.

The act called for assisting each participating state in establishing a competent and qualified water resources research institute at a Land Grant college or university or other suitable institution. The research may include, but is not limited to, the hydrologic cycle; supply and demand for water; conservation and best use of available supplies of water; methods of increasing water supplies; and economic, legal, social, engineering, recreational, biological, geographic, ecological, and other aspects of water problems. To implement the program, federal fundsld match state money on a dollar-for-dollar basis.

Title II of the act provided a special appropriation of funds to undertake research into aspects of water problems related to the mission of the Department of the Interior.

Title III of the act provided additional administrative direction, with an emphasis on interagency coordination, including:

1. Continuing review of the adequacy of the government-wide program in water resources research.
2. Identification of duplication and overlaps between two or more agency programs.
3. Identification of technical needs in various water resources research categories.
4. Recommendations with respect to allocation of technical effort among federal agencies.
5. Review of the technical staff needs and findings of the programs initiated by the act.
6. Recommendations concerning management plans to improve the quality of the government-wide research effort.
7. Facilitating interagency communication at management levels.

Amendments to this legislation include:

Public Law 89-404, April 19, 1966

Public Law 92-175, December 2, 1971

Public Law 93-608, January 2, 1975

Public Law 95-84, August 2, 1977

Water Resources Research Act of 1984 (Public Law 98-242, March 22, 1984)

The Water Resources Research Act of 1984 continued the program begun under the Water Resources Research Act of 1964 and the Water Research and Development Act of 1977. Congress found that:

1. The existence of an adequate supply of water of good quality for the production of food, materials, and energy for the nation's needs and for the efficient uses of the nation's energy and water resources is essential to national economic stability and growth, and to the well-being of the people.
2. The management of water resources is closely related to monitoring environmental quality and social well-being.
3. There is an increasing threat to impairment of the quantity and quality of surface and groundwater resources.
4. The nation's capabilities for technological assessment and planning and for policy formulation for water resources must be strengthened at the federal, state, and local levels.
5. There should be a continuing national investment in water and related research and technology commensurate with growing national needs.
6. It is necessary to provide for the research and development of technology for the conversion of saline and other impaired water to a quality suitable for municipal, industrial, agricultural, recreational, and other beneficial uses.
7. The nation must provide programs to strengthen research and associated graduate education because the pool of scientists, engineers, and technicians trained in fields related to water resources constitutes an invaluable natural resource which should be increased, fully utilized, and regularly replenished.

The purpose of the act was to assist the nation and the states in augmenting their water resources science and technology to:

1. Assure supplies of water sufficient in quantity and quality to meet the nation's expanding needs for the production of food, materials, and energy.

2. Discuss practical solutions of the nation's water problems, particularly those related to impaired water quality.
3. Assure the protection and enhancement of environmental and social order in connection with water resources management and utilization.
4. Promote the interest of state and local governments as well as private industry in research and the development of technology for reclaiming wastewater and to convert saline and other impaired water to a quality suitable for municipal, industrial, agricultural, recreational, and other beneficial uses.
5. Coordinate more effectively the nation's water resources research program.
6. Promote the development of a cadre of trained research scientists, engineers, and technicians to deal with future water resources problems.

The institutes' responsibilities were enlarged to include:

1. Plan, conduct, or arrange for competent research with respect to water resources, including investigation and experiments of either a basic or practical nature.
2. Promote the dissemination and application of the results of these efforts.
3. Provide for the training of scientists and engineers through research, investigation, and experiments.
4. Cooperate with other colleges and universities in the state that have demonstrated capabilities for research, information dissemination, and graduate training to develop a statewide program for resolving state and regional water and related land problems.

Water Facilities Act of 1937 (Chapter 870, August 28, 1937)

In the Water Facilities Act of 1937 Congress found that:

> The wastage and inadequate utilization of water resources on farm, grazing and forest lands in the arid and semiarid areas of the United States resulting from inadequate facilities for water storage and utilization contribute to the destruction of natural resources, injuries to public health and public lands, droughts, periodic floods, crop failures, decline in standards of living, and excessive dependence upon public relief, and thereby menace the national welfare. It is therefore hereby declared to be the

policy of Congress to assist in providing facilities for water storage and utilization in the arid and semiarid areas of the United States.

The Secretary of Agriculture was authorized to:

1. Formulate and keep current a program of projects for the construction and maintenance in the areas of ponds, reservoirs, wells, check dams, pumping installations, and other facilities for water storage or utilization, together with the appropriateness of such facilities. The facilities were to be located to promote the proper utilization of lands but not where they would encourage the cultivation of submarginal lands or those that should be devoted to other uses.
2. Construct and sell or lease water facilities.
3. Cooperate with all persons and groups in implementation.
4. Obtain options upon and acquire lands, or rights or interests therein, to rights to the use of water, by purchase, lease, gifts, exchange, condemnation, or otherwise.

Amendments to this legislation include:

Chapter 751, August 17, 1954

Public Law 85-748, August 25, 1958

Water Bank Act of 1970 (Public Law 91-559, December 13, 1970)

In passing the Water Bank Act of 1970, Congress found:

> It is in the public interest to preserve, restore and improve the wetlands of the Nation, and thereby to conserve surface water, to preserve and improve habitat for migrating waterfowl and other wildlife resources, to reduce runoff, soil and wind erosion, and contribute to flood control, to contribute to improved water quality and reduce stream sedimentation, to contribute to improved subsurface moisture, to reduce acres of new land coming into production and to retire land now in agricultural production, to enhance the national beauty of the landscape, and to promote comprehensive and total water management planning. The Secretary of Agriculture is authorized and directed to formulate and carry out a continuous program to prevent the serious loss of wetlands, and to preserve, restore, and improve such lands.

A major feature of this act was authorizing the Secretary of Agriculture to enter into agreements with landowners and operators in important migratory waterfowl nesting and breeding areas for the conservation of water and specified farm, ranch, or other wetlands identified in a conservation plan developed in cooperation with the Soil and Water Conservation District in which the land is located. These agreements were to extend for a period of ten years with provisions for renewal for additional periods of ten years each.

The landowner had to agree:

1. To place in the program for the period of the agreement eligible wetland areas he or she designates, which may include wetland covered by a federal or state government easement that permits agricultural use, together with adjacent appropriate areas.
2. Not to drain, burn, till, or otherwise destroy the wetland character of the area, nor to use the area for agricultural purposes.
3. To effectuate the wetland conservation and development plan for the land.
4. To forfeit all rights to further payments or grants under the agreement and refund to the United States all payments or grants received upon violation of the agreement.
5. Upon transfer of rights and interest in the lands subject to the agreement during the agreement period, to forfeit all rights to further payments or grants.
6. Not to adopt any practice specified by the secretary in the agreement as a practice which would tend to defeat the purposes of the agreement.
7. To such other provisions as the secretary determines are desirable and includes in the agreement to effectuate the purposes of the program or to facilitate its administration.

In return, the Secretary of Agriculture is to:

1. Make an annual payment to the owner or operator.
2. Bear part of the cost of establishing and maintaining conservation and development practices on the wetlands and adjacent areas.

The legislation was amended by:

Public Law 96-182, January 2, 1980.

Conservation in Arid and Semiarid Areas Act of 1937 (Public Law 399, August 28, 1937)

The purpose of Conservation in Arid and Semiarid Areas Act of 1937 was to promote water conservation in the arid and semiarid areas of the United States by aiding in the development of facilities for water storage and utilization. The law was based on the recognition that the wastage and inadequate utilization of water resources on farm, grazing, and forest lands in the arid and semiarid areas of the United States resulting from inadequate facilities for water storage and utilization contribute to the destruction of natural resources, injuries to public health and public lands, droughts, periodic floods, crop failures, declines in the standard of living, and excessive dependence upon public relief. It became the policy of Congress to assist in providing facilities for water storage and utilization in these semiarid and arid areas.

The act called for:

1. Developing a program of projects for the construction and maintenance of ponds, reservoirs, wells, dams, pumping installations, and other facilities for water storage and utilization.
2. Cooperating with other agencies to promote water conservation.
3. Constructing facilities to obtain conservation of water.

Disaster Assistance Act of 1988 (Public Law 100-387, August 11, 1988)

The Disaster Assistance Act of 1988 was a comprehensive measure to provide drought assistance to agricultural producers. Title IV provided water-related assistance, authorizing federal agencies to:

1. Conduct research and demonstration projects.
2. Provide technical assistance and extension services.
3. Make grants, loans, and loan guarantees.
4. Provide other forms of assistance for the purpose of helping rural areas make better and more efficient use of water resources.

Federal assistance included promotion or establishment of irrigation, watershed, and other water and drought management activities, such as water transmission, application, and regulation.

An Emergency Drought Authority was created under Title IV. Title IV is also cited as the Reclamation States Drought Assistance Act of 1988. To implement this act the Secretary of the Interior is to:

1. Perform studies to identify opportunities to augment, make use of, or conserve water supplies available to federal reclamation projects and Indian water resource developments.
2. Consistent with existing contractual arrangements and state law undertake construction, management, and conservation activities that will mitigate or can be expected to have an effect in mitigating losses and damages resulting from drought conditions in 1987, 1988, or 1989.
3. Assist willing buyers in their purchase of available water supplies from willing sellers and redistribute such water based upon priorities to be determined by the Secretary of the Interior consistent with state law, with the objective of minimizing losses and damages resulting from drought conditions in 1987, 1988, and 1989.

The act also authorized the release of water at existing federal reclamation projects to mitigate the drought conditions and loans to water users for management and conservation activities, or the acquisition or transportation of water. These activities were to be coordinated with existing federal and state agencies.

As a special project, the Secretary of the Interior was authorized to install a temperature control curtain as a demonstration project at Shasta Dam in the Central Valley Project in California. The purpose of the demonstration project was to determine the effectiveness of the temperature control curtain in controlling the temperature of water releases from Shasta Dam, so as to protect and enhance anadromous fisheries in the Sacramento River and San Francisco Bay/Sacramento-San Joaquin Delta and Estuary.

Lower Colorado Water Supply Act of 1986 (Public Law 99-655, November 14, 1986)

The Lower Colorado Water Supply Act of 1986 authorized the Secretary of the Interior to construct, operate, and maintain the Lower Colorado Water Supply Project in California to supply water for domestic, municipal, industrial, and recreational purposes. The project may be operated by nonfederal interests. The Secretary of the Interior is further authorized to facilitate a water exchange agreement between nonfederal interests in order to exchange a portion of water from the Colorado River for an equivalent quantity and quality of groundwater from a well field located in the Sand Hills area of Imperial County, California.

National Drinking Water Week Act of 1988 (Public Law 100-272, March 30, 1988)

Congress authorized the president to issue a proclamation calling upon the people of the United States to observe the period May 2 to May 8, 1988, with appropriate ceremonies, activities, and programs to enhance public awareness of drinking water issues and the importance of drinking water to health, safety, and quality of life.

Congress stated:

1. Whereas water itself is God-given, and the drinking water that flows dependably through our household taps results from the dedication of the men and women who operate the public water systems of collection, storage, treatment, testing, and distribution that insures that drinking water is available, affordable, and of unquestionable quality.
2. Whereas the advances in health effects due to research and water analysis and treatment technologies, in conjunction with the Safe Drinking Water Act Amendment of 1986 (Public Law 99-339), could create major changes in the production and distribution of drinking water.
3. Whereas this substance, which the public uses with confidence in so many productive ways, is without doubt the single most important product in the world and a significant issue for the future.
4. Whereas the public expects high quality drinking water to always be there when needed.
5. Whereas the public continues to increase the demand for drinking water of unquestionable quality.

Water Quality

Safe Drinking Water Act of 1974 (Public Law 95-523, December 16, 1974)

The Safe Drinking Water Act of 1974 amended the Public Health Service Act to assure, among other purposes, that the public is provided with safe drinking water. This act defines primary and secondary drinking water regulations.

Primary regulations:

1. Apply to public water systems.
2. Specify contaminants that adversely affect health.
3. Contain criteria and procedures to assure a supply of drinking water.

Secondary regulations apply to public water systems and specify the maximum contaminant levels requisite to protect public health. The act established primary drinking water regulations and gave the states enforcement responsibility for public water systems when they:

1. Have adopted water regulations.
2. Can implement adequate procedures.
3. Can keep records and make reports.
4. Have adopted and can implement an adequate plan to assure safe drinking water.

If a state fails to develop the necessary safe drinking water regulations, it is liable to federal prosecution.

The act also requires the protection of underground sources of drinking water by prohibiting underground injection of contaminants and calling for inspection, monitoring, record keeping, and reporting. The act prohibits, however, state underground injection control programs from prescribing "requirements which interfere with or impede the underground injection of brine or other fluids which are brought to the surface in connection with oil or natural gas production, or any underground injection for the secondary or tertiary recovery of oil or natural gas."

The act was amended by:

Public Law 95-190, November 16, 1977

Public Law 96-63, September 6, 1979

Public Law 100-572, October 31, 1988

Safe Drinking Water Act Amendment of 1986 (Public Law 99-339, June 19, 1986)

In 1986 Congress amended the Safe Drinking Water Act to establish that "each national interim or revised primary drinking water regulation promulgated under this section before such enactment shall be deemed to be a national primary drinking water regulation." It is the purpose of this act to protect all drinking water.

The act also covered a wide variety of topics in order to guarantee safe drinking water standards. The topics included were:

Public water systems

National primary drinking water regulations

Enforcement of regulations

Public notification

Variances exemptions

Monitoring for unregulated contaminants

Technical assistance for small systems

Tampering with public water systems

Lead-free drinking water

Water Pollution Control Act of 1948 (Chapter 758, June 30, 1948)

The Water Pollution Control Act of 1948 laid the foundation for the many acts and amendments that followed, establishing the responsibility of the federal government:

> To recognize, preserve, and protect the primary responsibilities and rights of the States in controlling water pollution, to support and aid technical research to devise and perfect methods of treatment of industrial wastes which are not susceptible to known effective methods of treatment, and to provide Federal technical services to State and interstate agencies and to industries, and financial aid to State and interstate agencies and to municipalities, in the formulation and execution of their stream pollution abatement programs.

The pollution control program was the responsibility of the Surgeon General, who was charged with developing comprehensive programs to:

1. Encourage cooperative activities by the states for the prevention and abatement of water pollution.
2. Encourage compacts between states for the prevention and abatement of water pollution.
3. Collect and disseminate information relating to water pollution and its abatement.
4. Support and aid technical research to devise and perfect methods of treatment of industrial wastes that are not susceptible to known effective methods of treatment.
5. Make available results of studies, research, investigations, and experiments relating to water pollution.

The act also established a Water Pollution Advisory Board to review the water policies and programs of the Public Health Service.

The act was subsequently amended by:

Chapter 927, July 17, 1952

Chapter 518, July 9, 1956

Public Law 94-273, April 21, 1976

Federal Water Pollution Control Act of 1977 (Clean Water Act, Public Law 95-217, December 17, 1977)

In 1977 Congress adopted comprehensive amendments of existing water quality legislation and restated congressional policy:

> The authority of each State to allocate quantities of water within the jurisdiction shall not be superseded, abrogated or otherwise impaired by this Act. It is the further policy of Congress that nothing in this Act shall be construed to supersede or abrogate rights to quantities of water which have been established by any State. Federal Agencies shall cooperate with State and local agencies to develop comprehensive solutions to prevent, reduce and eliminate pollution in concert with programs for managing water resources.

The continued development of information and the control and abatement of water pollution was a major objective of the new act. Measures and research topics included:

1. Training grants
2. Rural village study
3. Research and demonstration projects
4. Recreation
5. Water conservation
6. Allotment
7. State management assistance
8. Irrigation
9. Wastewater storage
10. Management practices
11. Toxic pollutants
12. Oil spills
13. Clean lakes
14. Aquaculture
15. Utilization of treated sewage
16. Seafood processing

A number of sections require elaboration:

Water Conservation To encourage water conservation, service charges to industrial users may be reduced if the total flow of sewage or unnecessary water consumption is reduced.

Toxic Pollutants All toxic pollutants listed on the Effluent Standards List are subject to effluent limitations, usually by the application of the best technology economically achievable.

Oil Spills Congress recognized the continued control of oil spills as determined under the Outer Continental Shelf Lands Act and the Deepwater Port Act of 1974. These acts provided guidance to prevent and control oil spills when drilling for oil in offshore waters and for the docking and unloading of oil tankers in ports.

Aquaculture Guidelines were established for the development of aquaculture.

This legislation was subsequently amended by:

Public Law 95-576, November 2, 1978

Public Law 96-478, October 21, 1980

Public Law 96-483, October 21, 1980

Public Law 97-35, August, 13, 1981

Public Law 97-119, December 29, 1981

Public Law 97-164, April 2, 1982

Public Law 97-357, October 19, 1982

Public Law 97-440, January 8, 1983

Public Law 100-4, February 4, 1987

Public Law 100-202, December 22, 1987

Public Law 100-236, January 8, 1988

Public Law 100-581, November 1, 1988

Public Law 100-653, November 14, 1988

Public Law 100-688, November 18, 1988

Public Law 101-380, August 18, 1990

Public Law 101-596, November 16, 1990

Water Quality Act of 1987 (Public Law 100-4, February 4, 1987)

The Water Quality Act of 1987 was specifically directed to water quality on the Chesapeake Bay and Great Lakes.

An existing Chesapeake Bay program was continued with the EPA providing an additional office to aid the program. In addition, the program must:

1. Coordinate and make available, through publications and other means, information pertaining to the environmental quality of the Chesapeake Bay.
2. Coordinate federal and state efforts to improve the water quality of the bay.
3. Determine the impact of sediment deposition in the bay and identify its sources, rates, routes, and distribution patterns.
4. Determine the impact of natural and artificially induced environmental changes on the living resources of the bay and the relationships among these changes, with particular emphasis on the impact of pollutant loadings of nutrients, chlorine, acid precipitation, dissolved oxygen, and toxic pollutants, including organic chemicals and heavy metals with special attention to be given to the impact of changes on striped bass.

This amendment also addressed the Great Lakes program, intending to achieve the goals embodied in the Great Lakes Water Quality Agreement of 1978 through improved organization and definition of purpose on the part of the EPA, funding of state grants for pollution control in the Great Lakes area, and improved accountability for implementation of such agreements. Congress found:

1. The Great Lakes are a valuable natural resource, continuously serving the people of the United States and other nations as an important source of food, fresh water, recreation, beauty and enjoyment.
2. The Environmental Protection Agency should take the lead in the effort to meet these goals, working with other federal and state agencies.

Congress also appropriated funds to implement the act and addressed:

1. The protection of underground sources of drinking water.
2. Restrictions on underground injection of hazardous waste.
3. State water programs.
4. Enforcement of water laws.
5. Aquifer demonstration programs.
6. Emergency power.
7. State programs to establish wellhead protection areas.
8. Indian water rights.
9. Judicial review of water laws and policies.

State Groundwater Control

About 1980 the states began to recognize that the subsurface was a major source of permanent water. As a result many states passed legislation to protect the availability and quality of their groundwater. These endeavors include:

1. Mapping aquifers, in cooperation with the U.S. Geological Survey, to determine quantity and characteristics of the aquifer.
2. Developing classification systems of aquifers based on their natural characteristics and use.

3. Establishing water quality standards.
4. Regulating the number of wells, well construction, and production of water from wells.
5. Identifying types of sources of groundwater from aquifers.

State groundwater quality protection was proposed by the Environmental Law Institute in 1984 (T. R. Henderson, J. Trauberman, and T. Gallagher, *Groundwater Strategies for State Action.* Washington, DC: Environmental Law Institute, 1984). There are three major steps to creating an effective overall groundwater protection program:

1. Adoption of a groundwater policy.
2. Drafting groundwater management strategies.
3. Implementing specific techniques for obtaining groundwater protection.

Reconciling groundwater policies with competing interests such as economic development has resulted in three major categories of state groundwater policies:

1. Nondegradation assurances to maintain groundwater at its highest quality for human consumption.
2. Limited degradation allowances in recognition that some degradation may have already occurred, but groundwater must be protected if future use is to be obtained.
3. Differential protection, aimed at satisfying the demand for groundwater by selecting only critical or high-quality aquifers for regulation.

State management strategies typically include one or more of the following:

1. Classification of aquifers as to present and future uses, or to natural conditions, with aquifers then treated according to their designated purpose and category.
2. Classification of contaminants and contaminant sources to determine seriousness of groundwater pollution problems and set priorities for control.
3. Uniform management, with controls that are applied equally statewide.
4. Recharge zone protection, recognizing the need for special protection of areas important for groundwater replenishment or recharge.

The third step of developing specific techniques for correcting or controlling groundwater contamination is the core of a groundwater program. These techniques usually include one or a combination of the following:

1. Groundwater quality standards, generally defining the maximum concentration of contaminants consistent with maintaining a desired level of quality.
2. Control of sources of contamination by permits, effluent limitations, discharge zones, facility design standards, and required management practices.
3. Restricting use of groundwater to protect quantity and quality.
4. Land use controls governing land overlying groundwater resources, including zoning, siting requirements, and public acquisition of land.
5. The removal of contaminants that are a threat to groundwater quality.

Supplementary measures include the consolidation of existing information and a system for collecting missing information, coordination of the various state agencies, developing a funding mechanism, and the adoption of groundwater monitoring and enforcement procedures.

Forty-eight states have developed laws and regulations regarding state groundwater protection; the following five programs illustrate the types of regulations that have been enacted.

Florida

The Florida Water Resources Act of 1972 provided authority for managing the state's water through five water management districts under the Florida Department of Environmental Regulation. The Water Quality Assurance Act of 1983 emphasized water quality and withdrawals from all sources. A statewide network with a centralized data base has been established to process information. Each of the districts currently has an ongoing cooperative program with the U.S. Geological Survey.

Florida has a classification system for both aquifers and injection wells. The Florida Department of Environmental Regulation controls the permitting, operating, and monitoring of these wells, as well as their plugging and abandonment. Injection wells are divided into five classes:

Class 1 Wells used to inject hazardous wastes beneath the lowermost rock formation containing water within a quarter mile of the well core or any underground source of drinking water.

Class 2 Wells used to reinject fluids brought to the surface in connection with conventional natural gas production, which may be commingled with wastewater from gas plants unless these waters are classified as hazardous wastes at the time of injection.

Class 3 Wells used to inject substances for extraction of minerals, including the mining of sulfur by the Frosche process and solution mining of minerals.

Class 4 Wells used to inject certain hazardous wastes or radioactive wastes into or above a formation that is within a quarter mile of wells containing either an underground source of drinking water or an exempted aquifer.

Class 5 Injection wells not in Class 1, 2, 3, or 4. These include air conditioning return flow wells used to return to an aquifer the water used for heating or cooling industrial or residential buildings; recharge wells used to replenish, augment, or store water in an aquifer to prevent salt intrusion; wells that are a part of the waste treatment system; dry wells used for the injection of wastes into a subsurface formation; and drainage wells used to drain surface fluids, primarily storm runoff or high lake levels.

Florida regulates construction of the various types of wells and imposes operating, monitoring, and reporting requirements. Permits are required for construction, operation, plugging, and abandonment of wells.

Pennsylvania

In Pennsylvania water management policies and practices are under the supervision of the Department of Environmental Resources (DER), guided by a comprehensive State Water Plan. At present Pennsylvania statutes provide that each landowner has a right to make reasonable use of groundwater below the surface, and no comprehensive program exists to allocate or provide for long-term management of groundwater. Research on groundwater is conducted by several DER offices and the U.S. Geological Survey.

Five specific statutes focus on groundwater aspects of water resource management. The Clean Stream Law is designed to control and prevent pollution of all state waters. It includes surface and underground water and prohibits the discharge of sewage or industrial wastes unless authorized by permit and the DER. The Water Well Drillers License Act controls the drilling of wells.

The Delaware River Basin and the Susquehanna River Basin commissions play a major role in managing the groundwater of the eastern two-thirds of Pennsylvania and portions of Virginia, New Jersey, Delaware, and Maryland. These two agencies review proposed groundwater uses that would "substantially affect development, conservation, management and control of water resources in the Basins." Both of these commissions limit their financial resources to projects withdrawing more than 100,000 gallons per day.

The Groundwater Protection Program is intended to improve management of a 1,500 square mile section of predominantly Triassic lowland formation in southeastern Pennsylvania. The withdrawal of groundwater is carefully regulated to accomplish the most effective utilization of the resource. The state also has a number of programs to control water quality, for example limiting concentrations of organisms and organic substances that may be dumped.

California

The California Department of Water Resources is the state's agency concerned with water resource development; it also provides data and information to other agencies. Water quality standards are established by the State Water Resources Control Board and nine regional boards. The Department of Health supervises the quality of drinking water. The U.S. Geological Survey has cooperative programs for data collection and hydrologic investigations. In 1987, pursuant to Proposition 65, an initiative measure passed by the voters, the state began to prepare a list of chemicals known to cause cancer or reproductive toxicity.

As the population and economy of California have grown, the availability of water has become a critical problem. Conflicts over water consumption have increased, particularly in southern California. The Water Resources Control Board has authority to restrict consumption and other practices such as groundwater overdraft, seawater intrusion, land subsidence, artificial recharge, and conjunctive use of groundwater.

The Department of Water Resources has surveyed the water resources of California and established 42 groundwater basins, 11 of which have experienced a major overdraft. In the coastal basins, sea-water intrusion has become a significant problem, and these areas are subject to management programs.

In spring 1991, water controls were imposed in southern California due to a five-year drought. Artificial recharge of groundwater has been used in southern California since the 1920s due to deficiencies of precipitation. An interesting variation on artificial recharge is "in-lieu replenishment," where imported water is used directly by the user in return for reduction of groundwater withdrawal by an equivalent amount.

New Mexico

The first groundwater regulations were enacted by the state legislature in 1851. In 1931 the state legislature established a permit system for the appropriation of groundwater. Groundwater use is regulated by the New Mexico state engineers in 31 water basins covering about 69 percent of the state's area. The quality of the water is under the protection of the Water Quality Control Commission. The various water programs are coordinated by the New Mexico Environmental Improvement Division.

New Mexico is particularly concerned with the total dissolved sediment (TDS). Groundwater with a TDS content greater than 10,000mg/l is considered nonpotable. New Mexico's management strategy is primarily aimed at the injection of wastes into wells and extraction from wells. There are controls over well drilling and abandonment.

Nebraska

The state of Nebraska has several agencies engaged in groundwater research, planning, regulation, and management. The Conservation and Survey Division of the University of Nebraska's Institute of Agriculture and Natural Resources has the responsibility of maintaining a natural resource data base and conducting research on natural resources. The State Water Research Institute is the Water Resource Center at the University of Nebraska at Lincoln. The Nebraska Natural Resources Commission is the state's agency for water planning and resource development. The Nebraska Department of Water Resources is responsible for regulating programs and the Department of Environmental Control (DEC) is responsible for the protection and improvement of water quality. Twenty-four natural resource districts coordinate land and water management programs with other government entities. Water conservation activities include monitoring water levels and groundwater quality and managing groundwater control areas.

Groundwater is the major source of water in the state, which is therefore concerned that the quality be maintained. DEC has wide-ranging regulatory authority to prevent groundwater pollution and protect water quality, including:

1. Developing comprehensive programs to prevent and control pollution.
2. Issuing orders prohibiting or abating groundwater pollution and requiring adoption of remedial measures to prevent, control, or abate pollution.
3. Issuing permits consistent with Environmental Control Council regulations to prevent, control, and abate pollution.
4. Requiring any person engaged in operations that may pollute groundwater to obtain a permit prior to continuing operations.
5. Litigation against, and fining of polluters to prevent or control groundwater pollution.
6. Delegating administration of groundwater pollution control programs to local governments with DEC control of the local program.

Irrigation

Desert Land Act of 1877 (Chapter 107, March 3, 1877)

The Desert Land Act of 1877 authorized citizens of the United States to purchase land, not exceeding one section, at 25 cents per acre in selected desert states and territories, on condition of conducting water upon it to develop agriculture. The law further stated that

> such right shall not exceed the amount of water actually appropriated, and necessarily used for the purpose of irrigation and reclamation; and all surplus water over and above such actual appropriation and use, together with the water of all lakes, rivers, and other sources of water supply upon the public lands and not navigable shall remain and be held free for the appropriation and use of the public for irrigation, mining, and manufacturing purposes subject to existing rights.

Carey Act of 1894 (An Act to Provide for the Sale of Desert Land in Certain States and Territories, Chapter 301, August 14, 1894)

The primary purpose of the Carey Act was to grant up to 1 million acres of land to the states for irrigation and reclamation. The intent of Congress was that these lands would be acquired by settlers in tracts not to exceed 160 acres to raise agricultural crops; it was hoped that desert lands within the states would be speedily reclaimed and made productive by irrigation. To secure the land, the state had to prepare a map of lands proposed to be irrigated and the plan of irrigation.

Reclamation Act of 1902 (An Act appropriating the receipts from the sale and disposal of public lands in certain States and Territories for the construction of irrigation works for the acclamation of arid lands, Chapter 1093, June 17, 1902)

The Reclamation Act of 1902 established a reclamation fund received from the sale of public land in 16 western states and territories. The funds were to be used to examine and survey areas for constructing and maintaining irrigation works for the purpose of storage, diversion, and development of water resources in order to reclaim arid and semiarid lands. Implementation was to be carried out by the Department of the Interior.

Act amending this act:

Chapter 192, May 20, 1920

Reclamation Act of 1911 (Chapter 32, February 2, 1911)

The 1911 Reclamation Act expanded the 1902 act to provide for the sale of irrigation works not needed for the purpose for which they were acquired; funds received from such sales were to be added to the reclamation fund.

The reclamation acts were further amended by:

Chapter 155, February 24, 1911 Chapter 247, August 14, 1914

Reclamation Project Act of 1939 (Chapter 418, August 4, 1939)

The Reclamation Project Act of 1939 called for feasible and comprehensive plans for economical and equitable treatment of repayment problems for construction charges for reclamation projects during periods of decline in agricultural income and unsatisfactory conditions of agriculture. "Project" was defined as any reclamation or irrigation project authorized by federal reclamation laws.

The reclamation acts were further amended by:

Chapter 94, August 24, 1945

Chapter 752, August 18, 1950

Chapter 256, May 10, 1956

Public Law 85-611, August 8, 1958

Public Law 86-308, September 21, 1959

Public Law 87-613, August 28, 1962

Public Law 93-608, January 2, 1975

Public Law 97-293, October 12, 1982

Reclamation Project Authorization Act of 1972 (Public Law 92-514, October 20, 1972)

The Reclamation Project Authorization Act of 1972 established the following projects:

1. Closed Basin Division, San Luis Valley Project, Colorado
2. Brantley Project, Pecos River Basin, New Mexico
3. Salmon Falls Division, Upper Snake River Project, Idaho
4. O'Neill Unit, Pick-Sloan Missouri Basin Program, Nebraska
5. North Loup Division, Pick-Sloan Missouri Basin Program, Nebraska

These projects were to develop the water resources of surrounding areas. The charge was to save water without flooding surrounding lands; to minimize loss of water through evaporation, transpiration, and seepage; to develop drainage projects; to furnish irrigation; to provide industrial and municipal water; to provide a wildlife refuge; to provide outdoor recreational opportunities; and to use the water for other useful purposes.

A further amendment of the reclamation laws was Public Law 93-608 (January 2, 1975).

The Reclamation Reform Act of 1982 (Public Law 97-293, October 12, 1982)

The Reclamation Reform Act of 1982 amended the original reclamation law to regulate the use of irrigation water in more detail. Under this legislation, water was not to be delivered to:

1. A qualified recipient for use in the irrigation of lands in excess of 960 acres of Class I lands or the equivalent.
2. A limited recipient for the use in the irrigation of lands in excess of 640 acres of Class I lands or the equivalent.

Any contract with an irrigation district entered into by the Secretary of the Interior had to provide for the delivery of irrigation water at field cost, and users of the water must certify to the Secretary that they are in compliance with the law.

Irrigation water made available in the operation of reclamation project facilities were not to be delivered for use in the irrigation of lands held in excess of the ownership limitations unless:

1. The disposal of the owner's interest in excess lands was required by an existing contract with the secretary.
2. The owners of the lands requested that a recordable contract be executed by the secretary.

The act also encouraged water conservation, and authorized the Secretary of the Interior to enter into agreements to develop plans, define goals, and provide for involvement of nonfederal agencies such as states, Indian tribes, and water use organizations to ensure full participation in water conservation efforts.

Further amendments of the reclamation laws included:

Public Law 97-570, October 30, 1984

Public Law 100-203, December 22, 1987

Public Law 100-516, October 24, 1988

Flood Control

Flood control acts were passed in 1917, 1928, 1936, 1937, 1938, 1941, 1944, 1946, 1948, 1950, 1954, 1958, 1960, 1962, 1965, 1966, 1968, and 1970. Altogether between 1917 and 1970, 18 flood control acts were passed, with 75 amendments. For example, the 1941 Act was amended 11 times between 1946 and 1987. Most of these acts concerned controlling floods on particular rivers or river basins.

Flood Control Act of 1917 (Chapter 144, March 1, 1917)

The initial flood control act passed in 1917 was specifically concerned with controlling floods on the Mississippi and Sacramento rivers and by continuing improvements to control floods on approval by the Mississippi River Commission and the California Debris Commission. The act required the preparation of surveys to show:

1. The extent and character of the area to be affected by proposed improvements.
2. The probable effect upon any navigable water or waterways.
3. The possible economical development and utilization of water power.
4. Other uses that might properly be related to or coordinated with the project.

Projects were to be submitted to the Board of Engineers for Rivers and Harbors for ranking as to importance and recommendations about construction. The board was to report its findings as to:

1. What federal interest, if any, was involved in the proposed improvement.
2. What share of the expenses, if any, should be borne by the United States and other sources such as state and local governments.
3. The advisability of adopting the project.

The act coordinated the work of the Board of Engineers for Rivers and Harbors, the Committee on Rivers and Harbors, and the Committee on Flood Control.

Flood Disaster Protection Act of 1973 (Public Law 93-234, December 31, 1973)

The Flood Disaster Protection Act of 1973 expanded the national flood insurance program by substantially increasing coverage limits and the total amount of insurance authorized to remain outstanding. It required known flood-prone communities to participate in the program.

In passing the act, Congress found:

1. Annual losses throughout the nation from floods and mud slides are increasing at an alarming rate, largely as a result of the accelerating development of, and concentration of, population in areas of flood and mud slide hazards.
2. The availability of federal loans, grants, guarantees, insurance, and other forms of financial assistance are often determining factors in the utilization of land and the location and construction of public and of private industrial, commercial, and residential facilities.
3. Property acquired or constructed with grants or other federal assistance may be exposed to risk of loss through floods, thus frustrating the purpose for which such assistance was extended.
4. Federal institutions insure or otherwise provide financial protection to banking and credit institutions whose assets include a substantial number of mortgage loans and other indebtedness secured by property exposed to loss and damage from floods and mud slides.
5. The nation cannot afford the tragic losses of life caused annually by flood occurrences, nor the increasing losses of property suffered by flood victims, most of whom are still inadequately covered by insurance.
6. It is in the public interest for persons already living in flood-prone areas to have both an opportunity to purchase flood insurance and access to more adequate coverage so that they will be indemnified for the losses in the event of future flood disasters.

The purpose of the act was, therefore, to:

1. Substantially increase the limits of coverage authorized under the national flood insurance program.
2. Provide for the expeditious identification of, and the dissemination of information concerning, flood-prone areas.

3. Require states or local communities, as a condition of future federal financial assistance, to participate in the flood insurance program and to adopt adequate flood plain ordinances with effective enforcement provisions consistent with federal standards to reduce or avoid future flood losses.
4. Require the purchase of flood insurance by property owners who are being assisted by federal programs or by federally supervised, regulated, or insured agencies or institutions in the acquisition or improvement of land or facilities located or to be located in identified areas having special flood hazards.

Title I, Expansion of National Flood Insurance Program, funding the National Flood Insurance Act of 1968, and Title II discussed disaster mitigation requirements.

This act was subsequently amended by:

Public Law 94-50, July 2, 1975

Public Law 94-198, December 31, 1975

Public Law 94-375, August 3, 1976

Public Law 95-128, October 12, 1977

Public Law 98-181, November 30, 1983

Public Law 98-479, October 17, 1984

Public Law 100-707, November 23, 1988

Watershed Protection and Flood Prevention Act of 1954 (Public Law 566, August 4, 1954)

The Watershed Protection and Flood Prevention Act of 1954 was passed by Congress because:

> Erosion, floodwater, and sediment damage in the watersheds of the rivers and streams of the United States, causing loss of life and damage to property, constitute a menace to the national welfare; and . . . it is the sense of Congress that the Federal Government should cooperate with States and their political subdivisions, soil and water conservation districts, flood prevention and control districts, and other local public agencies for the purpose of preventing such damages and of furthering the conservation, development, utilization and disposal of water and thereby of preserving and protecting the Nation's land and water resources.

To assist local organizations in preparing and carrying out plans for works of improvement the Secretary of Agriculture, working with state agencies, had the responsibility:

1. To conduct investigations and surveys necessary to prepare plans for works of improvement.
2. To make studies necessary for determining the physical and economic soundness of plans for works of improvement, including a determination as to whether benefits exceed costs.
3. To cooperate and enter into agreements with and to furnish financial and other assistance to local organizations, provided that for the land-treatment measures, federal assistance was not to exceed the rate of assistance for similar practices under existing national programs.
4. To obtain the cooperation and assistance of other federal agencies in carrying out the purposes of the act.

Federal assistance was to be provided after the Secretary of Agriculture has determined that the local organizations would:

1. Acquire without cost to the federal government land, easements, or rights-of-way needed in connection with works of improvement installed with federal assistance.
2. Assume a proportionate share of the cost of installing any works of improvement involving federal assistance as determined by the Secretary of Agriculture to be equitable in consideration of anticipated benefits.
3. Make arrangements satisfactory to the secretary for defraying costs of operating and monitoring works of improvement in accordance with regulations presented by the Secretary of Agriculture.
4. Acquire, or provide assurance that landowners have acquired, such water rights, pursuant to state law, or any water rights which may be needed in the installation and operation of the work of improvement.
5. Obtain agreements to carry out recommended soil conservation measures and proper farm plans from owners of not less than 50 percent of the lands situated in the drainage areas above each retention reservoir to be installed with federal assistance.

The act was subsequently amended by:

Public Law 639, July 19, 1956

Public Law 1027, August 7, 1956

Public Law 85-624, August 12, 1958

Public Law 85-865, September 2, 1958

Public Law 86-468, May 13, 1960

Public Law 86-545, June 29, 1960

Public Law 87-170, August 30, 1961

Public Law 87-703, September 27, 1962

Public Law 89-337, November 8, 1965

Public Law 90-361, June 27, 1968

Public Law 92-419, August 30, 1972

Public Law 95-113, September 19, 1977

Public Law 97-98, December 22, 1981

Public Law 99-662, November 17, 1986

Public Law 101-624, November 28, 1990

National Flood Insurance Act of 1968 (Public Law 90-448, August 1, 1968)

The National Flood Insurance Act of 1968 was Title XIII of the Housing and Urban Development Act of 1968. In passing this legislation, Congress found that (1) from time to time flood disasters have created personal hardships and economic distress which have required unforeseen disaster relief measures and have placed an increasing burden on the Nation's resources, (2) despite the installation of preventive and protective works and the adoption of other public programs designed to reduce losses caused by flood damage, these methods have not been sufficient to protect adequately against growing exposure to future flood losses, and, (3) as a matter of national policy, a reasonable method of sharing the risk of flood losses is through a program of flood insurance to complement and encourage preventive and protective measures.

Congress also found that many factors have made it uneconomic for the private insurance industry to make flood insurance available to those in need of protection on reasonable terms and conditions, but that a flood insurance program with large-scale participation of the federal government was feasible and could be initiated if carried out to the maximum extent practicable by the private insurance industry.

It was the purpose of this act to:

1. Authorize a flood insurance program by means of which flood insurance, over a period of time, could be made available nationwide through the cooperative efforts of the federal government and the private insurance industry.
2. Provide flexibility in the program so that flood insurance could be based on workable methods of pooling risks, minimizing costs, and distributing burdens equitably among those who would be protected by flood insurance and the general public.
3. Encourage state and local governments to make appropriate land use adjustments to prohibit the development of land exposed to flood damage and minimize damage caused by flood losses.
4. Guide the development of proposed future construction away from locations threatened by flood hazards.
5. Encourage lending and credit institutions, as a matter of national policy, to assist in furthering the objectives of the flood insurance program.
6. Assure that any federal assistance provided under the program be related closely to all other flood-related programs and activities of the federal government.
7. Authorize continuing studies of flood hazards in order to provide for a constant reappraisal of the flood insurance program and its effect on land use requirements.

Subsequent amendments included:

Public Law 92-213, December 22, 1971

Public Law 93-234, December 31, 1973

Public Law 94-173, December 23, 1975

Public Law 94-375, August 3, 1976

Public Law 95-60, June 30, 1977

Public Law 95-80, July 31, 1977

Public Law 95-128, October 12, 1977

Public Law 95-406, September 30, 1978

Public Law 95-557, October 31, 1978

Public Law 96-153, December 21, 1979

Public Law 96-399, October 8, 1980

Public Law 96-470, October 19, 1980

Public Law 97-35, August 13, 1981

Public Law 97-289, October 6, 1982

Public Law 97-348, October 15, 1982

Public Law 98-35, May 26, 1983

Public Law 98-109, October 1, 1983

Public Law 98-181, November 30, 1983

Public Law 98-479, October 17, 1984

Public Law 99-120, October 8, 1985

Public Law 99-156, November 15, 1985

Public Law 99-219, December 26, 1985

Public Law 99-267, March 27, 1986

Public Law 99-272, April 7, 1986

Public Law 99-345, June 24, 1986

Public Law 99-430, September 26, 1986

Public Law 99-450, October 6, 1986

Public Law 100-122, September 30, 1987

Public Law 100-154, November 5, 1987

Public Law 100-170, November 17, 1987

Public Law 100-200, December 21, 1987

Public Law 100-242, February 5, 1988

Public Law 100-628, November 7, 1988

Public Law 100-707, November 23, 1988

Public Law 101-137, November 3, 1989

Public Law 101-508, November 5, 1990

Public Law 101-591, November 16, 1990

Public Works

Water Resources Development Act of 1974 (Public Law 93-251, March 7, 1974)

The Water Resources Development Act of 1974 and its later amendments authorized the construction, repair, and preservation of selected public works on rivers and harbors for navigation, flood control, and other purposes. Scores of water resources development projects have been completed under these acts.

Flood control projects received special attention in these acts, with particular concern about the environmental quality of the regions involved, including but not limited to salinity intrusion, dispersion of pollutants, water quality, improvements for navigation, dredging, physical structures, bay fill, and other shoreline changes.

The act also authorized a national stream bank erosion prevention and control demonstration program. The program consisted of:

1. An evaluation of the extent of stream bank erosion on navigable rivers and their tributaries.
2. Development of new methods and techniques for bank protection.
3. Means for the prevention and correction of stream bank erosion.
4. Demonstration projects including bank protection works.

Demonstration projects were undertaken on streams selected to reflect a variety of geographical and environmental conditions, including streams with naturally occurring erosion problems and streams with erosion caused or increased by artificial structures.

In the development of these projects, nonfederal interests had to agree to provide lands, easements, and rights-of-way necessary for construction and subsequent operation of the works; hold the federal government free from damages due to construction, operation, and maintenance of the works; and operate and maintain the works upon completion.

In addition to flood control, the act authorized projects to:

1. Develop hurricane control.
2. Prevent loss of wildlife grazing areas.
3. Improve navigation.
4. Construct new bridges and relocate existing bridges.
5. Construct fish hatcheries.

Projects built pursuant to these acts were to be coordinated with work under the various flood control acts. Subsequent amendments to the 1974, 1976, 1986, 1988, and 1990 Water Resource Development Acts included:

Water Resources Development Act of 1976

Public Law 94-587, October 22, 1976

Public Law 99-88, August 15, 1985

Water Resources Development Act of 1986

Public Law 99-662, November 17, 1986

Public Law 100-71, July 11, 1987

Water Resources Development Act of 1988

Public Law 100-676, November 17, 1988

Public Law 100-707, November 23, 1988

Water Power

Federal Water Power Act of 1920 (Chapter 285, June 10, 1920)

The Federal Water Power Act of 1920 established the Federal Power Commission, composed of the secretaries of War, the Interior, and Agriculture. The commission was charged:

1. To make investigations and to collect and record data concerning the utilization of the water resources of any region to be developed. This included the water power industry and its

relation to other industries and to interstate or foreign commerce. Other responsibilities concerned the location, capacity, development costs, and relations to markets of power sites; whether the power from government dams could be advantageously used by the United States for its public purposes; and what was the fair value of the power.

2. To cooperate with the executive departments and other agencies of state and federal government in such investigations.
3. To make public from time to time such information as secured and provide for its publication.
4. To issue licenses to citizens or corporations for the purpose of constructing, operating, and maintaining dams, water conduits, reservoirs, power houses, transmission lines, and other project works necessary for the improvement of navigation and for the development, transmission, and utilization of power from the navigable waters of the United States, or for the purpose of utilizing the surplus water or water power from any federal dam.
5. To implement procedures for the development of the water power resources of the nation. The licenses issued under this act were limited to a period not exceeding 50 years.

When a license for a project was requested of the commission, the applicant for a license was to:

1. Prepare maps, plans, specifications, and estimates of cost as may be required for a full understanding of the proposed project.
2. Provide satisfactory evidence that the requirements of the laws of the states within which the project was to be located were fulfilled.
3. Provide information that the comprehensive scheme of development of water power and navigation was best adapted to the area.

The act required that if a dam or other project works were to be constructed near, along, or in navigable waters, the commission was to promote the present and future needs of navigation. The license could require the licensee to preserve and improve navigation facilities and construct, without expense to the United States, in connection with a dam, a lock or locks, booms, sluices, or other structures for navigation purposes, in accordance with the plans of the Chief of Engineers and the Secretary of War.

The act was later amended by:

Chapter 129, March 3, 1921

Chapter 572, June 23, 1930

Chapter 687, August 26, 1935

Chapter 351, May 28, 1948

Chapter 343, August 7, 1953

Chapter 351, June 4, 1956

Public Law 85-791, August 28, 1958

Public Law 86-619, July 12, 1960

Public Law 87-647, September 7, 1962

Public Law 90-451, August 3, 1968

Public Law 91-452, October 15, 1970

Public Law 95-617, November 9, 1978

Public Law 96-294, June 30, 1980

Public Law 97-375, December 21, 1982

Public Law 99-495, October 16, 1986

Public Law 99-546, October 27, 1986

Public Law 100-473, October 6, 1988

Public Law 101-575, November 15, 1990

Recreation

Federal Water Projects Recreation Act of 1965 (Public Law 89-72, July 9, 1965)

The Federal Water Projects Recreation Act of 1965 established uniform policies with respect to recreation and fish and wildlife benefits and costs of federal multiple-purpose water resource projects. It thus became federal policy that:

1. In investigating and planning for any federal navigation, flood control reclamation, hydroelectric, or multi-purpose water resources project, full consideration shall be given to the opportunities, if any, the project affords for outdoors recreation and for fish and wildlife enhancement, and that wherever a project can reasonably serve either or both of these purposes consistent with other provisions of this act, it shall be constructed, operated, and maintained accordingly.
2. Planning with respect to the development of recreation potential shall be based on the coordination of the recreational use of the project area with the use of existing and planned federal, state, and local public recreation developments.
3. Project construction agencies shall encourage nonfederal public bodies to administer project land and water areas for recreation and fish and wildlife enhancement purposes.

This act was subsequently amended by:

Public Law 93-251, March 7, 1974

Public Law 94-576, October 21, 1976

Ocean Pollution

National Ocean Pollution Research and Development and Monitoring Planning Act of 1978 (Public Law 95-273, May 8, 1978)

The National Ocean Pollution Research and Development and Monitoring Planning Act of 1978 was based on the findings of Congress that:

1. Human activities in the marine environment can have a profound short- and long-term impact on that environment and greatly affect ocean and coastal resources.
2. There is a need to establish a comprehensive federal plan for ocean pollution research and development and monitoring, with particular attention to the impacts, fates, and effects of pollutants in the marine environment.

3. The human population will increasingly need to rely on ocean and coastal resources as other resources are depleted. Our ability to protect, preserve, develop, and utilize these ocean and coastal resources is directly related to our understanding of the effects ocean pollution has upon the resources.
4. Numerous departments, agencies, and instrumentalities of the federal government sponsor and support or fund activities relating to ocean pollution research and development and monitoring, but these activities are often uncoordinated, resulting is unnecessary duplication.
5. Better planning and more effective use of available funds, personnel, vessels, facilities, and equipment is the key to effective federal action regarding ocean pollution research and development.

To implement an appropriate plan Congress:

1. Established a comprehensive plan (initially for five years) for federal ocean pollution research, development, and monitoring programs in order to provide planning, or coordination and dissemination of information, with respect to those programs in the federal government.
2. Developed the necessary base of information to support, and to provide for, the rational, efficient, and equitable utilization, conservation, and development of ocean and coastal resources.
3. Designated the National Oceanic and Atmospheric Administration (NOAA) as the lead federal agency to prepare the plan and carry out a comprehensive program of ocean pollution research, development and monitoring under the plan.

This act was subsequently amended by:

Public Law 96-17, June 4, 1979

Public Law 96-255, May 30, 1980

Public Law 97-375, December 21, 1982

Public Law 99-272, April 7, 1986

Public Law 100-636, November 8, 1988

Marine Protection, Research and Sanctuaries Act of 1972 (Public Law 92-532, October 23, 1972)

The Marine Protection, Research and Sanctuaries Act recognized that unregulated dumping of material into ocean waters endangers human health, welfare, and amenities; the marine environment; ecological systems; and economic potentialities. Congress declared:

1. It is the policy of the United States to regulate the dumping of all types of materials into ocean waters and to prevent or strictly limit the dumping into ocean waters of any material which would adversely affect human welfare of the marine environment.
2. It is the purpose of this act to regulate the transportation of material from the United States for dumping into ocean waters, and the dumping of material, transported from outside the United States, if the dumping occurs in ocean waters over which the United States has jurisdiction, or over which it may exercise control, under accepted principles of international law, in order to protect its territory or territorial seas.

Title I of the act covered ocean dumping: no person was to transport from the United States any radiological, chemical, or biological warfare agent or any high-level radioactive waste, except wastes authorized in a permit issued under this title. When a permit is issued the following items must be considered:

1. The need for the proposed dumping.
2. The effects of the dumping on human health and welfare including economic, aesthetic, and recreational values.
3. The effects on fishery resources, plankton, fish, shellfish, wildlife, shorelines, and beaches.
4. The effects on marine ecosystems.
5. The persistence and permanence of the effects of the dumping.
6. Effects of dumping particular volumes and concentrations of such materials.
7. Appropriate locations and methods of disposal or recycling, including land-based alternatives.
8. Effect on alternate uses of oceans, such as scientific study, fishing and other living resource exploitation, and nonliving resource exploitation.
9. In designating recommended sites, the NOAA shall utilize wherever feasible locations beyond the edge of the continental shelf.

Title II initiated a comprehensive and continuing program of monitoring and research regarding the effects of the dumping of material into ocean waters or other coastal waters where the tide ebbs and flows, or into the Great Lakes or their connecting waterways. The Secretary of Commerce has responsibility for the implementation.

Title III provided for marine sanctuaries. The Secretary of Commerce was authorized to designate as marine sanctuaries those areas of the ocean waters as far seaward as the outer edge of the continental shelf, other coastal waters where the tide ebbs and flows, and the Great Lakes and their connecting waters that he or she determines necessary for the purpose of preserving or restoring conservation, recreational, ecological, or aesthetic values. A marine sanctuary can be waters lying within the territorial limits of any state.

This act was amended by:

Public Law 93-254, March 22, 1974

Public Law 93-472, October 26, 1974

Public Law 94-62, July 25, 1975

Public Law 94-326, June 30, 1976

Public Law 95-153, November 4, 1977

Public Law 96-332, August 29, 1980

Public Law 96-381, October 6, 1980

Public Law 96-470, October 19, 1980

Public Law 96-572, December 22, 1980

Public Law 97-16, June 23, 1981

Public Law 97-106, December 26, 1981

Public Law 97-375, December 21, 1982

Public Law 97-424, January 6, 1983

Public Law 98-498, October 19, 1984

Public Law 99-272, April 7, 1986

Public Law 99-499, October 17, 1986

Public Law 99-662, November 17, 1986

Public Law 100-4, February 4, 1987

Public Law 100-17, April 2, 1987

Public Law 100-536, October 28, 1988

Public Law 100-627, November 7, 1988

Public Law 100-688, November 18, 1988 (The Ocean Dumping of Sewage Sludge and Industrial Wastes Act of 1988, often called he Ocean Dumping Ban Act of 1988)

Public Law 101-593, November 16, 1990

Public Law 101-596, November 16, 1990

Public Law 101-605, November 16, 1990

Oil Pollution

Oil Pollution Act of 1990 (Public Law 101-380, August 18, 1990)

In the Oil Pollution Act of 1990, Congress established limitations on liability for damages resulting from oil pollution by creating a fund for the payment of damages. These damages must be considered to result from an unanticipated grave natural disaster of an exceptional, inevitable, and irreversible character which could not have been averted by the exercise of due care or foresight.

Congress recognized the need for international oil pollution prevention and removal and specifically referred to Prince William Sound and the Trans-Alaska Pipeline System. The Secretary of Commerce was required to establish a Prince William Sound Oil Spill Recovery Institute to carry out educational and demonstration projects designed to:

1. Identify and develop the best available techniques, equipment, and materials for dealing with oil spills in the arctic and subarctic marine environment.
2. Complement federal and state damage assessment efforts and determine, document, assess, and understand the long-range effects of the *Exxon Valdez* oil spill on the natural resources of Prince William Sound and its adjacent waters, and the environment, economy, and life-style and well-being of the peoples who are dependent on them. Committees of the

institute include the Scientific and Technical Committee, the Oil Terminal Facilities and Oil Tanker Operations Association, the Regional Citizens' Advisory Council, the Committee for Terminal and Oil Tanker Operations and Environmental Monitoring, and the Committee for Oil Spill Prevention, Safety, and Emergency Response.

Title VIII, the Trans-Alaska Pipeline System Reform Act of 1990, besides liability provisions establishes a presidential task force with responsibilities to conduct a comprehensive review of the pipeline to determine whether:

1. The holder of the federal and state right-of-way is in full compliance with applicable laws, regulations, and agreements.
2. The laws, regulations, and agreements are sufficient to prevent the release of oil from the pipeline and prevent other damage or degradation to the environment.
3. Improvements are necessary to the pipeline to prevent release of oil.
4. Improvements are necessary for the offshore oil spill response capabilities for the pipeline.
5. Improvements are necessary for the security of the pipeline.

Possibly most important was the establishment of the Oil Pollution Research and Development Program, including the Interagency Coordinating Committee on Oil Pollution Research. This committee is to coordinate a comprehensive program of oil pollution research, technology development, and demonstration among federal agencies. The Oil Pollution Research and Technology Plan must:

1. Identify agency roles and responsibilities.
2. Assess the current state of knowledge on oil pollution prevention, response, and mitigation technologies and effects of oil pollution on the environment.
3. Identify significant oil pollution research gaps, including an assessment of major technological deficiencies in responses to past oil discharges.
4. Establish research priorities and goals for oil pollution technology related to prevention, response, mitigation, and environmental effects.
5. Estimate resources needed to conduct oil pollution research.

6. Identify regional oil pollution research needs and priorities for a coordinated multidisciplinary program of research at the regional level.

The program is to provide for research, development, and demonstration of new or improved technology effective in preventing or mitigating oil discharges and protecting the environment, including:

1. Development of improved design for vessels and facilities, and improved operational practices.
2. Research, development, and demonstration of improved technologies to measure the hulkage of a vessel tank, prevent discharges from tank vents, prevent discharges during lighting and bunking operations, contain discharges on the deck of a vessel, prevent discharges through the use of vacuums in tanks, and otherwise contain discharges of oil from vessels and facilities.
3. Research and demonstrate new or improved systems of mechanical, chemical, biological, and other methods (including the use of dispersants, solvents, and bioremediation) for the recovery, removal, and disposal of oil, including evaluation of the environmental effects of the use of such systems.
4. Research and training in consultation with the National Response Team to improve industry and government abilities to quickly and effectively remove an oil discharge, including the long-term use, as appropriate, by the National Spill Control School in Corpus Christi, Texas.
5. Research to improve information systems for decision making, including the use of data for coastal mapping, baseline data, and other data related to the environmental effects of oil discharges and cleanup technologies.
6. Development of technologies and methods to protect public health and safety from oil discharges, including the population directly exposed to any oil discharges.
7. Development of technologies, methods, and standards for protecting research personnel, including training, adequate supervision, protective equipment, and decontamination procedures.
8. Research and development of methods to restore and rehabilitate natural resources damaged by oil discharges.
9. Research to evaluate the relative effectiveness and environmental impacts of bioremediation technologies.
10. Demonstration of a satellite-surveillance vessel traffic system in Narragansett Bay to evaluate the utility of such systems in

reducing the risk of oil discharges from vessel collisions and groundings in unfamiliar waters.

Another aspect of the research program included oil pollution technology evaluation, including:

1. Evaluation and testing of technologies developed independently of the research and development program established under this law.
2. The establishment of standards and testing protocols based on national standards to measure the performance of oil pollution prevention or mitigation technologies.
3. Controlled field tests to evaluate real-world application to oil discharge prevention or mitigation technologies.

The Committee on Oil Pollution Effects established a research program to monitor and evaluate the environmental effects of oil discharges, including:

1. The development of improved models and capabilities for predicting the environmental fate, transport, and effects of oil discharges.
2. Development of methods, including economic methods, to assess damages to national resources resulting from oil discharges.
3. Identification of types of ecologically sensitive areas at particular risk of oil discharges and the preparation of scientific monitoring and evaluation plans, one for each of several types of ecological conditions, to be implemented in the event of major oil discharges in such areas.
4. The collection of environmental baseline data in ecologically sensitive areas at particular risk to oil discharges where such data are insufficient.

4

Directory of Organizations

ORGANIZATIONS HAVE DEVELOPED AND EVOLVED from the local to the federal level because water is such a fundamental resource. In the United States the Environmental Protection Agency (EPA) is the major governmental organization. There are also a large number of governmental advisory committees.

Private organizations vary, from those having a wide variety of functions to those with a single objective. The organizations listed here were functioning in 1990, but because the number of organizations have increased rapidly in recent decades, some may discontinue functioning when their objectives are completed.

There are also many important international organizations. Many of these are located in Europe and have a specific purpose.

The following directory is arranged in three sections: Private Organizations in the United States, U.S. Government Organizations, and International Organizations.

The following sources list water organizations.

Encyclopedia of Associations, Vol. 1, National Organizations of the U.S., Detroit, MI: Gale Research Co., 1991.

Encyclopedia of Associations, Vol. 4, International Organizations, Detroit, MI: Gale Research Co., 1991.

Encyclopedia of Governmental Advisory Organizations, 1992/1993, 8th ed., Detroit, MI: Gale Research Co., 1991.

The United States Government Manual, 1990/91, Washington, DC: Government Printing Office, 1991.

Private Organizations in the United States

American Institute of Hydrology (AIH)
3416 University Avenue, SE
Suite 200
Minneapolis, MN 55414

Established: In 1981. Members: 850. Staff: 3. State groups: 48. Meetings: Semiannual conference.

Purpose: To enhance and strengthen hydrology as a profession.

Activities: Sets standards and procedures to certify qualified hydrologists. Certifies and registers professionals in all fields of hydrological sciences. Grants awards in ground water hydrology and surface water hydrology for outstanding contributions.

Committees: Continuing Education; Institute Development; International; and Inter-Society.

Publications: AIH Bulletin (quarterly), *Hydrological Science and Technology* (quarterly), *Registry of Professional Hydrologists and Hydrogeologists* (annual), *Recent Advances in Ground Water Hydrology* (book), reports of workshops.

American Society of Irrigation Consultants (ASIC)
Four Union Square
Suite C
Union City, CA 94587

Established: In 1970. Members: 185. Regional groups: 5. Meetings: Annual.

Purpose: To advance the field of irrigation design and consultation and to promote ethical practices in the profession.

Activities: Monitors licensing and offers classes. Gives awards.

Committees: Uniform Standards.

Publications: ASIC Newsletter (quarterly), *Membership Roster* (annual), *Irrigation Design Standards.*

American Water Resources Association (AWRA)
5410 Grosvenor Lane
Suite 220
Bethesda, MD 20814

Established: In 1964. Members: 4,000. Staff: 1. State groups: 39. Members from 62 nations. Meetings: Annual.

Purpose: To advance water resources research, planning, development, and management.

Activities: Collects and disseminates ideas and information about water resources science and technology. Presents annual awards. Conducts symposia.

Committees: None.

Publications: Membership Directory—Who's Who in AWRA (annual), *Hydata—News and Views* (bimonthly), *Water Resources Bulletin* (bimonthly), *Coastal Water Resources, Groundwater Contamination and Reclamation, Water Resources Related to Mining and Energy—Preparing for the Future* (books), monographs.

American Water Works Association (AWWA)
6666 W. Quincy Avenue
Denver, CO 80235

Established: In 1881. Members: 45,000. Staff: 87. Regional groups: 10. State groups: 41. Meetings: Annual conference.

Purpose: To inform water utility mangers, boards of health, government officials, consultants, and others interested in public water supply of developments in the field.

Activities: Develops standards, supports research programs design, construction, operation, and management of waterworks, conducts in-service training schools, and prepares manuals for personnel. Maintains hall of fame and biographical archives. Operates placement service, sponsors competitions, and presents awards. Maintains a technical library and information center.

Committees: None.

Publications: AWWA Journal (monthly), *Mainstream* (monthly), *Membership Roster* (quadrennial): *Officers and Committee Directory* (annual), *OpFlow* (monthly), *Washington Report* (monthly), *Waterworld News* (bimonthly), catalogs, manuals, and reference books.

Association of Ground Water Scientists and Engineers (AGWSE)
6375 Riverside Drive
Dublin, OH 43017

Established: In 1963. Members: 9,000. Staff: 60. Meetings: Annual.

Purpose: To provide leadership and guidance for scientific, economic, and beneficial groundwater development and to promote the use, protection, and management of groundwater resources.

Activities: Holds educational seminars, short courses, symposia, programs, and field research projects. Offers a speakers' bureau. Operates a museum, sponsors competitions, compiles statistics. Maintains a library of 15,000 volumes.

Committees: Aquifer Protection; Certification; Regulatory.

Publications: Membership Directory (triennial), *Ground Water* (bimonthly), *Ground Water Monitoring Review* (quarterly), *Newsletter* (bimonthly).

Association of State and Interstate Water Pollution Control Administrators (ASIWPCA)
444 N. Capitol Street, NW
Suite 330
Washington, DC 20001

Established: In 1960. Members: 63. Meetings: Twice yearly, in August and February.

Purpose: To coordinate information regarding prevention and control of water pollution among state agencies and Congress, EPA, and other federal agencies.

Activities: Does research and maintains speakers' bureau.

Task Forces: Compliance, Groundwater, Municipal Assistance, Nonpoint Sources, Water Quality Management.

Publications: Annual Report, ASIWPCA position statements (periodic), *Membership Directory* (annual), brochures and conference proceedings.

Association of State Dam Safety Officials (ASDSO)
P.O. Box 55270
Lexington, KY 40555

Established: In 1984. Members: 401. Staff: 2. Meetings: Annual conference.

Purpose: To exchange ideas on dam safety problems and give assistance when needed. Provides state dam safety programs. Presents state interests before Congress and federal agencies. Promotes interstate cooperation.

Activities: Acts as a clearinghouse. Compiles statistics. Presents awards. Maintains a small collection of books. Holds technical training workshops.

Committees: Legal and Liability; Legislative Activities; Resolutions; Scholarship; Technical.

Publications: Proceedings (annual), *ASDSO Newsletter* (quarterly), *Membership Directory* (annual), Model State Dam Safety Program.

Clean Water Action (CWA)
317 Pennsylvania Avenue, SE
Suite 200
Washington, DC 20003

Established: In 1971. Meetings: Biennial.

Purpose: To ensure strong pollution controls and safe drinking water.

Activities: Works to ensure toxic protection for communities, preserve nation's wetlands, promote methods to recycle waste. CWA was influential in the passage of the 1986 Superfund for Toxic Cleanup (CERCLA).

Committees: None.

Publications: CWA News (quarterly).

Clean Water Fund (CWF)
c/o David Zwick
317 Pennsylvania Avenue, SE
Washington, DC 20003

Established: In 1974. Staff: 15. State groups: 17. Nonmembership.

Purpose: To promote public interest and involvement in problems related to water, toxic materials, and natural resources by protecting the environment, marine life, and groundwater by restricting use of pesticides.

Activities: Aims to help solve environmental problems. Provides research and assistance to environmental groups. Scrutinizes government agencies, public officials, and others for compliance with environmental protection laws and regulations. Compiles statistics.

Committees: None.

Publications: None.

Colorado River Association (CRA)
417 S. Hill Street
Room 1024
Los Angeles, CA 90013

Established: In 1947. Members: 25. Meetings: Annual.

Purpose: To conduct public education programs to protect the water and power rights of California in the Colorado River system.

Activities: Conducts public information and education programs through news releases and distribution of printed materials.

Committees: None.

Publications: Newsletter (quarterly), pamphlets and brochures.

Connecticut River Watershed Council (CRWC)
125 Combs Road
Easthampton, MA 01027

Established: In 1952. Members: 2,400. Staff: 7. State groups: 4. Meetings: Annual.

Purpose: To conserve and use wisely the natural resources in the Connecticut River Valley.

Activities: Bestows annual conservation awards for meritorious service to conservation. Has speakers' bureau. Conducts public education programs. Provides technical advisory service.

Committees: Conservation Education and Research; Land Acquisition; Development.

Publications: Current & Eddies (bimonthly), *Valley Newsletter* (quarterly), boating guide, and technical monographs.

Cooling Tower Institute (CTI)
P.O. Box 73383
Houston, TX 77273

Established: In 1950. Members: 300. Staff: 3. Meetings: Annual.

Purpose: To improve technology, design, and performance of water conservation apparatus.

Activities: Has developed standard specifications for cooling towers. Conducts research through technical subcommittees, sponsors workshops, seminars, and projects, maintains a speakers' bureau.

Committees: None.

Publications: Cooling Tower Institute—Bibliography of Technical Papers (periodic), *CTI News* (quarterly), *Newsletter, Journal of the Cooling Tower Institute* (semiannual), special publications and research reports.

Federal Water Quality Association (FWQA)
601 Wythe Street
Alexandria, VA 22314-1994

Established: In 1930. Members: 280. Meetings: Annual.

Purpose: Affiliated with Water Pollution Control Federation, aims to help prevent water pollution.

Activities: Seeks to advance knowledge concerning the collection, treatment, and disposal of wastewater. Presents scholarships to Washington, DC, students for wastewater operation training or environmental education. Holds annual awards dinner.

Committees: None.

Publications: None.

Ground Water Institute (GWI)
P.O. Box 981
Minneapolis, MN 55440

Established: In 1952. Members: 20. Meetings: Annual.

Purpose: To promote water production methods that will assure better use and management of natural groundwater.

Activities: Water supply contractors who conduct groundwater investigations, drill wells, construct pumping plants for municipal, industrial and domestic uses, and for irrigation.

Committees: None.

Publications: None.

Groundwater Management Caucus (GMC)
Box 637
White Deer, TX 79097

Established: In 1974. Members: 156. Regional groups: 133. State groups: 6. Meetings: Annual conference.

Purpose: To provide a forum for the exchange of information about groundwater management and conservation.

Activities: Does research on groundwater management and conservation technology. Encourages members to participate in legislative activities regarding the topic. Maintains small library.

Committees: None.

Publications: None.

Interstate Conference on Water Policy (ICWP)
955 L'Enfant Plaza, SW
Sixth Floor
Washington, DC 20024

Established: In 1959. Members: 70. Meetings: Annual conference.

Purpose: To inform state, intrastate, and interstate officials of new developments in the area of water quantity and quality.

Activities: Information exchange. Presents views to Congress and government agencies. Does research on water management, water energy, and water technology.

Committees: Drought; Groundwater; Legislative and Policy.

Publications: Annual Report, Membership Directory (annual), *Policy Statement* (annual), *Washington Report* (bimonthly), congressional testimony, papers and bulletins.

National Association of Conservation Districts (NACD)
509 Capital Court, NE
Washington, DC 20002

Established: In 1947. Members: 3,000. Staff: 16. Regional groups: 7. State groups: 54. Meetings: Annual.

Purpose: To direct and coordinate, through local self-government efforts, the conservation of soil, water, and related natural resources.

Activities: Establishes soil and water conservation districts organized by the citizens of watersheds, counties, or communities under provision of state laws.

Committees: Business Advisory; Coastal and Urban; Conservation Research and Technology; Cropland Conservation; District Operations; Education and Youth; Forestry; Great Plains; Information and Communications; Public Lands; Pasture and Range; Resource Policy Planning and Development; Soil Stewardship; Water Resources and Programs.

Publications: Conservation Bits and Bytes (quarterly), *District Leader* (monthly), *NACD Directory* (annual), *Proceedings of Annual Convention, Tuesday Letter* (monthly), guides and catalogs, films, videos, and slide sets.

National Association of Flood and Storm Water Management Agencies (NAFSMA)
1225 I Street, NW
Suite 300
Washington, DC 20005

Established: In 1977. Members: 44. Staff: 5. Meetings: Annual.

Purpose: To provide proper management of water resources.

Activities: Works to eliminate flooding, improve storm water management, and conserve watersheds.

Committees: Corps Liaison; FEMA Liaison; Policy; Legislative; Stormwater Management.

Publications: Legislative Report (periodic), *Newsletter* (monthly), *Technical Bulletin* (periodic).

National Association of Water Companies (NAWC)
1725 K Street, NW
Suite 1212
Washington, DC 20006

Established: In 1895. Members: 500. Staff: 7. Regional groups: 12. Meetings: Annual conference.

Purpose: To keep members informed of economic, legal, and regulatory developments, to encourage communication between investor-owned water companies and regulatory agencies, and to improve members' service to the public.

Activities: Conducts seminars, compiles statistics, presents the J. J. Barr Annual Scholarship.

Committees: Customer Service; Employee Relations; Government Relations; Management Information Systems; Public Information; Rates and Revenue; Regulatory Law; Regulatory Relations; Small Companies; Taxation; Water Technology.

Publications: Financial and Operating Data for Investor Owned Water Utilities (annual), *Financial Summary for Investor-Owned Water Utilities* (annual), *Water* (quarterly).

National Association of Water Institute Directors (NAWID)
c/o Paul Godfrey
Water Resources Research Center
Blairsdell House
University of Massachusetts
Amherst, MA 01003

Established: In 1973. Members: 54. Meetings: Annual conference.

Purpose: To provide a forum for communication among members and their institutes.

Activities: Promotes programs and activities of member institutes. Participates in technical meetings of related organizations.

Committees: None.

Publications: Handbook (biennial), *Newsletter* (bimonthly).

National Environmental Development Association/Ground Water Project (NEDA/Ground)
1440 New York Avenue, NW
Suite 300
Washington, DC 20005

Established: In 1985. Members: 14. Staff: 3. Meetings: Annual conference.

Purpose: To ensure supply of groundwater.

Activities: Aims to present methods to be sure there will always be groundwater for industry, agriculture, and domestic use.

Committees: None.

Publications: None

National Water Center (NWC)
P.O. Box 264
Eureka Springs, AR 72632

Established: In 1979. Members: 300.

Purpose: To make the public aware of the need to protect sources of unpolluted water. To disseminate information on conservation and research.

Activities: Offers consulting services to aid in detoxifying wastes and protecting water resources. Coordinates National Water Week. Conducts the Ellen Swallow Richard Retreat and presents the awards. Maintains a 300 volume library and speakers bureau.

Committees: None.

Publications: Aquaterra: Water Concepts for the Ecological Society (occasional), *We All Live Downstream* (book, 1986), *Heal the Waters, Wastewater Blues,* and *We All Live Downstream* (video documentaries), and audiocassettes.

National Water Resources Association (NWRA)
955 L'Enfant Plaza, SW
Suite 1202 N
Washington, DC 20024

Established: In 1932. Members: 4,800. Staff: 4. State groups: 17. Meetings: Annual conference.

Purpose: To develop, control, conserve, and use water resources in the reclamation states (17 western states).

Activities: Conducts legislative tracking and updates. Grants awards. Holds seminars.

Committees: Energy Issues; Environmental; Groundwater; Project Development and Finance; Resolutions; Small Projects; Water Quality; Water Rights.

Publications: Directory (biennial), *National Waterline* (monthly), *Water Report* (biweekly).

National Water Supply Improvement Association (NWSIA)
P.O. Box 102
St. Leonard, MD 20865

Established: In 1985. Members: 400. Staff: 1. Regional groups: 1. Meetings: Biennial conference.

Purpose: To promote research and development programs in water sciences such as desalination and wastewater reclamation.

Activities: Provides programs of water supply and urban environment improvement . Trains water treatment plant operators. Presents William E. Warne Man of the Year Award and Lifetime Achievement Award. Holds workshops and seminars.

Committees: Legislative Political Action; Standardization of Design; Technology Transfer.

Publications: Proceedings (biennial), *Newsletter* (monthly), papers.

National Water Well Association (NWWA)
6375 Riverside Drive
Dublin, OH 43017

Established: In 1948. Members: 16,000. Staff: 70. State groups: 49. Meetings: Annual.

Purpose: To help solve problems of locating, developing, preserving, and using underground water supplies.

Activities: Conducts seminars and continuing education programs. Encourages scientific education, research, and the development of standards. Compiles statistics. Maintains a speakers' bureau and 50,000 volume library. Maintains museum and software clearinghouse. Bestows awards.

Committees: Aquifer Protection; International; Regulatory Officials; Safety; Specifications.

Publications: Groundwater Monitoring Review (quarterly), *Journal of Ground Water* (bimonthly), *Membership Directory* (triennial), *National Well Association Briefings* (quarterly), *Newsletter* (bimonthly), *Proceedings* (annual), *Water Well Journal* (monthly), *Well Log* (monthly), special topic newsletters, quarterly, textbooks, and manuals.

National Watershed Congress (NWC)
c/o National Association of Conservation Districts
509 Capital Court, NE
Washington, DC 20002

Established: In 1954. Members: 28. Meetings: Periodic.

Purpose: To foster development of natural resources conservation through upstream watershed programs.

Activities: Deals with river basin planning, land treatment, and national water policy. Bestows Outstanding Watershed Management Award. Presently inactive.

Committees: None.

Publications: Proceedings.

North American Lake Management Society (NALMS)
1000 Connecticut Avenue, NW
Suite 300
Washington, DC 20036

Established: In 1980. Members: 1,640. Staff: 4. Regional groups: 10. Meetings: Annual conference.

Purpose: To promote and understand needs for management of lakes and their watersheds. Exchange information on technical and administrative aspects of management. Supports national, state, and local conservation programs.

Activities: Encourages development of lake coalitions. Sponsors surveys. Monitors water pollution and water standards. Reports on legislative action affecting lake management. Gives slide and tape presentations. Makes list of states' needs for the Environmental Protection Agency. Sponsors workshops and seminars. Holds student photo, paper, and poster competitions. Grants awards.

Committees: Fund Raising; Government Affairs; Industrial Relations; Technology Transfer.

Publications: Lake Line (bimonthly), *Lake and Reservoir Management* (semiannual), *NALMS Membership Directory* (annual), brochures, booklets, bibliographies, and monographs.

Passaic River Coalition (PRC)
246 Madisonville Road
Basking Ridge, NJ 07920

Established: In 1971. Members: 2,500. Staff: 7. Regional groups: 60. State groups: 30. Meetings: Twice yearly, June and fall.

Purpose: To resolve problems of an urban river system with attention given to population growth, water pollution and supply, flood control, sewage and garbage disposal, and urban decay.

Activities: Does research on land use, water supply and quality, flood control, wildlife and vegetation, water testing, solid waste recovery, and historic preservation. Offers environmental education at elementary through university levels, and citizens awareness education. Provides information to interested persons or groups. Receive grants for special studies and projects. Offers college interns professional experience. Maintains a speakers' bureau and library of 9,000 volumes. Keeps statistics. Grants annual award for outstanding contribution to the improvement of the Passaic River Watershed.

Committees: Education; Flood Plain Watch; Passaic River Restoration; Passaic Valley Ground Water Protection; Recreation; Water Resource; Watershed Preservation.

Publications: Goals and Strategies (quarterly), *Groundwater Monitor* (quarterly), *Passaic River Restoration Newsletter* (quarterly), *Vibes From the Libe* (monthly), *Watershed News* (bimonthly), *Citizen Alerts,* reports, articles, pamphlets, and inventory.

Rural Community Assistance Program (RCAP)
602 S. King Street
Suite 402
Leesburg, VA 22075

Established: In 1973. Members: 13. Staff: 10. Regional groups: 6. Meetings: Quarterly.

Purpose: To provide adequate and affordable drinking water and wastewater disposal to low-income rural areas in the United States and developing countries by developing water and wastewater systems.

Activities: Conducts research. Provides technical assistance.

Committees: None.

Publications: Directory of Network Organizations (semiannual), *Rural Water News* (bimonthly), policy and technical manuals, management guides.

Soil and Water Conservation Society (SWCS)
7515 N.E. Ankeny Road
Ankeny, IA 50021

Established: In 1945. Members: 13,000. Staff: 16. Local groups: 100. Meetings: Annual.

Purpose: To advance the proper use of land and water, and management of natural resources.

Activities: Grants awards, offers scholarships.

Committees: More than 25 committees, including College and University Relations, Conservation History, International Affairs, and Student Services.

Publications: Conservogram (bimonthly), *Journal of Soil and Water Conservation* (bimonthly), *Technical Monographs* (periodic), booklets and glossary.

Submersible Wastewater Pump Association (SWPA)
600 S. Federal Street
Suite 400
Chicago, IL 60605

Established: In 1976. Members: 26. Staff: 2. Meetings: Semiannual.

Purpose: To promote use of the submersible wastewater pump in handling waste.

Activities: Offers educational slide programs. Compiles statistics on shipment of pumps.

Committees: None.

Publications: SWPA Membership Directory (annual), *SWPA News* (annual), handbooks, and brochures.

Universities Council on Water Resources (UCOWR)
4543 Faner Hall
Southern Illinois University
Carbondale, IL 62901

Established: In 1962. Members: 83. Meetings: Annual conference.

Purpose: To expand and strengthen programs in water resources education and research.

Activities: Explores many aspects in the field of water resources. Conducts internship and visiting scientists programs. Grants awards. Holds seminars and workshops.

Committees: Education; Engineering and Physical Sciences; Environmental and Biological Sciences; International; Policy, Legislation, and Administration; Public Service; Research; Social and Behavioral Sciences.

Publications: Proceedings (annual), *Water Resources Update* (quarterly), brochures, reports, and proceedings of workshops and seminars.

U.S. Committee on Irrigation and Drainage (USCID)
P.O. Box 15326
Denver, CO 80215

Established: In 1952. Members: 700. Meetings: Triennial international congress.

Purpose: To exchange information with other engineers, scientists, and anyone interested in irrigation and drainage.

Activities: Affiliated with International Commission on Irrigation and Drainage and works with the commission in managing water and land resources for irrigation and flood control methods. Does research in the field. Studies planning, financing, and design for river control.

Committees: None.

Publications: USCID Newsletter (quarterly).

U.S. Committee on Large Dams (USCOLD)
P.O. Box 15103
Denver, CO 80215

Established: In 1928. Members: 1,150. Meetings: Annual, triennial international congress.

Purpose: To inform interested people of the planning, design, construction, and maintenance of large dams.

Activities: Takes part in the International Commission of Large Dams and provides opportunities for discussion, study tours, and publication of technical research.

Committees: Concrete; Construction; Dam Safety; Earthquakes; Education and Training; Environmental Effects; Hydraulics of Dams; Materials for Embankment Dams; Measurements; Methods of Numeric Analysis of Dams; Register of Dams; Tailings Dams; Technical Activities.

Publications: Membership Directory, Newsletter (3 times yearly), books, and technical bulletins.

Water and Wastewater Equipment Manufacturers Association (WWEMA)
P.O. Box 17402
Dulles International Airport
Washington, DC 20041

Established: In 1908. Members: 85. Meetings: Annual.

Purpose: To promote equipment for waterworks, wastewater, and industrial wastes disposal plants.

Activities: Sponsors public relations, public affairs, and marketing programs. Does research. Compiles statistics.

Committees: Equipment and Services; Marketing; Political Action; Public Affairs.

Publications: Membership Directory (annual), *News* (monthly), *Washington Analysis* (monthly).

Water Pollution Control Federation (WPCF)
601 Wythe Street
Alexandria, VA 22314-1994

Established: In 1928. Members: 32,000. Staff: 62. State groups: 44. Meetings: Annual.

Purpose: To advance knowledge concerning collection and disposal of domestic and industrial wastewater and management of facilities used.

Activities: Disseminates information to promote proper water pollution control and improve working conditions in this field. Holds educational programs. Sponsors High School Science Fair National Award. Maintains library of 6,000 volumes.

Committees: Groundwater; Hazardous Wastes; Industrial Wastes; Personnel Advancement; Public Relations; Research; Safety; Standard Methods; Technical Practice; Toxic Substances; Water Reuse.

Publications: The Bench Sheet (bimonthly), *Highlights* (monthly), *Joe Bank* (biweekly), *Literature Review* (annual), *Operations* (monthly), *Research Journal* (bimonthly), *Safety and Health Bulletin* (quarterly), *Washington Bulletin* (monthly), *Water, Environment, and Technology* (monthly), *Manuals of Practice,* brochures and water quality curriculum for schoolchildren.

Water Quality Association (WQA)
4151 Naperville Road
Lisle, IL 60532

Established: In 1974. Members: 1,550. Staff: 18. Regional groups: 3. State groups: 35. Meetings: Annual.

Purpose: To promote the acceptance and use of industry equipment, products, and services for water treatment.

Activities: Has programs and activities to help improve economy and efficiency in the industry. Conducts symposia, seminars, and expositions. Maintains library. Compiles statistics and presents annual hall of fame award for greatest contribution to the industry.

Committees: Technical Standards.

Publications: National Directory of Leading Water Conditioning Dealers, Manufacturers, and Suppliers (annual), *Newsletter* (bimonthly), *Point-of-Use Report* (quarterly), glossary.

Water Quality Research Council (WQRC)
4151 Naperville Road
Lisle, IL 60532

Established: In 1950. Members: 1,465. Staff: 18. Meetings: Annual.

Purpose: To improve economy and efficiency within the industry dealing with water quality by performing research.

Activities: Does research. Sponsors technical and marketing research and symposia on water quality.

Committees: None.

Publications: Materials for schools on water quality and water project ideas.

Water Resources Congress (WRC)
3800 N. Fairfax Drive
Suite 5A
Arlington, VA 22203

Established: In 1971. Members: 300. Regional groups: 9. Meetings: Annual conference.

Purpose: To inform agencies, associations, businesses, and organizations about recent developments regarding water resources.

Activities: Works for development, improvement, and utilization of the nation's rivers, lakes, and water and land resources.

Committees: Policy and Projects; Resolutions and Coordinating.

Publications: Hotline (periodic), *Platform* (every 5–6 years), *Water Resources Congress—Washington Report* (bimonthly), brochure.

Water Systems Council (WSC)
600 S. Federal Street
Suite 400
Chicago, IL 60605

Established: In 1932. Members: 40. Staff: 4. Meetings: Semiannual.

Purpose: To disseminate information regarding water systems markets, sales, installation, servicing, and engineering.

Activities: Sponsors workshops for engineering companies, county agents, public health officials, and others. Compiles statistics on employee compensation and training practices.

Committees: None.

Publications: None.

Western Snow Conference (WSC)
P.O. Box 2646
Portland, OR 97208

Established: In 1939. Members: 600. Regional groups: 4. Meetings: Annual.

Purpose: To share developments and research in the nonrecreational uses of snow.

Activities: Studies ways to use snow for irrigation and hydroelectric power. Studies ways to prevent floods and avalanches and to forecast water supplies.

Committees: Wilderness.

Publications: Proceedings (annual).

U. S. Government Organizations

Government Agency

Environmental Protection Agency
401 M Street, SW
Washington, DC 20460

Established: On December 2, 1970, as an independent agency in the executive branch pursuant to Reorganization Plan No. 3 of 1970.

Purpose: To protect and enhance our environment today and for future generations, and to control and abate pollution in air, water, solid waste, pesticides, radiation, and toxic substances.

Activities: EPA water quality activities represent a coordinated effort to restore the nation's waters, including:

1. Development of national programs, technical policies, and regulations for water pollution control and water supply.
2. Groundwater protection.
3. Marine and estuarine protection.
4. Development of water quality standards and efficiency guidelines.
5. Technical direction, support, and evaluation of regional water activities.
6. Development of programs for technical assistance and technology transfer.
7. Training in the field of water quality.

Committees: Other EPA divisions include Air and Radiation, Solid Wastes and Emergency Response, and Pesticides and Toxic Substances.

The EPA Office of Research and Development is responsible for a national research program in pursuit of technological controls of all forms of pollution. The office directly supervises the research activities of EPA's national laboratories. Close coordination of the various research programs is designed to yield a synthesis of knowledge from the biological, physical, and social sciences that can be interpreted in terms of total human and environmental needs.

Regional Offices: Ten regional offices are EPA's principal contact and relationships with federal, state, interstate, and local agencies, industry and academic institutions, and other public and private groups. These offices disseminate national program objectives and propose, develop, and implement regional programs for comprehensive and integrated environmental protection activities.

Publications: EPA Journal (bimonthly), a wide variety of special reports.

Government Advisory Committees

Advisory Committee on Water Data for Public Use
Office of Water Data Coordination
Geological Survey
Department of the Interior
417 National Center
Reston, VA 22092

Established: In 1964 under the authority of Office of Management and Budget Circular A67, dated August 28, 1964. Members: 29 representatives of national, state, and regional water-oriented organizations. Meetings: At least once yearly.

Purpose: To act as the public advisory committee of the Geological Survey, Department of the Interior.

Activities: Advises the Department of the Interior on plans, policies, and procedures regarding water data acquisition programs and their effectiveness in meeting national water data needs.

Committees: None.

Publications: None.

Colorado River Basin Salinity Control Advisory Council
Colorado River Water Quality Office
U.S. Bureau of Reclamation
P.O. Box 25007 Code: D-1000
Denver, CO 80225

Established: On February 25, 1976, by Public Law 93-320, the Colorado River Basin Salinity Control Act. Members: No more than three members from each of the states in the Colorado River Basin, appointed by the governor of the state. Staff: 1. Meetings: Annual.

Purpose: To act as a public advisory council under the Bureau of Reclamation, Department of the Interior.

Activities: Acts as liaison between the departments of the Interior and Agriculture, EPA, and the Colorado River Basin states (AZ, CA, CO, NM, NV, UT, WY) to protect the quality of water in the Colorado River upstream of Imperial Dam. Reviews reports on progress of salinity control program and recommends ways to improve the program.

Committees: None.

Publications: None.

Colorado River Floodway Task Force
Bureau of Reclamation
Department of the Interior
Nevada Highway and Park Street
P.O. Box 427
Boulder City, NV 89005

Established: On October 8, 1986 by Public Law 99-450, the Colorado River Floodway Protection Act. Members: One representative from each state (appointed by the governor), Indian reservation, county and district in the floodway, representatives from cities affected by the floodway, the Colorado River Wildlife Council, Army Corps of Engineers, Federal Emergency Management Agency, and the departments of Agriculture, the Interior, and State.

Purpose: To act as a public advisory task force of the Department of the Interior.

Activities: Makes recommendations to the Department of the Interior and Congress about the restoration, maintenance, and management of the Colorado River floodway. Reviews its boundaries and the hardship caused by the 1983 flood outside its boundaries. Studies legislation and regulations needed to achieve the purpose of the floodway and considers its effect on Indian lands.

Committees: None.

Publications: None.

Committee to Review the Potomac Estuary Experimental Water Treatment Project
Water Science and Technology Board
Commission on Engineering and Technical Systems
National Research Council
2101 Constitution Avenue, NW
Washington, DC 20418

Established: In 1976, and functions under the Water Science and Technology Board, Commission on Engineering and Technical Systems and the National Research Council. Members: 10.

Purpose: To study the future water resources needs of the Washington, DC, area.

Activities: Investigates using the estuary of the Potomac River as a water supply source. Reviews and reports on conclusions reached by the Army Corps of Engineers on the construction and testing of a pilot water treatment plant.

Committees: None.

Publications: None.

Committee to Review the Washington Metropolitan Area Water Supply Study
Water Science and Technology Board
Commission on Engineering and Technical Systems
National Research Council
2101 Constitution Avenue, NW
Washington, DC 20418

Established: In 1977, and functions under the Water Science and Technology Board, Commission on Engineering and Technical Systems, and the National Research Council. Members: 12.

Purpose: To study the future water resources needs of the metropolitan Washington, DC, area.

Activities: Reviews and comments on scientific findings made by the Army Corps of Engineers.

Committees: None.

Publications: None.

Interagency Advisory Committee on Water Data
Geological Survey
Department of the Interior
417 National Center
Reston, VA 22092

Established: In 1964, as the Federal Advisory Committee on Water Data under authority of Office of Management and Budget Circular A-67, August 28, 1964. Its name was changed in 1973. Members: Representatives of federal departments, their water-data acquisition and user bureaus, and independent agencies concerned with water data. Staff: Administrative services provided by the Office of Water Data Coordination, Geological Survey. Meetings: At least once yearly.

Purpose: To advise on water policies as an interagency advisory committee of the Geological Survey.

Activities: Provides advice on the design and operation of the national water data network. Offers advice on appropriate policies and procedures for getting statements, data, and need for data from federal agencies.

Committees: None.

Publications: None.

International Boundary and Water Commission, United States and Mexico, U.S. Section
4171 North Mesa
The Commons Building C
Suite 310
El Paso, TX 79902

Established: Pursuant to a treaty of March 1, 1889 as the International Boundary Commission. Its jurisdiction was extended by subsequent treaties and reconstituted under its present name by the Water Treaty of 1944. Members: United States and Mexico.

Purpose: To monitor treaties between the United States and Mexico dealing with boundary and water issues affecting both countries.

Activities: Provides flood control projects for urban and irrigation developments along the border. Helps control distribution of Rio Grande

River and the Colorado River waters. Provides storage, conservation, and regulation of the Rio Grande River for usage by the United States and Mexico. Authorizes structures in the two rivers for diversion of flood and irrigation waters. Helps solve salinity problems. Preserves international boundary line between the two countries. Monitors the boundary section for joint hydroelectric power development.

Committees: None.

Publications: None.

Interstate Commission on the Potomac River Basin
6100 Executive Boulevard
Suite 300
Rockville, MD 20852

Established: On July 11, 1940, by 54 Stat 78 as amended by Public Law 91-407, September 25, 1970, and Public Law 97-88, December 4, 1981. Members: Three members from each state in the Potomac Valley Conservancy District and three members appointed by the President. Staff: 1.

Purpose: To serve as an interstate commission on water problems.

Activities: Conducts studies and sponsors research on pollution and water problems of the Potomac Valley Conservancy District (MD, VA, WV, PA, DC) drained by the Potomac River and its tributaries. Disseminates information and serves as liaison with public and non-public agencies on pollution, utilization, conservation, and development of water in the district. Establishes standards of water quality for the district.

Committees: None.

Publications: None.

Management Advisory Group to the Construction Grants Program
Office of Water
Office of Municipal Pollution Control
Room 1219G
Washington, DC 20460

Established: On January 31, 1972, the Administrator of the EPA, under authority of Section 5 of the Federal Water Pollution Control Act, as amended, established the Technical Advisory Group for Municipal Waste Water Systems. It was restructured and name changed September 1975 and to Management Advisory Group and to Municipal Construction

Division in 1977; changed to current name in 1982. Members: 16. Staff: 1. Meetings: Two or three times yearly.

Purpose: To act as a public advisory group of the Office of Water, EPA.

Activities: Advises in implementation and review of federal laws and regulations concerning municipal wastewater treatment plants. Advises municipalities so they can deal effectively with wastewater problems. Comments on changes in federal regulations. Communicates with public agencies, contractors, engineers, and manufacturers of supplies. Advises on program objectives and how they can be achieved.

Committees: None.

Publications: None.

National Drinking Water Advisory Council
Office of Drinking Water
Environmental Protection Agency
Washington, DC 20460

Established: On February 26, 1975, by Section 1146 of P.L. 93-523, the Safe Drinking Water Act. Members: 15. Staff: 1. Meetings: Twice yearly.

Purpose: To provide advice to EPA on matters and policies concerning drinking water quality and hygiene.

Activities: Keeps aware of new issues regarding problems related to water quality. Advises the EPA on regulations and guidelines required by the Safe Drinking Water Act. Recommends special studies and research. Studies drinking water standards for improvement. Helps to identify environmental or health problems related to hazardous drinking water conditions.

Committees: None.

Publications: None.

Water Research Committee
Office of Water
Office of Research and Development, P.D. 674
Environmental Protection Agency
401 M Street, SW
Washington, DC 20460

Established: In March 1987 by the assistant administrator for the Office of Research and Development. Members: Managers from the research program (headquarters and laboratory), regulatory and enforcement program office staff, and representatives from regional offices of EPA.

Purpose: Serves as internal research committee of the Office of Research and Development, EPA.

Activities: Communicates research needs and ways to meet the needs. Advises on matters regarding air and radiation.

Committees: None.

Publications: None.

International Organizations

Abwassertechnische Vereinigung (ATV)
Markt 71
D-5205 St. Augustin 1
Federal Republic of Germany

Established: In 1965. Members: 6,100. Meetings: Periodic conferences.

Purpose: To promote and develop new methods to combat water pollution in Germany,

Activities: Works for cooperation among members and constituent organizations. Conducts research in the area of water pollution. Trains operators in sewage works and sponsors seminars.

Committees: Maintains numerous committees.

Publications: Annual Report, ATV Journal (monthly), scientific and technical manuals.

Baltic Marine Environment Protection Commission—Helsinki Commission (HELCOM)
Mannerheimintie 12A
SF-00100 Helsinki
Finland

Established: In 1980. Members: 7. Staff: 9. Language: English. Meetings: Annual conference.

Purpose: To develop methods to protect the Baltic Sea area from industrial pollution in accordance with the Helsinki Convention.

Activities: Conducts programs on pollution control to help the ecological condition of the Baltic area. Does scientific research on pollution control. Disseminates scientific data on pollution. Establishes guidelines for members in violation of regulations. Cooperates with international groups working on pollution. Has several working groups.

Committees: None.

Publications: Baltic Sea Environment Proceedings (periodic).

Canadian Water Resources Association (CWRA)
334 13th Street N
Lethbridge, AB T1H 2R8
Canada

Established: In 1947. Members: 800. Languages: French, corresponds in English.

Purpose: To increase public understanding of water resources and proper use of water policies.

Activities: Exchanges information on water use and management within Canada.

Committees: None.

Publications: Canadian Water Resources Journal (quarterly), *ICID Bulletin* (semiannual), *Water News* (quarterly), makes available technical dictionaries, books, notes, bulletins, and conference translations.

European Desalination Association (EDA)
University of Glasgow
Department of Mechanical Engineering
Glasgow G12 8QQ
Scotland

Established: In 1985. Members: 150. Language: English.

Purpose: To study the process of desalination and ways to reuse water.

Activities: Holds seminars on desalination, pumps and energy recovery for RO plants, and corrosion.

Committees: None.

Publications: None.

European Institute for Water (EIW)
23 Boulevard d'Anvers
F-67000 Strasbourg
France

Established: In 1983. Languages: French, Italian, corresponds in English.

Purpose: To encourage exchange of ideas between legislators and managers of water resources in Europe.

Activities: Reviews adequacy, value, and implementation of policies of the Commission of European Communities. Helps develop new directives. Examines needs for research that would help in cooperation between countries. Implements water related programs. Holds training courses and seminars on wastewater treatment.

Committees: None.

Publications: None.

European Mediterranean Commission on Water Planning (EMCWP)
Via Cimarosa 10
I-95 128 Catania
Italy

Established: In 1970. Members: 93. Languages: French, corresponds in English. Meetings: Semiannual symposium.

Purpose: To coordinate experimental work and research in the field of water studies.

Activities: Organizes scientific meetings and congresses on hydrology and water protection. Affiliated with FAO and UNESCO.

Committees: None.

Publications: Proceedings of Symposium (semiannual), monographs.

European Water Pollution Control Association (EWPCA)
Markt 71
D-5205 St. Augustin 1
Federal Republic of Germany

Established: In 1981. Members: 17. Languages: French, German, corresponds in English. Meetings: Triennial.

Purpose: To promote the practice of water pollution control, including sewerage, surface water drainage, sewage and waste treatment, and management in Germany.

Activities: Works to make the profession of water pollution control appealing to others. Bestows triennial Dunbar Award. Holds periodic seminars.

Committees: Artificial Wetlands; Education and Training; Radioactivity Due to Accident; Technical and Scientific; Technology Transfer; Urban Run-Off Water Quality Data in Member Countries.

Publications: Membership List (periodic), *Newsletter* (periodic), *Euro Water,* proceedings of seminars and symposia (triennial).

Institution of Water and Environmental Management (IWEM)
15 John Street
London WC1N 2EB
England

Established: In 1987. Members: 10,600. Staff: 14. Regional groups: 14. National groups: 2. Language: English. Meetings: Annual conference.

Purpose: To study water and environmental management issues worldwide.

Activities: Promotes water and environmental management, including treatment and disposal of drinking water and domestic and industrial wastewater. Holds symposia. Has library of 3,000 volumes. Sponsors study tours and competition.

Committees: None.

Publications: Journal of Institution of Water and Environmental Management (bimonthly), *Newsletter* (bimonthly), *Year Book,* manuals and symposia proceedings.

Inter-African Committee for Hydraulic Studies (ICHS)
Boite Postale 369
Ouagadougou 01
Burkina Faso

Established: In 1960. Regional groups: 2. Language: French. Meetings: Biennial council.

Purpose: To promote research and exchange of information regarding water resources in Africa.

Activities: Gives professional training and consulting services in water management, irrigation, soil and water conservation. Has a 12,000 volume library of books on water resources and a biographical archive. Compiles statistics and gives awards.

Committees: None.

Publications: Bulletin de Liaison du CIEH (quarterly), study reports.

Intergovernmental Council for the International Hydrological Programme (IHP)
c/o UNESCO
Maison de l'UNESCO
Division of Water Sciences
7, Place de Fontenoy
F-75700 Paris
France

Established: In 1974. Members: 30. Meetings: Biennial council session.

Purpose: To help countries cope with water problems by increasing education in hydrology and water sciences.

Activities: Investigates hydrological and ecological effects of human activities, especially pollution. Has interest in groundwater hydrology. Provides cooperation and technical assistance in countries sharing rivers or groundwater basins. Has ad hoc groups and panels.

Committees: None.

Publications: Nature and Resources (quarterly), *Bulletin,* technical documents in hydrology, hydrological studies, reports and technical papers.

Interim Committee for Coordination of Investigations of the Lower Mekong Basin (CCILMB)
Mekong Secretariat
Pibultham Ville
Kasatsuk Bridge, Rama I Road
Bangkok 10330
Thailand

Established: In 1957. Members: 3. Languages: French, corresponds in English. Meetings: Semiannual conference.

Purpose: To develop water resources in the Lower Mekong River Basin and its tributaries.

Activities: Coordinates improvements in irrigation, hydroelectric power, flood control and watershed management in the basin. Established the Lower Mekong Information System and hydrologic and meteorologic stations. Uses data collected to forecast floods and droughts. Collects documents in the field of water resources. Holds training courses, seminars, and workshops. Has library of 8,000 volumes.

Committees: None.

Publications: Annual Report, Mekong News (periodic), *Work Programme* (annual).

International Association of Hydrological Sciences (IAHS)
TNO Commission on Hydrological Res.
Postbus 297
NL-2501 BD The Hague
Netherlands

Established: In 1922. Members: 1,600. Languages: French, corresponds in English. Meetings: Biennial assembly.

Purpose: To promote the study of hydrology worldwide.

Activities: Studies the hydrological cycle of land and waters of the continents. Coordinates research in water research problems internationally. Discusses, compares, and publishes results of research. Bestows annual prize and the Tison Award.

Committees: Remote Sensing and Data Transmission Committee.

Publications: Catalogue of IAHS Publications (biennial), *Hydrological Sciences Journal* (bimonthly), *IAHS Newsletter* (3 times yearly), *IAHS Yearbook* (quadrennial), *Proceedings of Symposia* (6–12 times yearly), monographs and reports series.

International Association on Water Pollution Research and Control (IAWPRC)
1 Queen Anne's Gate
London SW1H 9BT
England

Established: In 1965. Members: 2,400. Staff: 8. Language: English. Meetings: Biennial.

Purpose: To contribute to advancement of research, development, and proper use in water pollution control and to work for communication and better cooperation among those who are involved in water pollution and quality management.

Activities: Exchanges information on water pollution control and its application. Bestows awards. Holds seminars and symposia.

Committees: Biennial Conference Programme; Editorial; Honours and Awards; Scientific and Technical; Technology Transfer.

Affiliated National Committees: Argentina; Australia; Austria; Belgium; Brazil; Canada; Chile; China; Cyprus; Czechoslovakia; Denmark; Egypt; Finland; France; Hungary; Israel; Japan; Korea; Kuwait; Malaysia; Netherlands; New Zealand; Norway; Philippines; Poland; Portugal; Italy; Singapore; South Africa; Spain; Sweden; Switzerland; Thailand; Turkey; United Kingdom; Uruguay; Venezuela; Yugoslavia.

Publications: IAWPRC Yearbook, Water Quality International (quarterly), *Water Research* (monthly), *Water Science and Technology* (monthly), *Advances in Water Pollution Control* (book series), scientific and technical reports.

International Commission for the Protection of the Rhine Against Pollution (ICPRAP)
Hohenzollernstrasse 18
Postfach 309
D-5400 Koblenz
Federal Republic of Germany

Established: In 1963. Members: 6. Staff: 8. Languages: French and German. Meetings: Annual.

Purpose: To determine origin, nature, and extent of pollution of the Rhine River and take measures to protect the Rhine from pollution.

Activities: Prepares antipollution treaties and agreements between members. Sponsors Action Program Rhine that deals with chemical, chloride, and thermal pollution. Maintains monitoring program.

Committees: Current Research; Hydrological; Legal.

Publications: Activity Report (annual), *Statistical Report on Physio-Chemical Analysis* (annual).

International Commission on Irrigation and Drainage (ICID)
48 Nyaya Marg
Chanakyapuri
New Delhi 110 021
Delhi, India

Established: In 1950. Members: 69. Staff: 28. Languages: French, corresponds in English. Meetings: Triennial congress.

Purpose: To promote use of engineering, economics, ecology and social science in managing water and land resources for irrigation, flood control, and research.

Activities: Makes plans for irrigation and drainage projects, and flood control methods to reclaim land. Does research on design and planning for canals and channels. Studies methods to control rivers. Sponsors symposia, seminars, and workshops. Maintains library of 25,000 volumes and 180 journals.

Committees: None.

Publications: Bibliography on Irrigation, Drainage and Flood Control (annual), *Bulletin* (semiannual), *Newsletter* (quarterly), *Proceedings of Special Sessions and Symposia* (triennial), *Transactions of Congress* (triennial), posters, pamphlets, and magazines.

International Commission on Large Dams (ICOLD)
151 Boulevard Haussman
F-75008 Paris
France

Established: —. Members: 78. Regional groups: 5. Languages: French, corresponds in English. Meetings: Triennial congress.

Purpose: To promote improvement in design and construction of dams.

Activities: Conducts tours for the study of dams. Operates bulletin archives. Holds symposia. Acts as a clearinghouse.

Committees: Analysis and Design of Dams; Dam Aging; Dam Safety; Design Flood; Environment; Glossary and Dictionary of Dams; Hydraulics for Dams; Materials for Concrete Dams; Materials for Fill Dams; Monitoring of Dams and Their Foundations; Technology of Dam Construction; World Registry of Dams.

Publications: Abstracts of ICOLD Publications (periodic), *Bulletin* (periodic), *Congress Proceedings and Transactions* (triennial), *ICOLD Directory* (triennial), reports, abstracts, documents, and technical dictionaries.

International Desalination Association (IDA)
P.O. Box 387
Topsfield, MA 10983

Established: In 1985. Members: 400. Staff: 1. Meetings: Biennial.

Purpose: To develop and promote desalination in maintaining water supplies, to control water pollution, and to promote the purification, treatment, and reusing of water.

Activities: Provides information on desalination and water reuse. Encourages establishment of standards and specifications and efficient use of water for energy. Offers scholarships. Conducts workshops and seminars. Maintains library and bestows awards.

Committees: France.

Publications: Proceedings (biennial), *Membership Directory,* Newsletter, inventories, monographs.

International Hydrological Programme (IHP)
c/o UNESCO
Division of Water Sciences
7 Place de Fontenoy
F-75700 Paries
France

Established: In 1965. Members: 138. Staff: 20. Regional groups: 5. Languages: French, corresponds in English. Meetings: Biennial council session.

Purpose: To create scientific and technological program for improving management of world water resources.

Activities: Studies economic, political, social, and industrial factors affecting management of water resources. Develops regional water management programs. Offers technical assistance to developing countries. Give technical training and sponsors research. Holds symposia and workshops. Conducts projects in Africa, Latin America, Middle East, and the Caribbean.

Committees: None.

Publications: Studies and Reports in Hydrology (periodic), *Technical Documents in Hydrology* (periodic), *Technical Reports in Hydrology* (periodic), *Long-Term Programme in the Field of Hydrology.*

International Mine Water Association (IMWA)
Department of Hydrogeology
Madrid School of Mines
Rios Rosas 21
E-28003 Madrid
Spain

Established: In 1979. Members: 200. Staff: 8. Language: English. Meetings: Quadrennial congress.

Purpose: To promote the development of scientific research and technology dealing with water in mines, to encourage research and training, to exchange scientific and engineering information, and to protect the environment against mine drainage.

Activities: Assists developing countries in solving mine water problems. Experts visit these countries and offer special courses and symposia. Maintains commissions and working groups.

Committees: None.

Publications: Directory (annual), *International Journal of Mine Water* (quarterly), *Proceedings of International Congress* (quadrennial).

International Ocean Pollution Symposium (IOPS)
c/o Dr. Iver W. Duedall
Department of Chemical and Environmental Engineering
Florida Institute of Technology
Melbourne, FL 32901

Established: In 1978. Language: English. Meetings: Annual.

Purpose: To meet and exchange ideas regarding waste disposal in the ocean.

Activities: Drafts guidelines for study of ocean disposal practices. Collects symposia information for books. Sponsors bibliographic seminars. Offers stipends for graduate students to attend symposia. Maintains a 2,300 volume library on ocean waste disposal.

Committees: None.

Publications: Marine Pollution Bulletin (annual), *Wastes in the Ocean and Oceanic Processes in Marine Pollution.*

International River Network (IRN)
301 Broadway
Suite B
San Francisco, CA 94118

Established: In 1985. Members: 1,000. Staff: 2.

Purpose: Communicates with members to update and encourage development of water policy worldwide.

Activities: Publicizes problems of large water projects. Aims to prevent building of destructive dams. Performs research and disseminates information about solutions to environmental problems. Objects to World Bank providing funds for projects that are destructive to the environment. Operates speakers' bureau. Maintains library.

Committees: None.

Publications: World Rivers Review (bimonthly), mail alerts and flyers.

International Society for the Prevention of Water Pollution (ISPWP)
Little Orchard
Bentworth
Alton, Hants
England

Established: In 1980. Members: 460. National groups: 3. Languages: French, Italian, corresponds in English.

Purpose: Water experts from 18 countries unite to prevent water pollution throughout the world.

Activities: Provides funds for research on water pollution problems. Gets national and regional governments and municipalities involved in control of water pollution. Maintains speakers' bureau. Offers children's services. Maintains small library.

Committees: English; Italian.

Publications: None.

International Water Resources Association (IWRA)
205 N. Mathews Avenue
University of Illinois
Urbana, IL 61801

Established: In 1972. Members: 1,400. Meetings: Annual conference and triennial world congress.

Purpose: To promote water resources planning, development, management, research, and education on an international level.

Activities: Encourages international programs and an international forum. Bestows Crystal Drop Award and Best Paper Awards. Offers scholarships and lectureships.

Committees: Geographical; Water Reuse; Water for the 21st Century.

Publications: Newsletter (quarterly), *Water International* (quarterly), *Proceedings,* books.

International Water Supply Association (IWSA)
1 Queen Anne's Gate
London SW14 9BT
England

Established: In 1947. Members: 1,200. Staff: 5. Languages: French, corresponds in English. Meetings: Biennial congress.

Purpose: To form an international body interested in public supply of water through pipes for domestic, agricultural, and industrial uses. To become more knowledgeable about public water supplies.

Activities: Exchanges information on water supply methods. Encourages cooperation among those involved in water supply. Holds workshops. Gives awards.

Committees: Corrosion; Desalination and Reuse of Wastewater; Instrumentation, Control and Automation; Management and Training; Public Relations; Statistics; Water Distribution and Engineering; Water Quality and Treatment; Water Resources.

Publications: Aqua (6 times yearly), *IWSA, Yearbook, Directory, Water Supply* (quarterly), journal and proceedings of conferences.

International Water Tribunal (IWT)
Damrak 83-1
NL-1012 LN Amsterdam
Netherlands

Established: In 1983. Staff: 4. Nonmembership. Languages: Dutch, French, Spanish, corresponds in English. Meetings: Periodic.

Purpose: To organize a second tribunal on water problems, focusing on Africa, Asia, and Latin America and their industrial activities.

Activities: Does scientific research. Endorses economic use of water and also management along with exchange of information on water. Aims to develop new legislation and carry out old legislation on environmental issues. Maintains working and advisory groups.

Committees: None.

Publications: IWT Newsletter (periodic), *Manual* (periodic), tribunal results and brochures.

International Working Association for Danube Research (IWADR)
Schiffmuhlenstrasse 120
A-1223 Vienna
Austria

Established: In 1956. Members: 300. Language: German. Meetings: Annual.

Purpose: To study the water quality of the Danube River.

Activities: Conducts research regarding hydrology, limnology, chemistry, and water quality of the Danube. Collects data. Has a 300 volume library.

Committees: None.

Publications: Archiv für Hydrobiologie (periodic), *Fachgruppenbericht* (annual), *Papers of the Symposium* (annual), *Proceedings* (annual), *Win Salvesbericht* (annual), monographs and bibliographies.

IRC International Water and Sanitation Centre (IRC)
Postbus 93190
NL-2509 The Hague
Netherlands

Established: In 1968. Staff: 20. Languages: Dutch, French, Spanish, corresponds in English. Meetings: Semiannual.

Purpose: To promote and provide means for potable water and sanitation facilities in developing countries.

Activities: Concentrates work on rural and semi-urban areas of Africa, Asia, and Latin America and cooperates with United Nations efforts to improve water. Focuses on information, support, services, technology, operation, and maintenance. Has training courses and community education programs. Studies role of women, hygiene, and education of population and maintains 8,000 volume library on sanitation and water supply.

Committees: None.

Publications: Accession List (bimonthly), *Current Awareness Bulletin* (quarterly), *IRC Newsletter* (in French, bimonthly), *IRC Newsletter* (in English, 10 times yearly), technical papers, training series, and occasional paper series.

Joint Group of Experts on the Scientific Aspects of Marine Pollution (GESAMP)
c/o International Maritime Organization
4 Albert Embankment
London SE1 7SR
England

Established: In 1969. Members: 25. Languages: French, Russian, Spanish, corresponds in English. Meetings: Annual.

Purpose: To assess effects of marine pollutants.

Activities: Provides scientific bases for research. Monitors programs. Internationally exchanges information on control of marine pollution. Scientific control and management of marine pollution sources. Control over scientific bases and legal measures for prevention, control or abatement of marine pollution. Has several working groups.

Committees: None.

Publications: GESAMP reports and periodic studies.

Oslo Commission (OSCOM)
New Court
48 Carey Street
London WC2A 2JE
England

Established: In 1972. Members: 13. Staff: 5. Languages: French, corresponds in English. Meetings: Annual.

Purpose: To control the disposal of waste in the North Sea and northeast Atlantic by regulating dumping activities.

Activities: Receives and reviews dumping permits and approvals. Studies the general condition of the waters within its sphere of interest and controls the measures adopted. Has a monitoring program with the Paris Commission.

Committees: Standing Advisory Committee for Scientific Advice.

Publications: Annual Report, The Oslo and Paris Commissions—The First Decade, Review of Sewage Sludge Disposal.

Regional Commission on Land and Water Use in the Near East (RCLWUNE)
Food and Agriculture Organization of the United Nations
Via delle Terme di Caracalla
I-00100 Rome
Italy

Established: In 1967. Members: 23. Languages: Arabic, corresponds in English. Meetings: Biennial conference.

Purpose: To ensure efficient use of land and water in the Near East.

Activities: Develops ways to improve land and water management and makes suggestions to FAO member states. Exchanges information among members states. Compiles data on conservation of water resources. Holds seminars, workshops, and training courses.

Committees: None.

Publications: None.

Regional Organization for the Protection of the Marine Environment (ROPME)
P.O. Box 26388
Safat 13124
Kuwait

Established: In 1979. Members: 8. Languages: Arabic, Persian, corresponds in English. Meetings: Biennial council and periodic meetings.

Purpose: To implement the Kuwait Regional Convention on the Protection of the Marine Environment and Pollution ratified in 1979.

Activities: Organizes scientific programs and projects as suggested in the Kuwait Action Plan. Operates the Marine Emergency Mutual Aid Centre to coordinate and cooperate activities in emergency cases such as oil spills and dumping of waste products. Works with United Nations organizations. Holds training programs, seminars, and workshops in marine science. Maintains a 500 volume library.

Committees: None.

Publications: ROPME Newsletter (quarterly), directories, and children's books (annual).

Southern Africa Society of Aquatic Scientists (SASAS)
Town Planning Department
P.O. Box 1694
Cape Town 1699
South Africa

Established: In 1963. Members: 400. Regional groups: 8. Language: English. Meetings: Annual congress.

Purpose: To promote the study of limnology and water resources in South Africa.

Activities: Holds seminars on water environment topics. Has symposia on water law and management.

Committees: None.

Publications: Limnological Bibliography of Africa South of the Sahara (annual), *Southern African Journal of Aquatic Sciences* (semiannual).

Union of Water Associations from Countries of the European Communities (EUREAU)
255 Chaussee de Waterloo
Boite 6
B-1060 Brussels
Belgium

Established: In 1975. Members: 11. Staff: 2. Languages: French, corresponds in English. Meetings: Periodic conference.

Purpose: To provide information on water sources and supplies.

Activities: Assesses and monitors water supply problems. Reviews and discusses legislation regarding the industry. Suggests solutions to problems in member states. Works with technical associations.

Committees: None.

Publications: Report (periodic).

5

Bibliography

SINCE THE PUBLICATION OF *Silent Spring* by Rachel Carson in the early 1960s there has been an outpouring of literature on a wide variety of environmental problems including the quality and quantity of water. The literature varies from complex, scientific articles to popular accounts. The selection of bibliographical items in this chapter provides a wide perspective on this worldwide problem. The first section presents reference sources, followed by annotated entries on books, journal articles, and governmental publications, with a final section listing journals that publish articles on water related topics.

Reference Works

Cohen, Phyllis. *The Public's Role in Water Resource Planning: Selected Bibliography.* Public Administration Series: Bibl. no. P772. Monticello, IL: Vance Bibliographies, 1981. 7p. No ISBN.

This short bibliography of key studies in the current literature on public participation in water resource planning covers books, journal articles, and government documents. It will be of value to planners and the public interested in water resource planning.

Cohen, Sanford F. *Cost-Benefit Analysis for Water Project Evaluation: A Selected Bibliography.* Public Administration Series: Bibl. no. P578. Monticello, IL: Vance Bibliographies, 1981. 7p. No ISBN.

This bibliography covers books, articles, and studies dealing with the economic impact of water resources development projects. It should be useful to environmental planners working with water project evaluation.

———. *Pollution from Land Activities in the Great Lakes Basin: A Selected Bibliography.* Public Administration Series: Bibl. no. P577. Monticello, IL: Vance Bibliographies, 1980. 10p. No ISBN.

This bibliography lists important books, papers, articles, and studies dealing with pollution of land in the Great Lakes Basin in alphabetical order. It is of interest to environmental planners.

Duensing, Edward E., and Lenore B. Duensing. *The Environmental Effects of Acid Rain.* Public Administration Series: Bibl. no. P624. Monticello, IL: Vance Bibliographies, 1980. 13p. No ISBN.

The authors include a two-page introduction relating to acid rain to set the stage for the user. The bibliography includes scholarly materials published since 1975 and materials available in the Rutgers University Library system. The material is listed alphabetically.

Everson, Curtis A., and Rodney L. Sharp. *Great Plains Irrigation, 1975–80; A Literature Review.* Bibliographies and Literature of Agriculture, no. 22. Washington, DC: U.S. Department of Agriculture, Economic Research Service, 1981. 69p. No ISBN.

This bibliography covers literature on irrigation in the Great Plains, including such topics as water distribution, pump selection, management techniques and production costs for irrigated crops, efficiency and economics of irrigation, water quality, and conservation. Each topic begins with a short introduction followed by bibliographic items listed by state. There are 467 bibliographic items.

Feeser, Sally A., and Mary Ann Filler. *Acid Rain: A Pennsylvania Problem?* Bibliographical Series no. 10. University Park, PA: Pennsylvania State University Libraries, 1984. 59p. No ISBN.

This bibliography is for those interested in the problems acid rain causes in our environment. It includes annotated citations for articles, books, monographs, theses, and reports for the period 1948 through 1983 with both an author and description index arranged alphabetically.

Gray, Elaine. *Groundwater Overview of Issues, a Partially Annotated Bibliography.* CPL Bibl. no. 261. Chicago, IL: Council of Planning Librarians, 1990. 22p. ISBN 0-86602-261-9.

This bibliography covers articles and monographs published 1985–1990. Some of the articles are annotated. It gives sources for basic information about groundwater, its contamination, and methods and regulations to protect it. The bibliography is helpful to planners, environmentalists, teachers, and anyone interested in protecting groundwater.

Hawkins, Donald T. *Physical and Chemical Properties of Water: A Bibliography: 1957–1974.* New York: IFI/Plenum, 1976. 556p. ISBN 0-306-65164-5.

This bibliography emphasizes writings on the properties of liquid water, including the heavy isotopic forms, and ice. Fundamental properties of water as a substance are included in the bibliography, but not explicit industrial properties or properties of natural waters, mineral waters, etc. Some works on the molecular nature of pure water are included.

Huck, Peter, and Peter Toft. *Treatment of Drinking Water for Organic Contaminants.* New York: Pergamon Press, 1986. 383p. ISBN 0-08-031876-2.

This volume contains the proceedings of the Second National Conference on Drinking Water held in Edmonton, Alberta, April 7–8, 1986. "Treatment for Organic Contaminants" was the theme of the conference, which was attended by practitioners, researchers, and people concerned with the treatment of contaminants in drinking water. Invited papers from Europe, the United States, and Canada were presented. The conference stressed that people must be educated regarding why, what, and how organic contaminants are found in drinking water. Figures, diagrams, and bibliographic references are found within chapters.

Hyman, Eric, et al. *The Theory and Practice of Environmental Quality Analysis: Water Resources Management, Land Suitability Analysis, Economics, and Aesthetics.* CPL Bibl. no. 27. Chicago, IL: Council of Planning Librarians, 1980. 89p. No ISBN.

This bibliography gives references that show changes that have taken place in water resources planning since the passage of the National Environmental Policy Act of 1969 and guidelines adopted by the

Water Resources Council in 1973. Entries are divided into nine categories and cross-referenced where appropriate. A list of acronyms is also given.

Jones, E. A. *Irrigation and Human Adaptation: Annotated Bibliography*, Commonwealth Bureau of Agricultural Economics no. 19. Farnham Royal, UK: Commonwealth Agricultural Bureau, 1973. 8p. No ISBN.

References in this bibliography were selected from *World Agricultural Economics and Rural Sociology Abstracts* for the period 1971–1973, covering irrigation facilities. Some references deal with problems people have in taking advantage of new methods. The bibliography was compiled in response to a request from the Dambarawa Peasant Colonization Scheme in Sri Lanka. It covers Africa, America, Asia, and Australia.

Knight, Allen W., and Mary Ann Simmons. *Water Pollution: A Guide to Information Sources.* Man and the Environment Information Guide Series, vol. 9. Detroit, MI: Gale Research Company, 1980. 278p. ISBN 0-8103-1346-4.

This guide covers reference material, case studies, and methodology ideas in journal articles, government documents, and professional journals on water pollution. All material, except for journal articles, are annotated. One chapter lists educational programs in water pollution, another chapter gives addresses of federal, regional, and state water agencies. The volume includes a glossary of terms and a few selected readings. There is a good chapter on water pollutants, listing references to pollution by pesticides, detergents, oil, metals, radioactivity, and thermal pollution. The book has author, title, and subject indexes.

Lockerby, Robert W. *Desalination Technology*. Public Administration Series: Bibl. no. P668. Monticello, IL: Vance Bibliographies, 1981. 12p. No ISBN.

This 12-page bibliography begins with a discussion of methods of desalination. An alphabetical list of books and articles pertaining to desalination follows.

Moe, Christine. *Environmental Impacts of Spillway Openings and Flooding in the Lower Mississippi River*. Public Administration Series: Bibl. no 376. Monticello, IL: Vance Bibliographies, 1979. 17p. No ISBN.

This short bibliography offers a brief introduction about the effects of spillways on the Mississippi River. Sections on manuscript materials, government documents, and books and articles are arranged alphabetically.

———. *Rio Grande Flood Control and Drainage.* Public Administration Series: Bibl. no 769. Monticello, IL: Vance Bibliographies, 1981. 27p. No ISBN.

This bibliography covers government documents, books, and articles, with entries arranged in alphabetical order.

Northeast Regional Center for Rural Development. *Community Guide to Groundwater Protection and Management: An Annotated Bibliography.* 2d ed. Publication no. 38. University Park, PA: Pennsylvania State University, 1985. 65p. No ISBN.

This bibliography includes sources of technical advice and assistance to protecting and managing groundwater in the United States with some emphasis on the northeastern states.

Paylore, Patricia, ed. *Desertification: A World Bibliography.* Tucson: University of Arizona, Office of Arid Lands Studies, 1976. 644p. No ISBN.

Prepared for the 23d International Geographical Congress, Moscow, 1976, preconference meeting of the IGU Working Group on Desertification, Desert Research Institute, Ashkhabad, Turkmen SSR, July 20–26, 1976. This mammoth work is a compilation of information showing the causes of environmental changes extending deserts into marginal areas or intensifying desert conditions in arid regions. It begins with general references on desertification worldwide and then follows with entries on Africa, the Middle East, the Soviet Union, Pakistan, India, China, Australia, South America, and finally North America. Each section begins with a general discussion. Articles are arranged alphabetically within each section.

Rinaldi, Amalia. *Irrigation in Southern Africa: An Annotated Bibliography.* MSU International Development Papers. Working Paper no. 19. East Lansing: Michigan State University, Department of Agricultural Economics, 1985. 49p. No ISBN.

This annotated bibliography is arranged alphabetically by country and then by articles dealing with irrigation in that country. The countries included are: Angola, Botswana, Lesotho, Malawi, Mozambique, Swaziland, Tanzania, Zambia, Zimbabwe, and the South African Department

Coordination Conference (SADCC) countries in southern Africa. A few references are given for the Republic of South Africa; the literature on South Africa is in Afrikaans, English, and German. Some of the literature on Angola and Mozambique is in Portuguese. The remaining sources are in English. Annotations are in English. A list of acronyms used in the bibliography is given.

Salter, Christopher L. *Doing Battle with Nature: Landscape Modifications and Resources Utilization in the People's Republic of China, 1960–1972.* Occasional Paper no. 1. Eugene, OR: Asian Studies Committee, University of Oregon, 1973. 96p. No ISBN.

This annotated bibliography is arranged by year, beginning with a pre-1960 section and ending with 1973. Within each section the works are listed alphabetically. Irrigation and water conservancy is the largest category in the bibliography, with emphasis placed on construction of small dams, canals, sluices, and wells. Land reclamation is the other major category, including citations on drought and reclaiming riverbanks and land from the sea.

Seabrooke, Allan K., and John S. Marsh. *The Environmental Impact of Water-Based Recreation: An Annotated Bibliography.* Public Administration Series: Bibl. no. P777. Monticello, IL: Vance Bibliographies, 1981. 15p. No ISBN.

The authors set the stage for their bibliography in a four-page introduction before listing books, journal articles, and government documents that show the impact of water recreation activities on the environment. They exclude the impact of fishing and the shooting of waterfowl. The volume is in alphabetical order and provides guidelines for water recreation management.

Stine, Jeffrey K., and Michael C. Robinson. *The U.S. Army Corps of Engineers and Environmental Issues in the Twentieth Century: A Bibliography.* Environmental History Series. Washington, DC: U.S. Army Corps of Engineers, Office of the Chief of Engineers, 1984. 110p. No ISBN.

This bibliography was compiled by the Historical Division Office of Administrative Services with the support of the Public Works Historical Society. It deals with environmental aspects of the Corps civil works and regulatory programs, emphasizing the period after 1969.

Stopp, G. Harry, Jr. *Acid Rain: A Bibliography of Research Annotated for Easy Access.* Metuchen, NJ: Scarecrow Press, 1985. 174p. ISBN 0-8108-1822-1.

This bibliography will help those interested in acid rain locate information. Entries are arranged alphabetically by author and each entry is annotated. References include journal articles, books, and some government documents. The book lists acronyms and members of the National Atmospheric Deposition Program 1983–1984 and National Atmospheric Deposition Program/National Trend-Network Sites.

Summers, W. K., and Zane Spiegel. *Ground Water Pollution: A Bibliography.* Ann Arbor, MI: Ann Arbor Science Publishers, 1975. 83p. ISBN 0-250-40048-0.

This bibliography offers more than 400 entries covering groundwater contamination by nitrates, heavy metals, pesticides, and herbicides. It also covers the impact of urbanization and the effects of animal wastes, petroleum products, and other solid waste disposal on groundwater quality.

Toft, Peter, Richard S. Tobin, and James Sharp, eds. *Drinking Water Treatment: Small System Alternatives.* New York: Pergamon, 1988. 346p. ISBN 0-08-036936-7.

This volume gives the proceedings of the Third National Conference on Drinking Water held in Saint John's, Newfoundland, June 12–14, 1988. "Small System Alternatives" was the theme of the conference, which was attended by researchers, practitioners, and personnel concerned with design, construction, and operation of small systems providing treatment of drinking water. There were 233 participants from Canada and five other countries. References are given at the end of chapters. Many diagrams and figures are included. Many authors contributed to this volume. Bibliographical information is found on the editors along with a list of reviewers.

University of California, Water Resources Center, Archives. *Dictionary Catalog of the Water Resources Center Archives, University of California.* Boston, MA: G. K. Hall, 1970. 5 vols. No ISBN.

This publication contains more than 80,000 items covering water as a natural resource and its utilization: municipal and industrial water uses and problems, flood control, reclamation, waste disposal, water quality, water pollution, water law, and water resources development and management. The main body of the center's holdings is located on the Berkeley campus with a smaller branch on the Los Angeles campus. The emphasis of the collection is on report literature, books and journals, technical and scientific reports, government publications, publications of societies and associations, symposia, and manuscripts. The bibliography covers the period 1890 to 1970 in five volumes.

U.S. National Fertilizer Development Center, Muscle Shoals, Alabama. *Effects of Fertilizers on Water Quality.* Muscle Shoals, AL: 1969. 107p. No ISBN.

This bibliography is a collection of references dealing with the relationships between fertilization of agricultural lands and chemical composition of surface and groundwater. Abstracts are arranged alphabetically by author according to subject headings and subheadings. Each abstract is numbered to facilitate cross-indexing. At the end of each category is a list of all other abstract numbers relating to the subject under another category. A supplemental list of unabstracted references is found at the end of most major subjects. At the end of the book there is an author index, geographic location index, and soil classification index.

Vance, Mary. *Drinking Water Standards: A Bibliography.* Public Administration Series: Bibl. no. P2796. Monticello, IL: Vance Bibliographies, 1989. 13p. ISBN 0-7920-0406-X.

This short bibliography covers journal articles, some books, and government publications which address the quality of drinking water.

———. *Groundwater Protection: A Bibliography.* Public Administration Series: Bibl. no. P2639. Monticello, IL: Vance Bibliographies, 1989. 26p. ISBN 0-7920-0159-1.

This bibliography covers journal articles, books, and government documents addressing the protection of groundwater.

———. *Utilization of Sewage: A Bibliography.* Public Administration Series: Bibl. no. P878. Monticello, IL: Vance Bibliographies, 1982. 33p. No. ISBN.

The author categorized 348 works under specific catagories: general references, sewage irrigation, sewage fuel, and sewage as fertilizer. The book includes an author index and a journal index.

———. *Water Purification: A Bibliography.* Public Administration Series: Bibl. no. P933. Monticello, IL: Vance Bibliographies, 1982. 39p. No ISBN.

Entries in two sections covering books and journal articles make up this 39 page bibliography, with each part arranged alphabetically. No introduction is given.

———. *Watersheds: A Bibliography.* Public Administration Series: Bibl. no. P935. Monticello, IL: Vance Bibliographies, 1982. 18p. No. ISBN.

Bibliographic material on watersheds is arranged alphabetically within two sections on books and journal articles. No introduction is given.

White, Anthony G. *Administration, Construction, and Operation of a River Master Project—The Colorado River: A Selected Bibliography.* Public Administration Series: Bibl. no. P537. Monticello, IL: Vance Bibliographies, 1980. 6p. No ISBN.

This six-page bibliography deals with dams, reservoirs, and diversions built on the Colorado River to improve irrigation projects, control floods and drainage, and generate electric power. It gives a good cross-section of issues that come up when a government project is undertaken. It covers books, articles, and a few government documents within an alphabetical listing.

———. *Arid Lands: Government Impact, Water and the Public: A Selected Bibliography.* Public Administration Series: Bibl. no. P585. Monticello, IL: Vance Bibliographies, 1980. 7p. No ISBN.

In this bibliography the author lists books and articles that address the absence of sufficient quantities of potable water necessary to support human life. He sets the background for the bibliography in an introduction.

———. *Major Irrigation Projects, with an Emphasis on Public Administration Impacts: A Selected Bibliography.* Public Administration Series: Bibl. no. P604. Monticello, IL: Vance Bibliographies, 1980. 18p. No ISBN.

After a short introduction the author lists books, journal articles, and government documents on irrigation projects in alphabetical order with no topical breakdown.

White, Bernard. *Septic Tank Density and Groundwater Contamination: An Annotated Bibliography.* CPL Bibl. no. 234. Chicago, IL: Council of Planning Librarians, 1989. 15p. ISBN 0-86602-234-1.

This bibliography contains more than 60 annotated items showing the impact on the environment caused by septic system effluent leaching into groundwater. The purpose of this work is to make local planning and zoning officials aware of groundwater contamination.

Books

General

Aucoin, James. *Water in Nebraska.* Lincoln: University of Nebraska Press, 1984. 157p. ISBN 0-8032-1013-2.

Nebraska is one of the many states that will face serious water problems unless patterns of overuse and abuse of water are corrected. The rivers and streams of Nebraska have shrunk with the development of dams and reservoirs, and groundwater deposits are being depleted and polluted. State water policy is frequently based on political matters. Some of the issues facing Nebraskans are the allocation of water, in-stream flows, groundwater management, use of surface and groundwater, water quality, transbasin diversion, and financing water development. The book includes figures, photos, and notes.

Bocking, Richard C. *Canada's Water: For Sale?* Toronto: James Lewis & Samuel, 1972. 188p. ISBN 0-88862-028-4.

This book was written as the result of a film by the same name produced by the Canadian Broadcasting Corporation. The first three chapters discuss the water "crisis" in the United States and the possible need for Canadian water. The next few chapters discuss the impact of water development on humans and the environment, and methods to deal with water shortages. The final chapters discuss water policy and development in Canada. The volume includes a few sources and references.

Canada Water Year Book, 1985. Canada Water Year Book Series. Ottawa: Canadian Government Publishing Centre, 1986. 98p. ISBN 0-660-11997-8.

This is the sixth edition in the series and deals with freshwater resources. Data used in this edition were derived from surveys of agriculture, industry, municipalities, etc. and corrects information given in earlier year books. The volume contains many tables and photos.

Clark, Ira G. *Water in New Mexico.* Albuquerque: University of New Mexico Press, 1987. 839p. ISBN 0-8263-0923-2.

This book is a historical narration of the management and use of water in the state of New Mexico. Clark begins by sketching the Indian tradition of water management predating North America's Christian era and follows with the laws and customs developed during Spanish occupation. Chapters 3 through 10 cover New Mexico's territorial period with a few chapters devoted to early statehood. Several chapters address water use and management after World War II. The state's groundwater statute, enacted in 1931, did not fully take effect until the postwar years. The book concludes with a discussion on the current water problems and the probable future of the state water policy. There are several illustrations, many pages of notes for the individual chapters and also a table of cases and a selective bibliography.

Davis, Kenneth, and John A. Day. *Water: The Mirror of Science.* Science Study Series no. S18. Garden City, NY: Anchor Books, 1961. 195p. No ISBN.

This short but enlightening essay on water begins by addressing water as an unusual substance, and then moves on to discuss water's peculiar properties. After a section dwelling on the geological origin of water a chapter is devoted to the physical causes of its unusual properties.

Eisenberg, David S., and Walter Kauzmann. *The Structure and Properties of Water.* New York: Oxford University Press, 1969. 296p. No ISBN.

Water, the principal component of all living organisms, is held in the rocks of the Earth's crust in the form of hydration. The authors discuss the water molecule in detail: its composition, properties, density, and distribution. One chapter deals with the structure of ice and another chapter gives the properties of liquid water, followed by specific models. An addendum lists articles that were overlooked in the writing of the book according to the sections of the book to which they correspond. The book includes a lengthy bibliography and both author and subject indexes.

Fox, Cyril S. *Water: A Study of Its Properties, Its Constitution, Its Circulation on the Earth, and Its Utilization By Man.* London: Technical Press, 1951. 148p. No ISBN.

This treatise on water is divided into three parts. Part I, the natural history of water, deals with the physical properties, chemical constitution, circulation, and distribution of water on the Earth. Part II describes the work done by water, such as erosion, leaching, sedimentation, and the movement of underground water. Part III discusses the problems of collecting data, obtaining water supplies from springs, wells, and rivers, artisan water storage reservoirs, purification of water, and obtaining power from waterfalls. The book includes many photographs.

Franks, Felix. *Water.* rev. ed. London: Royal Society of Chemistry, 1984. 96p. ISBN 0-85186-473-2.

This small volume presents a condensed review of the present state of knowledge on liquid water. It discusses the remarkable physical properties of water, how these properties give rise to a unique liquid "structure," the influence of water on the interactions between dissolved solutes, the role of water in maintaining biologically active molecular structures, its involvement in chemical reactions, and the problems posed by management and provision of adequate amounts of water of acceptable quality. The author covers the nature of water, its eccentric physical properties, the manner it which it renders the Earth habitable, and its involvement in life processes.

Furon, Raymond. *The Problems of Water: A World Study.* Translated by Paul Barnes. New York: American Elsevier, 1967. 207p. No ISBN.

This interesting work, translated from French, begins with a definition of water and then discusses such topics as freshwater, water in agriculture and industry, floods, water pollution, and development problems in arid zones. The book includes references for each chapter and a list of illustrations.

King, Thomson. *Water: Miracle of Nature.* New York: Macmillan, 1960. 238p. No ISBN.

The author gives an excellent description of what water is and how it is found in solid, liquid, and gaseous form depending on temperature and pressure. In the second part of the book the author describes how human life depends on water for irrigation, power, drinking, and transportation. He concludes by showing that to conserve water is to conserve life.

Kuenen, P. H. *Realms of Water: Some Aspects of Its Cycle in Nature.* New York: John Wiley, 1955. 327p. No ISBN.

Water is a unique and remarkable product of nature; it is the only inorganic fluid that occurs in great quantities and the only substance found in the natural state as a solid, a liquid, and a vapor. The author discusses the water cycle, which begins with the evaporation of water from the seas and ends with the return there by way of discharging rivers. This is the "short" cycle. A still shorter cycle involves condensation in the atmosphere and evaporation before reaching the Earth's surface. A "long" cycle includes evaporation from the ocean with the moisture being precipitated upon dry land. If the route includes a glacier, it might be called the longest cycle. The author uses 16 plates and 190 figures to illustrate the book. There is a short list of references.

Leopold, Luna B. *Water: A Primer.* San Francisco, CA: W. H. Freeman, 1974. 172p. ISBN 0-7167-0264-9.

This book is an expansion of a manuscript entitled *A Primer on Water* written by the author and W. B. Langbein for the U.S. Geological Survey. Environmental problems dealing with water are discussed in this volume, including such topics as surface water and groundwater, surface runoff and storage, flood plains, sediment load of rivers, water in the world, the amount of water available, and its use. The book includes many figures and tables as well as a glossary.

Leopold, Luna B., and Kenneth S. Davis. *Water.* Life Science Library. New York: Time Inc., 1966. 200p. No ISBN.

The nature of water, what it can do, and how humans use it is discussed in this book. The book not only addresses consumption of water, but its uses in power, transportation, and irrigation as well. A picture essay supplements each chapter. A table giving the chemical content of water and a dictionary of a hydrologist is found at the end of the book.

Majumdar, Shyamal, E. Willard Miller, and R. R. Parizek, eds. *Water Resources in Pennsylvania: Availability, Quality and Management.* Easton, PA: Pennsylvania Academy of Science, 1990. 580p. ISBN 0-945809-02-6.

This volume presents a comprehensive review of Pennsylvania water resources and the availability of water within the atmosphere, surface, and subsurface. The diversity of uses and water quality problems associated with these uses are discussed. Major consideration is given to climatic trends that serve as long-term driving forces determining water renewal and changes in quality. Specific aspects include problems faced by utilities, water-based recreation, transportation, and groundwater heat pumps. Other water quality management concerns include the environmental consequences of agriculture, domestic water disposal, pollution by organic chemicals, and acid mine drainage. Water management considerations include review of nonpoint source area control of pollutants, regional planning, importance of computer models, legal basis for management, drought measurement options, artificial groundwater recharge, and international agreements aimed at protecting water quality and availability. This volume presents a broad insight into Pennsylvania's water resource opportunities, limitations, major problems, and present and future challenges.

Marston, E., ed. *Western Water Made Simple.* Washington, DC: Island Press, 1987. 237p. ISBN 0-933280-39-4.

This book contains information found in special issues of *High Country News* that dealt with water in the West. The book begins with a map showing the locations of the three major river basins—the Columbia, the Missouri, and the Colorado. A discussion of the size, the flow, the runoff, the land irrigated, and the reservoir storage of each of the basins begins the main section of the book. The major themes and issues discussed in the book are: the importance of understanding the western water situation by the nonspecialist, the reformation of the Columbia River, the uncertain fate of the Missouri River and its surrounding land, and the change that has taken place on the Colorado River as a result of various projects. The book includes many maps and photographs along with a selected reading list. A list of contributors is found at the beginning.

Swank, Wayne T., and D. A. Crossley, Jr., eds. *Forest Hydrology and Ecology at Coweeta.* Ecological Studies, vol. 66. New York: Springer-Verlag, 1987. 469p. ISBN 3-540-96547-5.

Research on the Coweeta watersheds in North Carolina is the longest continuous environmental study on any landscape in North America. This book is based on papers presented at a three-day symposium held in Athens, Georgia, in 1984 to celebrate 50 years of research at the Coweeta Hydrologic Laboratory. The book begins with a description of the laboratory and its research accomplishments and then discusses human management of forested watersheds, including: effects of pesticides, stream flow changes, acid precipitation, trace metals in the atmosphere, forest floor soil, and vegetation. The Coweeta Hydrologic Laboratory has played a major role in the programs created through ecosystem science. The book includes several pages of references as well as tables, graphs, and charts.

What Is the Most Effective Water Policy for the United States? Washington, DC: Government Printing Office, 1985. 957p. No ISBN.

The title of this book was the National Debate topic for high schools, 1985–1986 as a result of the passage of Public Law 88-246. The book was compiled by the Congressional Research Service of the Library of Congress. It begins by giving background materials and references, followed by a discussion on the control of the allocation of water, the quality of water, and the supply of potable water, with references for each topic. It ends with a selected bibliography, additional source material, and government publications on the topics.

Wheeler, William B., and Michael J. McDonald. *TVA and Tellico Dam, 1936–1979: A Bureaucratic Crisis in Post-Industrial America.* Knoxville: University of Tennessee Press, 1986. 290p. ISBN 0-87049-492-9.

The Tennessee Valley Authority (TVA) closed the gates to Tellico Dam, November 20, 1979, impounding a free-flowing stretch of the Little Tennessee River and creating Tellico Lake, a human-constructed lake. This ended a controversy that had gone on for years. The backdoor approval of the Tellico Project was received by the public with a mixture of outrage, anger, and bitterness. The Tellico project was only one phase of TVA's development, serving as a pilot program and a catalyst for future programs, as well as a watershed for the region. The book has an appendix on land purchases by TVA at Tellico, 1967–1976, a key to TVA document collections, and several pages of notes.

Wollman, Nathaniel, and Gilbert W. Bonem. *The Outlook for Water: Quality, Quantity and National Growth.* Published for Resources for the Future. Baltimore, MD: Johns Hopkins University Press, 1971. 286p. ISBN 0-8018-1260-7.

Water has been overused and misused. Economists term it a "common property resource," owned by everyone and therefore not owned by anyone. The authors describe the future outlook for water in the United States, concluding that new technologies will need to be used if high-quality water is to be available in the future, that the capital needed to start and operate facilities, primarily in waste treatment plants, will continue to grow, that the Southwest will remain an area of water shortage, and that the quality of pure water will become costly. The book has many tables, figures, and appendices for explanatory purposes.

Groundwater

Agricultural Chemicals and Groundwater Protection: Emerging Management and Policy. Navarre, MN: Freshwater Foundation, 1988. 235p. No ISBN.

This volume is a compilation of papers presented at a conference held October 22–23, 1987 in St. Paul, Minnesota. The main issue of the conference was the growing concern over the increased use of agricultural chemicals and how to minimize their effects on groundwater. They concluded that improved water management and increased cooperation between federal, state and local governments are needed to ease the problem. The book includes some tables, diagrams and references.

Balek, Jaroslav. *Groundwater Resources Assessment.* Developments in Water Science, no. 38. New York: Elsevier, 1989. 249p. ISBN 0-444-98895-5.

Groundwater is one of the most important natural resources. In arid regions it is often the only source of water and in humid regions it is highly important as surface water is often polluted, scarce and inadequately developed. Groundwater levels decline because of excessive pumping, thereby increasing the cost of pumping. Groundwater can be recharged by condensation of atmospheric humidity, rainfall, and infiltration of surface water from streams and lakes. Urban areas are concerned about ensuring an adequate supply of water and how to dispose of used water. Recycling has become an economic issue. Industrial and mining activities also have a tremendous effect on groundwater resources. Water resources, and water resource management, are two factors examined in detail by industries before developing in an area. The book has bibliographic references and many figures.

Fujimura, Faith N. and Williamson B. C. Chang, eds. *Groundwater in Hawaii.* Honolulu, HI: University Press of Hawaii, 1981. 257p. ISBN 0-8248-0788-X.

This volume contains the proceedings of the Artesian Water Centennial Symposium, held to celebrate the centennial of the first well drilled in the Hawaiian Islands by James Campbell. The symposium was sponsored by the estate of James Campbell and the Water Resources Research Center of the University of Hawaii at Manoa.

Gordon, Wendy. *A Citizen's Handbook on Groundwater Protection.* New York: Natural Resources Defense Council, 1984. 208p. No ISBN.

This handbook begins by discussing the nature of groundwater and cites specific examples of areas where groundwater contamination occured, the health risks encountered, and methods used to prevent and slow down contamination. Part II discusses citizen participation in groundwater management. Part III advises the testing of water. Part IV gives laws and regulations protecting groundwater and Part V is a list of questions to be asked when the protection of groundwater is considered. The book has six appendices, a glossary and references, and resource list.

Ground Water Quality Protection: State and Local Strategies. Committee on Ground Water Quality Protection; Water Science and Technology Board; Commission on Physical Sciences, Mathematics and Resources; National Research Council. Washington, DC: National Academy Press, 1986. 309p. ISBN 0-309-03685-2.

The Committee on Ground Water Quality Protection was established in 1984 to identify and review several state and local ground water protection programs with special attention given to prevention of ground water contamination. The committee included experts in such areas as water supply management, toxicology, environmental management, ground water quality and protection, environmental law, and hydrogeology. They studied activities that cause contamination of groundwater, such as solid-waste disposal, storage and management of hazardous materials, septic systems, agricultural chemicals applied to land, land application of waste waters, and production and storage of oil and gas. The committee conducted interviews, led discussions, reviewed literature, and presented their conclusions and recommendations for improvement of ground water quality within this volume. This book has many references, tables, figures, and appendices.

Groundwater Protection. Washington, DC: Conservation Foundation, 1987. 240p. ISBN 0-89164-102-5.

This book is divided into two parts. Part I, Groundwater: Saving the Unseen Resource, gives the recommendations and conclusions reached by the National Groundwater Policy Forum. Public hearings were held in December, 1985 in Austin, Texas; Milwaukee and Wausau, Wisconsin; and Hauppauge, New York, where government officials, industrialists, environmentalists, and interested citizens testified. Part II, A Guide to Groundwater Pollution, discusses the problems of, causes for, and government responses to groundwater pollution. A lengthy chapter is devoted to the sources of contamination, including waste disposal, mining and drilling methods, storage tanks, agricultural fertilizers, pesticides, and livestock. This book has many figures and references for each chapter.

Groundwater Protection: Groundwater: Saving the Unseen Resource. Washington, DC: The Conservation Foundation, 1987. 240p. ISBN 0-89164-102-5.

This book gives the final report of the National Groundwater Policy Forum. It is a guide to groundwater pollution, examining the problems of, causes for, and government responses to groundwater pollution. This book has many figures, a list of forum participants, staff, and references for each of the five chapters.

Lloyd, J. W. ed. *Case-Studies in Groundwater Resources Evaluation.* New York: Oxford University Press, 1981. 206p. ISBN 0-19-854530-4.

Over the past two decades, the knowledge and understanding of groundwater conditions have improved. Probably the most important developments have been concerned with investigating the possibility of groundwater resources to support freshwater supplies. Industrial countries have become aware of the need to conserve and protect water resources to ensure economical supplies of water. Planning programs have been initiated, often showing that groundwater resources are preferred to surface water resources, as they are easier to protect from pollution, more dependable during drought periods, and also more economical. In nonindustrialized countries, groundwater resources are emphasized as a source for drinking and irrigation. In tropical areas, groundwater is important as a dry season supplementary supply for water for agricultural production and human consumption. In semiarid and arid regions, groundwater investigations are being pursued to help solve the scarcity of water. This book has references for each chapter, tables, and figures.

Mumme, Stephen P. *Apportioning Groundwater Beneath the U.S.–Mexico Border: Obstacles and Alternatives.* Research Report Series 45. La Jolla,

CA: Center for U.S.–Mexican Studies, University of California, San Diego, 1988. 54p. ISBN 0-935391-79-7.

This small book discusses the possibile alternatives to resolve this international natural resource controversy.

Page, G. William, ed. *Planning for Groundwater Protection.* Orlando, FL: Academic Press, 1987. 387p. ISBN 0-12-543615-7.

Remedial attempts must be made not only to contend with the contamination of groundwater, but also to develop and use methods to prevent contamination. Specific case studies are used to show how to plan for groundwater protection. Each case contains and details different circumstances, in order to portray a variety of approaches. Case studies cited in the book are: Long Island, New York; Dade County, Florida; Wausau, Wisconsin; Austin, Texas; Perth Amboy, New Jersey; Bedford, Massachusetts; South Brunswick, New Jersey; Silicon Valley, California. This book has many figures and tables. References are found at the end of each chapter.

Schmidt, Kenneth D., ed. *Symposium on Groundwater Contamination and Reclamation.* American Water Resources Association Technical Publication Series, TPS-85-2. Bethesda, MD: American Water Resources Association, 1985. 175p. No ISBN.

This volume contains the proceedings of a symposium held in Tucson, Arizona, August 14–15, 1985, sponsored by the American Water Resources Association. Studies in groundwater contamination have intensified over the years as it has become necessary to investigate and discover the cause of contamination while attempting to solve the problem. The papers are listed under four major topics: monitoring, legal and political issues, protection, and reclamation. Problems and solutions are discussed. Many figures and tables, are found throughout the volume. This book contains many bibliographical references.

Smerdon, Ernest T., and Wayne R. Jordan, eds. *Issues in Groundwater Management.* Water Resources Symposium no. 12. Austin: University of Texas, Center for Research in Water Resources, 1985. 498p. No ISBN.

This volume contains the proceedings of a symposium on groundwater management sponsored jointly by the Water Resources Institute and the Texas A & M University System held at San Antonio, October 29–31, 1984. The symposium, entitled "Groundwater—Crisis or

Opportunity," primarily addressed groundwater issues in Texas and the Southwest. Included were topics such as conservation of water, improving water management, and increasing research to improve the water situation. Work accomplished at the symposium should be helpful in remedying existing water problems in the near future. Most of the papers include illustrations and figures.

Smith, Zachary A. *Groundwater in the West.* San Diego, CA: Academic Press, 1989. 308p. ISBN 0-12-652995-7.

For many sections of the West groundwater is the only dependable water resource, so proper groundwater management remains one of the major concerns. This book discusses the use, management, laws, and politics of groundwater in 19 western states. The opening chapter gives the history of groundwater use and the legal, political, and economic trends related to groundwater. Chapters 2 through 20 review groundwater conditions in each of the states, with the last chapter concluding the discussion. The book has a lengthy bibliography, a glossary, and a map showing the states discussed.

———. *Groundwater Policy in the Southwest.* Southwestern Studies Monograph no. 76. El Paso, TX: Texas Western Press, 1985. 57p. ISBN 0-87404-152-X.

This small volume examines groundwater conditions and policies in Arizona, California, and New Mexico. These southwestern states depend on groundwater as a major water resource and have felt the impact of decreased groundwater levels. Water is also a vital element in the future growth of the Southwest and must be managed better. Pollution from toxic substances and salt water is a serious threat to the water quality in the region. Groundwater mining, in which water is removed from an aquifer and not replaced, and overdrafting are other serious problems, especially in California and Texas. Individual case studies point out how serious the need for good groundwater management has become. The book includes many excellent references.

———. *Interest Group Interaction and Groundwater Policy Formation in the Southwest.* Lanham, MD: University Press of America, 1985. 216p. ISBN 0-8191-4456-8.

Groundwater has become an important source of water in the United States. Some places, such as Tucson, Arizona, depend entirely upon groundwater for their water supply. The demand for groundwater continues to rise with growing populations and increased use by industry. This book discusses management of groundwater in Arizona, New Mexico, and California as well as offering reasons on why particular interest groups may or may not join coalitions to study and address groundwater problems. Each chapter includes footnotes.

van der Heijde, Paul, et al. *Groundwater Management: The Use of Numerical Models.* 2d ed. Water Resources Monograph no. 5. Washington, DC: American Geophysical Union, 1985. 180p. ISBN 0-87590-314-2.

This monograph examines the intrinsic strength and deficiencies of existing groundwater models as well as factors not directly related to the models that affect model use. Groundwater systems and groundwater management along with models for management are discussed. The authors point out models that water managers need but do not have, and why they do not use certain models available. This second edition is based on information collected at the International Ground Water Modeling Center from its inception in 1978 to December 1983. Because of the introduction of many new mathematical models since the publication of the first edition in 1980, several chapters in the second edition were completely rewritten. A lengthy appendix, bibliographic references, and a list of model contact addresses are given.

Ward, C. H., W. Giger, andP. L. McCarty, eds. *Ground Water Quality.* Environmental Science and Technology Series. New York: John Wiley, 1985. 547p. ISBN 0-471-81597-7.

This volume discusses current research geared toward solving the problems of groundwater contamination, including waste disposal practices and waste treatment facilities. Methods are being developed to make accurate damage assessments and to improve waste disposal technology. This volume brings together many authors and has chapters on sources, types, and quantities of contaminants in groundwater; methods for groundwater quality research; subsurface character in relation to groundwater pollution; and the transportation and fate of subsurface contaminants. The book is illustrated with figures and tables. Each chapter gives a conclusion and many bibliographical references.

Supply

Brown, F. Lee, and Helen M. Ingram. *Water and Poverty in the Southwest.* Tucson: University of Arizona Press, 1987. 226p. ISBN 0-8165-1038-5.

This book examines the relationship between water and poverty within the "Four Corners" area of the Southwest (Colorado, Utah, Arizona, and New Mexico). Control over scarce water resources within this arid region has caused extensive conflict. This book discusses the importance of water strategies for improving living conditions of the rural poor through two case studies: a Hispanic community in the Upper Rio Grande and the Tohono O'odham Nation. The poor can gain influence in society by participating in decisions concerning the control of water. The book includes notes for each chapter, maps, and tables.

El-Ashry, Mohamed T., and Diana C. Gibbons, eds. *Water and Arid Lands of the Western United States.* A World Resources Institute Book. New York: Cambridge University Press, 1988. 415p. ISBN 0-521-35040-9.

This book, based on a two-year study, primarily covers agricultural areas—the Central Valley of California, the High Plains of Texas, and the Upper Colorado River Basin—but also discusses the problems of metropolitan Denver, Tucson, and the southern California megalopolis. In the West, about 90 percent of the water is used for agricultural irrigation. Despite the water shortage, the West is one of the fastest growing regions in the United States. Each chapter has bibliographical references. A few figures are interspersed throughout the book.

Enge, Kjell I., and Scott Whiteford. *The Keepers of Water and Earth: Mexican Rural Social Organization and Irrigation.* Austin:University of Texas Press, 1989. 222p. ISBN 0-292-74309-2.

Irrigation is a necessity for people who occupy arid lands. The authors have worked in and studied the water situation in the Tehuacán Valley of Mexico over a period of several years. They studied the valley's irrigation systems in great detail, noting the importance of these systems to the area. In Mexico, most irrigated agricultural lands use water that is federally owned, but in the Tehuacán Valley, the water resources are controlled privately. Private ownership has been equated by the people of the valley with local control. The people of the valley feel that state ownership does not favor interests of the peasants, the landless, or the indigenous population.The book includes notes on each chapter, tables, figures, and a lengthy list of bibliographic references.

Engelbert, Ernest A., and Ann Foley Scheuring, eds. *Water Scarcity: Impacts on Western Agriculture.* Berkeley: University of California Press, 1984. 489p. ISBN 0-520-05300-1.

This book is a result of an interdisciplinary conference on water problems in western United States held in Monterey, California, September 1982. The primary purpose of the conference was to assess the impacts of limited water supplies on agriculture in the semiarid West for local, state, national, and international areas. The West is faced with major decisions about managing declining water supplies in order to maintain a prosperous agricultural economy. The summary, written by Gilbert F. White, suggests that careful management is necessary to avoid making the water situation worse in the West, consequently affecting agricultural irrigation and the economy. A few illustrations are found throughout the book. Some of the papers included have references and all of the papers have an abstract. A list of contributors is also given.

Foster, Harold D., and W. R. Derrick Sewell. *Water: The Emerging Crisis in Canada.* Canadian Institute for Economic Policy Series. Toronto: James Lorimer in association with the Canadian Institute for Economic Policy, 1981. 117p. ISBN 0-88862-447-6.

A water crisis in Canada is apparent. There are deficiencies in the management of water, general failure to maintain the quality of rivers and lakes, and an inability to control flood and drought losses, to list a few of the most basic reasons. Both the availability and quality of water are threatened by attempts to solve problems related to water use. The United States is pressuring Canada to enter water agreements in order to supply water for the growing migratory population and industry of the Sunbelt. Six tables are found in the appendix. There are bibliographic notes for each chapter.

Francko, David A., and Robert G. Wetzel. *To Quench Our Thirst: The Present and Future Status of Freshwater Resources of the United States.* Ann Arbor: University of Michigan Press, 1983. 148p. ISBN 0-472-10032-7.

Water resource crises have existed as long as humans have banded together in settlements for survival. We use more water—in our homes, in agriculture, and in industry—than is being replenished through natural processes to usable water sources. To continue our uncontrolled water consumption and contamination patterns will directly cause a water crisis. Our response to the existing and developing water quality and availability problems will determine the future of our water resources. Bibliographic notes are found with each chapter. The book includes a glossary.

Frederick, Kenneth D., ed., with assistance of Diana C. Gibbons. *Scarce Water and Institutional Change.* Washington, DC: Resources for the Future, 1986. 207p. ISBN 0-915707-21-7.

The authors discuss alternative ways to meet long-term water needs and to solve conflicts among competing water users within five study regions: the Columbia River Basin; Kern County, California; southern California; Virginia Beach, Virginia; and northeastern Colorado. The authors contend that our water supply problem is due to the policy of underpricing water, failure to manage water resources properly, and bad water use habits. Each chapter has a list of bibliographical references. The book is illustrated with tables and figures.

Kneese, Allen V., and F. Lee Brown. *The Southwest Under Stress: National Resource Development Issues in a Regional Setting.* Baltimore, MD: Johns Hopkins University Press, 1981. 268p. ISBN 0-8018-2707-8.

This volume is a report on the Southwest Under Stress Project covering the "Four Corners" states of Arizona, Colorado, New Mexico, and Utah. Part I gives the background of the area. Part II details water supply and use in the area, emphasizing the upper Colorado River Basin and the development of energy resources. The Colorado River is being called upon to meet out-of-basin demands by diverting water to other basins, such as the eastern slope of the Rocky Mountains, southern California, and the Rio Grande Basin in New Mexico. Also discussed are the potential effects on the freshwater ecosystem and the implications for water management within the region. Many important issues about water in the West are examined, including equity, management, environmental quality, and water development. Part III deals with air quality and other environmental issues, and Part IV examines who benefits from economic development. The book has several tables and a list of other works published as a result of the Southwest Under Stress Project.

Moss, Frank E. *The Water Crisis.* New York: Praeger, 1967. 305p. No ISBN.

This book begins by giving the history and politics of water use in the United States. The author then discusses problems such as water pollution, water shortages, water variabilities, water depletions, water wastes, and solutions to these problems. The appendix contains information on water resources legislation, civil works projects, and water development projects. Bibliographic references are given.

Overman, Michael. *Water: Solutions to a Problem of Supply and Demand.* Doubleday Science Series. New York: Doubleday, 1969. 192p. No ISBN.

This interesting volume has chapters discussing the water crisis, water budgets, irrigation and dams, hydroelectricity, desalination, water purification, and future water resources. The book is beautifully illustrated with color photographs, but includes only a sketchy bibliography.

Page, G. William, ed. *Planning for Groundwater Protection.* Orlando, FL: Academic Press, 1987. 387p. ISBN 0-12-543615-7.

Toxic substances contaminating groundwater is one of the most important environmental problems today. Groundwater conditions are so different in each locality that there is no single set of groundwater protection measures which can be used universally. This book describes methods to protect the groundwater from toxic contaminants with programs developed to fit special case studies. The book includes figures and tables. References are found at the end of each chapter.

Perrault, Pierre. *On the Origin of Springs.* Translated by Aurele La Rocque. New York: Hafner, 1967. 209p. No ISBN.

This is the translation of Perrault's *Treatise on the Origin of Springs,* published in Paris in 1674. This text explains how Perrault was the first person to show that rainfall is more than enough to account for the flow of rivers and springs. This treatise was a turning point in the study of water as a geologic agent and the understanding of the hydrologic cycle. Opinions of several philosophers concerning the origin of springs are discussed before the author scientifically refutes their beliefs. The book includes references and many footnotes.

Powledge, Fred. *Water: The Nature, Uses, and Future of Our Most Precious and Abused Resource.* New York: Farrar, Straus & Giroux, 1982. 423p. ISBN 0-374-28660-4.

Written in a style that is easy to read, this book discusses water contaminants such as toxic chemicals, acid rain, pesticides, and fertilizers. The role played by the Bureau of Reclamation, the Army Corps of Engineers, and the Tennessee Valley Authority in the conservation and use of water is also discussed. The last section of the book discusses the future of this most precious and abused natural resource. The book lacks photos and tables.

Reisner, Marc, and Sarah Bates. *Overtapped Oasis: Reform or Revolution for Western Water.* Washington, DC: Island Press, 1990. 200p. ISBN 0-933280-76-9.

The authors wrote this book as a "sequel" to Reisner's book *Cadillac Desert.* They begin by giving a brief history of western water, discussing the dams that have been constructed in the West and the role played by the Bureau of Reclamation. In the second part of the book they examine western water law, and how water is allocated, used, and wasted. In the final section they suggest ways to modernize water management in the West. While the first part of the book has several diagrams, there are few in the last two sections. Appendix A is a case study of the Imperial Irrigation District, Appendix B is the Department of the Interior's water transfer policy, and Appendix C is the directory of Bureau of Reclamation. The book includes a selected bibliography.

Sheaffer, John R., and Leonard A. Stevens. *Future Water: An Exciting Solution to America's Most Serious Resource Crisis.* New York: Morrow, 1983. 269p. ISBN 0-688-01575-1.

The authors cite specific examples to show that the national water crisis is not necessarily due to the shortage of water, but to the mismanagement of current water supplies. Technological solutions are now being attempted to remedy water pollution problems such as sewage disposal into lakes, rivers, and oceans. This book is beautifully illustrated with photographs and diagrams. Bibliographical references are found for each chapter.

Storper, Michael, and Richard A. Walker. *The Price of Water: Surplus and Subsidy in the California State Water Project.* Berkeley: University of California, Institute of Governmental Studies, 1984. 30p. ISBN 0-08772-293-5.

This small monograph deals with the distribution of the California State Water Project's costs and benefits to agricultural interests in Kern County and to consumers in the Metropolitan Water District of Southern California. The authors conclude that costs are shifted from the Kern County Water Agency to the people in southern California. They argue that water agencies and the legislature should reconsider the financing and management of the water supply in California and devise equitable arrangements for water allocation and cost repayment. The volume includes many notes and an appendix on the evidence for transfers between contractors.

Williams, Robert B., and Gordon L. Culp, eds. *Handbook of Public Water Systems.* Van Nostrand Reinhold Environmental Engineering Series. New York: Van Nostrand Reinhold, 1986. 1113p. ISBN 0-442-21597-5.

This excellent handbook contains a wealth of information usable by public water systems managers to help solve the many challenges of water management from the point of source development up to the delivery of the final product to the consumer. Appendix A gives a list of conversion factors, Appendix B lists chemicals used in treatment of water and wastewater, and Appendix C has several miscellaneous tables. The handbook has a very useful author-subject index.

Drought

Dregne, H. E. *Desertification of Arid Lands.* Advances in Desert and Arid Land Technology and Development, vol. 3. New York: Harwood Academic Publishers, 1983. 242p. ISBN 3-7186-0168-0.

Desertification is defined as the impoverishment of terrestrial ecosystems under the impact of human activities—deterioration in the ecosystems and the degradation of the land. The author discusses desertification processes such as grazing, dryland cropping, mining, irrigated agriculture, and recreation. He then discusses the causes and what steps can be taken to prevent desertification. Many tables are found throughout the book. Each chapter has a conclusion and several bibliographic references.

Page, Jake, and the editors of Time-Life Books. *Arid Lands.* Planet Earth Series. Alexandria, VA: Time-Life Books, 1984. 176p. ISBN 0-8094-4512-3.

This book has four major chapters: "Discovering the Demons of Noon," "Studies in Deprivation," "Survivors in a Sere World," and "Stemming the Tides of Sand." Several essays discuss the beauty of the desert, migration in arid lands, adaptation of animals and plants to the desert, and the progress being made in reclaiming wastelands for production through irrigation. The book is beautifully illustrated with colored photographs of the desert and many maps.

Russell, Clifford S., David G. Arey, and Robert W. Kates. *Drought and Water Supply: Implications of the Massachusetts Experience for Municipal Planning.* Baltimore, MD: Johns Hopkins University Press, 1970. 232p. No ISBN.

The authors examined five years of very low rainfall in New England during the mid-1960s to show how different communities respond to water shortages. The demand for water throughout the state was met with very little foresight or planning: not one of the water managers mentioned or provided for periodic water restrictions, increases in rates, or the installation of meters. The book includes many figures and tables with eight appendices giving specific information. The chapters have footnotes.

Somerville, Carolyn M. *Drought and Aid in the Sahel: A Decade of Development Cooperation.* Westview Special Studies on Africa. Boulder, CO: Westview, 1986. 306p. ISBN 0-8133-7087-6.

The author spent several months in the Sahel in western Africa doing research on drought to write this book. She examines the problem of drought as well as the relief efforts, discussing in detail the work done by the Comité Permanent Inter-états de Lutte contre la Sécheresse dans le Sahel (CILSS) and the Club du Sahel to combat drought. The author also analyzes the support given to CILSS projects by the Sahelians. The book includes many tables, bibliographic references, and a list of countries providing financial aid to alleviate the drought problems. Abbreviations used in the book are included.

Spooner, Brian, and H. S. Mann, eds. *Desertification and Development: Dryland Ecology in Social Perspective.* New York: Academic Press, 1982. 407p. ISBN 0-12-658050-2.

This book focuses on human society as the causative factor in all aspects relating to desertification. The book is divided into two parts—the first part addresses the global perspective of desertification and the other section discusses regional programs. The text is written in a readable but scholarly style. Specific case studies show how types of human activities—canal irrigation, pastoralism, and resettlement—are related to desertification. The book includes tables, figures, and a lengthy bibliography.

Timberlake, Lloyd. *Africa in Crisis: The Causes, the Cures of Environmental Bankruptcy.* An Earthscan Book published in cooperation with the International Institute for Environment and Development. Philadelphia, PA: New Society Publishers, 1986. 232p. ISBN 0-86571-082-1.

For someone needing information on Africa, this book will answer many questions. A drought can be a disaster, depending on how land has been managed previous to the drought. Africa needs efficient irrigation as the little rain that falls only comes during a few months. Many African river basins have great irrigation potential. Although the Aswân Dam is used mainly for irrigation, many more dams are needed, even though their development will mean the displacement of people. The book includes many references and photographs.

Floods

Handmer, John, ed. *Flood Hazard Management: British and International Perspectives.* Norwich, UK: Geo Books, 1987. 297p. ISBN 0-86094-208-2.

This volume is the result of research and policy development at the Middlesex Polytechnic Flood Hazard Research Centre in Great Britain. It concentrates on policy-related issues of coastal and river flooding rather than on the physical aspects of flooding. The book discusses the British flood problem, the institutional and policy context of British flood hazard management, implementation of land use management, and the communication of flood warnings. The book concludes with an overview of the issues and recommendations for changes and research priorities. There are only a few tables and maps. Bibliographic references are given with each chapter.

Water Use

Ballard, Steven C., et al. *Water and Western Energy:Impacts, Issues, and Choices.* Studies in Water Policy and Management no. 1. Boulder, CO: Westview, 1982. 321p. ISBN 0-86531-332-6.

This book is a study sponsored by the Environmental Protection Agency on the relationship between water resources and energy development. It discusses the availability and quality of water, pollution from energy facilities, and salinity control. Water conservation, increasing water supply, protecting water quality, and water management by the states and federal government are also evaluated. The book includes many tables and figures and bibliographic references.

Frederick, Kenneth D., and James C. Hanson. *Water for Western Agriculture.* Series: RFF Research Paper. Washington, DC: Resources for the Future, 1982. 241p. ISBN 0-8018-2832-5.

Because rainfall in most of the West is insufficient to support agriculture, irrigation must be used. This book discusses the availability and demand for water, water laws and institutions, impacts of increasing energy costs and pumping depths on water costs, environmental problems associated with irrigation, and the expensive costs of transporting water. The book includes many tables, figures, and footnotes.

Kindler, J., and C. S. Russell, eds. *Modeling Water Demands,* Orlando, FL: Academic Press, 1984. 248p. ISBN 0-12-407380-8.

This book, written by several authors knowledgeable in their fields, describes ways to analyze water demands for industry, agriculture, and urban settlements. The book can be of value to hydrologists, water resources planners, water technologists, economists, and anyone working in the area of water management. The book begins by defining the problems of water demands, using case studies to explain these problems, and discussing some of the approaches that can be used in addressing water demand issues at the regional level. The book concludes with some thoughts on the existing limitations and the additional research needed on the role of water demand modeling. There is a lengthy bibliography and notes about the authors. A few tables and figures are found throughout the book.

Nolte, Byron H., ed. *Water Resources for Agricultural Production in the United States.* Public Policies Issues Report Developed by Water Issues Task Force. St. Joseph, MI: American Society for Agricultural Engineers, 1985. 39p. ISBN 0-916150-72-0.

This small monograph discusses the water quality and supply, managing excess water, storage, competing uses of water, and institutional and administrative water issues related to agriculture.

Postel, Sandra. *Conserving Water: The Untapped Alternative.* Worldwatch Paper no. 67. Washington, DC: Worldwatch Institute,1985. 66p. ISBN 0-916468-67-4.

This small monograph contains a wealth of information on water conservation methods. Only by managing water demand can there be hope for a secure water supply in the future. Irrigation systems must work more efficiently, which will also help produce a greater crop yield and more fertile soil. Recycled wastewater will be more extensively used in the future. Many homes have already installed low-flow showerheads—which also save electricity—and water-saving toilets to help conserve water. The book includes bibliographic references and several tables.

———. *Water for Agriculture: Facing the Limits.* Worldwatch Paper no. 93. Washington, DC: Worldwatch Institute, 1989. 54p. ISBN 0-916468-94-1.

Supplying water for irrigation is putting a strain on rivers, streams, and underground aquifers. The demand for water is approaching its limits and there will be shortages in areas such as northern China, Africa, parts of India, Mexico, and the western United States. In competing for water supplies, it seems that the farmer usually loses to cities. Watersheds need to be restored to stabilize water cycles that have been disrupted by deforestation and the denuding of land. Irrigation presents a hazard in that it washes a greater quantity of dangerous chemicals out of the soil than natural rainfall. Many areas in the West have overpumped groundwater for irrigation. This small volume has many excellent tables depicting such matters as water rights sales in the western United States, overpumped groundwater, and water scarcity in certain countries. Several pages of bibliographic references are included.

Willey, Zach. *Economic Development and Environmental Quality in California's Water System.* Berkeley: University of California, Institute of Governmental Studies, 1985. 73p. ISBN 0-87772-302-8.

This small volume examines California's water shortage. Among recent problems, the pollution of surface and groundwater by toxic substances is addressed. Higher rates are being charged for water from long-distance pumping and construction stations. The author concludes that many of the problems are due to the absence of a clearly defined objective in California water allocation and project planning. The monograph shows ways to reform the state water policy and benefits that might be made by the reform. The book has five appendices and several figures and tables.

Agriculture and Irrigation

Arid Zone Hydrology for Agricultural Development. FAO Irrigation and Drainage Paper no. 37. Rome: Food and Agriculture Organization of the United Nations, 1981. 367p. ISBN 92-5-101079-X.

This book is based on the work of hydrologist K. R. Jones and the FAO staff. The aim of the book is to describe the characteristics of important hydrological elements such as rainfall, evaporation, runoff, sedimentation, and groundwater, and how these elements function within arid zones. For instance, the annual flow of a stream in an arid area may take place in a few hours, whereas a stream in a humid area might be perennial. The book includes 271 pages of text and references and 96 pages of appendices.

Braden, John B., and Stephen B. Lovejoy, eds. *Agriculture and Water Quality: International Perspectives.* Boulder, CO: Lynne Rienner, 1990. 224p. ISBN 1-55587-183-6.

Several authors contributed to this book. It is divided into three parts. Part I discusses issues and policy options of water quality and incentives for reducing agricultural pollution. Part II studies specific cases of agricultural pollution in Sweden, Denmark, California, and Australia. Part III deals with agricultural competitiveness and environmental policies in international trade relations, concluding that nations must be aggressive in solving environmental problems in order to compete in international agricultural markets. The book includes many tables and figures along with references at the end of each chapter.

Chambers, Robert. *Managing Canal Irrigation: Practical Analysis from South Asia.* Wye Studies in Agricultural and Rural Development. New York: Cambridge University Press, 1988. 279p. ISBN 0-521-34554-5.

The author stresses the challenges of canal irrigation management in South Asia, and the various ways to confront these challenges. Part I gives background on the scale of canal irrigation within South Asia and how this type of irrigation might help the rural poor. Part II is concerned with the learning and mislearning surrounding canal irrigation. Part III addresses gaps or blind spots where water is missing, such as canal irrigation at night, farmers' activities above the outlet, assuring better water supply, and the environment and motivation of managers of canal irrigation systems. Part IV analyzes the problems and studies ways to improve the canal irrigation systems. The book includes a glossary, figures and tables, bibliographic references, and a list of abbreviations used. The appendix is an interesting poem titled "How to Succeed with Irrigation Action Research."

Hazlewood, Arthur, and Ian Livingstone. *Irrigation Economics in Poor Countries.* Elmsford, NY: Pergamon, 1982. 144p. ISBN 0-08-027451-X.

Irrigation, crucial to agricultural progress, can be expensive if errors are made in planning and construction. Engineers, hydrologists, agriculturalists, and others with proper training are required to carry out effective irrigation. This book begins by defining the concept of irrigation potential, followed by a discussion of programming methods, water supply, the economics of scale in irrigation projects, and the importance of economics and other social studies in the planning of irrigation. This book is based on a study of the Usangu Plains of Tanzania. There are many figures and tables and references are included for each chapter.

Korten, Frances F., and Robert Y. Siy, Jr., eds. *Transforming a Bureaucracy: The Experience of the Philippine National Irrigation Administration.* Case Study Series. West Hartford, CT: Kumarian, 1988. 175p. ISBN 0-931816-73-4.

This volume shows the role played by the farmer in planning and managing irrigation projects in the Philippines.

Uphoff, Norman. *Improving International Irrigation Management with Farmer Participation: Getting the Process Right.* Studies in Water Policy and Management no. 11. Boulder, CO: Westview, 1986. 215p. ISBN 0-8133-7330-1.

Irrigation is necessary to sustain agriculture in areas that have inadequate rainfall, but irrigation systems require correct technical planning and design. The farmer must play a greater role in irrigation management. This study has three objectives: (1) to make farmer organization and participation more comprehensible, (2) to assemble and evaluate experience with farmer organization and participation in irrigation management, and (3) to generate suggestions for establishing farmer organization and participation that can help improve irrigation performance. The following information can be found in the appendices: criteria for irrigation management performance, analysis of irrigation activities, and irrigation management roles. The book has several pages of bibliographic references and both a country/case and a subject index. It has only two tables and three figures.

Walker, Wynn R., and Gaylord V. Skogerbee. *Surface Irrigation: Theory and Practice.* Englewood Cliffs, NJ: Prentice-Hall, 1987. 386p. ISBN 0-13-877929-5.

The book is divided into two major parts. Part I, including chapters 1–10, covers the fundamentals of volume balance evaluation and design, and discusses measures in surface irrigation maintenance and operation. The second part deals with the theoretical contributions of pioneers of modern surface irrigation in the 1960s, the solutions developed in the mid-1970s, and the computer solutions developed in the 1980s. Surface irrigation systems are usually classed as basin irrigation, border irrigation, furrow irrigation, and wild flooding. The book includes photographs, figures, tables, and references for each chapter.

Weatherford, Gary D., et al., eds. *Water and Agriculture in the Western U.S.: Conservation, Reallocation and Markets.* Studies in Water Policy and Management no. 2. Boulder, CO: Westview, 1982. 269p. ISBN 0-86531-367-9.

This book is the product of two years of research funded by the National Science Foundation and conducted by a team of 18 persons sponsored by the John Muir Institute during the late 1970s. The project's main concern was to study conservation of agricultural water and reallocation. The book gives an overall view of water management policy, case studies in three agricultural settings—the Navajo Indian Irrigation Project, Central Arizona Project, and the Tulare Basin project—and water rights transfer investigations in Arizona, New Mexico, Colorado, and Utah. The appendix is in the form of a questionnaire. The book contains tables, figures, and notes.

Technology

Boyle, Robert H., John Graves, and T. H. Watkins. *The Water Hustlers.* San Francisco: Sierra Club, 1971. 253p. ISBN 0-87156-053-4.

This interesting book deals with the manipulation of large streams. In some cases, dams built to prevent floods often cause damage while in other cases stream manipulation is a waste of water. Yet developers—the hustlers—believe they are doing the greatest good for a large number of people. Flood control is only part of the issue, which includes the harnessing of rivers to supply water for urban and rural areas, to produce electric power and even to shuttle water from one watershed to another in interbasin water transfers. The authors examine the water problems in Texas, California, and New York. A chapter on manipulation shows what has been done on the Susquehanna, Hudson, Delaware, Raritan, and Passaic rivers, as well as Long Island Sound. The book shows the water features for each state discussed.

Canter, Larry W. Water Resources Assessment: Methodology and Technology Sourcebook. Ann Arbor, MI: Ann Arbor Science Publishers, 1979. 529p. ISBN 0-250-40320-X.

A total of 254 references were examined and studied for this book in a comprehensive review and evaluation of methodologies and technologies that have potential application to environmental impact assessment for water resources programs. The literature reviewed covers the period 1960–1978, subdivided into six chronological periods: 1960–1970, 1971–1973, 1974, 1975, 1976, and 1977–1978. Literature for the periods was examined with 12 criteria in mind: interdisciplinary team, assessment variables, baseline studies, impact identification, critical impacts, importance weighting, scaling or ranking, impact summarization, documentation, public participation and conflict management, and resolution. The book is written for those working on environmental impact studies and is oriented to water resources. It includes many figures, tables, and references.

Fahim, Hussein M. *Dams, People, and Development: The Aswân High Dam Case.* Pergamon Policy Studies on International Development. New York: Pergamon, 1981. 187p. ISBN 0-08-026307-0.

In this book the author has compiled what he learned from the study of the Aswân Dam since 1963 as a member of the Academy of Michigan project which was composed of the Egyptian National Academy of Scientific Research and Technology and the School of Public Health at the University of Michigan. The purpose of the project was to show the present and future effects of the dam. The book is divided into four parts covering cost/benefit variables, human implications, development potentials and constraints, and concluding remarks. The book has many figures, tables, and bibliographic references.

Haimes, Yacov Y., ed. *Scientific, Technological and Institutional Aspects of Water Resource Policy.* AAAS Selected Symposium no. 49. Boulder, CO: Westview, 1980. 128p. ISBN 0-89158-842-6.

This book collects papers presented at a symposium held at the AAAS annual meeting in Houston, Texas, January 3–8, 1979. Individual papers deal with water policy issues related to water resources research, groundwater, conservation, urban water systems, water resource planning, water supply and demand, costs and benefits, and institutional aspects of local, state, and federal policies. President

Carter's *Federal Water Policy Initiatives* is included in the book, though not a part of the symposium.

Hanada, O. P. *Water Well Technology*. Rotterdam, The Netherlands: A. A. Balkema, 1989. 311p. ISBN 90-6191-969-X.

This book examines the various drilling techniques and the difficulties encountered in drilling wells for water. It has an excellent chapter on water-lifting devices and well failure and maintenance. Testing procedures for the quality of water are also discussed. Several case histories are incorporated in the book to show how problems can be solved. The book contains a glossary of water well drilling terms, tables and formulas, conversion factors, and a bibliography.

Lawson, Michael L. *Dammed Indians: The Pick Sloan Plan and the Missouri River Sioux, 1944–1980*. Norman: University of Oklahoma Press, 1982. 261p. ISBN 0-8061-1657-9.

This is a very interesting book telling how dam construction on the Missouri River and its major tributaries resulted in the flooding of five Indian reservations. The Pick-Sloan Plan is said to have caused more damage to Indian land than any other public works project in America. The author goes on to show that the damages suffered by the Missouri River tribes outweigh the benefits provided to them. The book includes several illustrations and bibliographic references. Notes are given for each chapter.

Miller, Taylor O., et al. *The Salty Colorado*. Washington, DC: Conservation Foundation, 1986. 102p. ISBN 0-89164-093-2.

This book, published jointly with the John Muir Institute, recommends that the Colorado River Basin states and the federal government consider new methods for solving the salt content problem in the Colorado River. It also suggest ways to fund the control of salinity.

Porteous, Andrew, ed. *Desalination Technology: Developments and Practice*. New York: Applied Science Publishers, 1983. 271p. ISBN 0-85334-175-3.

The purpose of this book is to give the processes necessary to change saline or brackish water to water suitable for human consumption. The book is of value to anyone wishing to know more about the field of desalination. It includes many tables, figures, and references with each chapter.

Waldram, James B. *As Long As the Rivers Run: Hydroelectric Development and Native Communities in Western Canada.* Winnipeg: University of Manitoba Press, 1988. 253p. ISBN 0-88755-143-2.

This book deals with the environmental, social, and economic impact the building of dams has on the Indian and Metis people and details their struggle against hydro development. The author discusses the treaties made for their protection and several case studies, including the Cumberland House and Squaw Rapids Dam, Easterville and the Grand Rapids Dam, and the South Indian Lake and the Churchill River Diversion Project. In each of these cases the Indians were shoved off the land they lived on and placed on reservations. The early treaties were to remain in force "as long as the rivers run" but the native people find this promise difficult to believe and they feel that the situation is dominated by politics. Appendix 1 treats Treaty No. Five; Appendix 2 treats the Forebay Agreement; and Appendix 3 treats the Manitoba Hydro 1969 Compensation Proposal for South Indian Lake. The book includes many bibliographical references and notes.

Quality

Gabler, Raymond, and the editors of Consumer Report Books. *Is Your Water Safe to Drink?* Current Affairs/Environment/Health. Mount Vernon, NY: Consumers Union, 1988. 390p. ISBN 0-89043-041-1.

The availability and quality of drinking water affects our daily lives. Water problems are found everywhere. In the East, groundwater is contaminated by synthetic chemicals, and the West faces shortages. People are forced to boil water because of the parasite *Giardia lamblia.* The solutions to drinking water problems, whether they are political, technological, social, or economic, are as complex as the problems themselves. The purpose of this book is to define the water problem and to help people solve it. The book has many references and a glossary.

Greenley, Douglas A., Richard G. Walsh, and Robert A. Young. *Economic Benefits of Improved Water Quality: Public Perceptions of Option and Preservation Values.* Studies in Water Policy and Management no. 3. Boulder, CO: Westview, 1982. 164p. ISBN 0-86531-414-4.

This study of the South Platte River Basin was made to develop and apply a procedure for measuring the benefits of improved water quality. Benefits included enhanced enjoyment of water-based recreation, choice of recreation use through avoidance of pollution by mineral development, and preservation of a natural ecosystem for the future. Other benefits would include improved health for people, less expensive municipal and industrial water treatment, and improved water for irrigation. The book includes many figures and tables, three appendices, and a lengthy list of bibliographic references. Chapters also have bibliographic notes.

The Integrity of Water. Washington, DC: U.S. Environmental Protection Agency, Office of Water and Hazardous Materials, 1975.230p. No ISBN.

This volume covers the proceedings of a symposium held in Washington, DC, March 10–12, 1975. The program dealt with integrity from a physical, chemical, and biological point of view. It ended with an interpretation of integrity as viewed by the government, the conservationist, the industrialist, and the public. Included are many bibliographical references.

Kneese, Allen V., and Blair T. Bower. *Managing Water Quality: Economics, Technology, Institutions*. Baltimore, MD: Johns Hopkins University Press, 1968. 328p. No ISBN.

Much of the research for this volume, a revision of *The Economics of Regional Water Quality Management* published in 1964, was conducted under the auspices of Resources for the Future. The book aims to shed light on water quality issues such as how water quality is determined, what the best system of management measures are for achieving specified water quality, and what the most effective institutional or organizational arrangements for managing water quality are. The book is divided into five parts. Part I provides background, Part II outlines the theory of economic resource allocation as it relates to waste disposal problems, Part III focuses on a broader concept of regional water quality management, Part IV deals with organizational approaches to regional water quality management, and Part V summarizes what was said about water quality management. Selected references are given for the chapters along with tables and figures.

National Research Council, Committee on Ground Water Quality Protection. *Ground Water Quality Protection: State and Local Strategies*. Washington, DC: National Academy Press, 1986. 309p. ISBN 0-309-03685-2.

This book identifies technical and institutional features of programs used in protecting groundwater.

Pettyjohn, Wayne A. *Water Quality in a Stressed Environment.* Minneapolis, MN: Burgess, 1972. 309p. No ISBN.

Material in this book provides excellent supplementary readings for anyone interested in water. Topics discussed include drinking water and sources of surface and groundwater pollution, trace elements in pollution, as well as an examination of legal approaches to control pollution. Each reading has figures, tables, and bibliographical references. The epilogue looks into future environmental monitoring.

Ward, C. H., W. Giger, and P. L. McCarty, eds. *Ground Water Quality.* A Wiley-Interscience Publication. New York: John Wiley,1985. 547p. ISBN 0-471-81597-7.

This volume includes papers presented at the First International Conference on Ground Water Quality Research, October 7–10, 1981, at Rice University in Houston, Texas, which was sponsored by the National Center for Ground Water Research. Papers were presented on such topics as sources, types, and quantities of contaminants in groundwater, methods for groundwater investigation, the relation of subsurface characterization to groundwater pollution, and the subsurface transport and fate of pollutants and chemical contamination. The book includes figures, tables, and references.

Contamination

Barcelona, Michael, et al. *Contamination of Ground Water: Prevention, Assessment, Restoration.* Pollution Technology Review no. 184. Park Ridge, NJ: Noyes Data, 1990. 213p. ISBN 0-8155-1243-0.

As a resource document, this book brings together technical information on groundwater management that can be used at all levels of government as well as in the private sector. It contains information on preventing contamination, assessing the extent of contamination, and restoring groundwater quality. Bibliographic references and figures are found throughout the book. An appendix gives names and addresses by state for sources of information about groundwater contamination, the EPA Office of Groundwater Protection and the Federal Agency Groundwater Protection Committee.

Canter, Larry. *Environmental Impact of Water Resources Projects.* Chelsea, MI: Lewis, 1985. 352p. ISBN 0-87371-015-0.

This volume is divided into five chapters, the first serving as an introduction. Other chapters cover the environmental impact of dam and reservoir projects, channelization projects, dredging projects, and irrigation and shoreline projects. The book contains 16 appendices with 434 annotated references on such topics as: impoundment, channelization and dredging, nonpoint sources of pollutants, the transport and fate of pollutants in the water environment, water quantity/quality impact prediction, groundwater noise, cultural and visual socioeconomic impact prediction, and impact mitigation measures. The book includes tables and figures.

Churchill, Peter, and Ruth Patrick, eds. *Ground Water Contamination—Sources, Effects and Options to Deal with the Problem.* Third National Water Conference, 1987. Philadelphia, PA: Academy of Natural Sciences, 1987. 453p. No ISBN.

This volume contains the proceedings of the Third National Water Conference held in Philadelphia, January 13–15, 1987, and sponsored by the Academy of Natural Sciences. Papers covered such topics as contamination versus health effects, what happens to contaminants, how to determine cleanliness, remedial technology, and options for preventing contamination in the environment. The book includes figures and tables along with bibliographic references.

Ciaccio, Leonard L., ed. *Water and Water Pollution Handbook.* 4 vols. New York: Marcel Dekker, 1971–1973. Vol. 1: 449p. ISBN 0-8247-1104-1; Vol. 2: 800p. ISBN 0-8247-1116-5; Vol. 3: 1313p.ISBN 0-8247-1117-3; Vol. 4: 1945p. ISBN 0-8247-1118-1.

Many authors contributed to this four volume handbook on water and water pollution, which covers a multidisciplinary area of chemical, physical, and biological definitions of the water environment and environmental problems. Chapters are devoted to the chemical, physical, and biological characteristics of estuaries, irrigation and soil waters, wastes and waste effluents, and the effects of pollution, self-purification, water purification, and waste treatment. Each volume has illustrations and references.

Clark, Edwin H., II, Jennifer A. Haverkamp, and William Chapman. *Eroding Soils: The Off-Farm Impacts.* Washington, DC: Conservation Foundation, 1985. 252p. ISBN 0-69164-086-X.

Significant steps have been taken to clean up conventional pollutants such as dissolved solids, fecal coliform, and biochemical oxygen, but much more has to be done with nonpoint source pollution problems such as runoff from agricultural lands, construction sites, mining operations, and city streets. Runoff accounts for one-third to one-half of the pollution load in rivers, lakes, and estuaries. This book examines the problems caused by soil erosion from farms, which is the main cause of nonpoint source pollution. The book focuses on water pollution caused by sediments and other contaminants that are carried off farms with cropland erosion. Each chapter has a list of references. The book contains many figures.

Coffel, Steve. *But Not a Drop To Drink! The Lifesaving Guide to Good Water.* New York: Rawson, 1989. 323p. ISBN 0-89256-328-1.

Water shortages and poor water quality are bound to affect each of us. To ensure that we do not have polluted water supplies, we must take steps to avoid contamination. The author discusses the threat of toxic elements in drinking water and the ways we can protect water in our homes. He discusses the cause and effects of the 1980s drought. The book concludes with chapters ranking states by volume of hazardous waste generation, and the quality of the hazardous waste program. It also lists, by state, water information sources, health agencies, and Superfund toxic waste sites

Culp, Gordon L., and Russell L. Culp. *New Concepts in Water Purification.* Van Nostrand Reinhold Environmental Engineering Series. New York: Van Nostrand Reinhold, 1974. 305p. ISBN 0-442-2178-1.

The authors discuss the problems of water purification which arise because of pollution by domestic wastes, as well as chemical waste pollution from industry and agriculture. They assess treatment techniques for heavy metals, herbicides, and pesticides found in drinking water. These techniques can be used in existing treatment facilities to increase capacity and efficiency with little cost increase. Examples of full-scale use of the new concepts are found throughout the book. The authors indicate it is possible to solve virtually any water quality problem. References are found at the end of each chapter.

Dee, Norbert, William F. McTernan, and Edward Kaplan, eds. *Detection, Control, and Renovation of Contaminated Ground Water.* New York: American Society of Civil Engineers, 1987. 213p. ISBN 0-87262-595-8.

The volume contains the proceedings of a symposium sponsored by the Committee on Water Pollution Management of the Environmental Engineering Division of the American Society of Civil Engineers in conjunction with the ASCE Convention in Atlantic City, New Jersey, April 27–28, 1987. It was cosponsored by the EPA Office of

GroundWater Protection. The symposium covered such topics as groundwater contamination, monitoring and detection, control of contamination, and renovating contaminated groundwater. The volume contains many figures, tables, and bibliographic references.

D'Itri, Frank M., and Lois G. Wolfson, eds. *Rural Groundwater Contamination.* Chelsea, MI: Lewis, 1987. 416p. ISBN 0-87371-100-9.

The quality of groundwater is declining because of contaminants from natural and human generated sources. The authors give an overview of the sources, impacts, assessments, methods, and health risk implications of groundwater pollution along with remedial programs to help protect the groundwater. The public must be educated on the causes and prevention of groundwater contaminants in order to protect the nation's groundwater resources. Many authors contributed to write this volume. Bibliographic references are given with each chapter. Figures are found in most chapters.

Ground Water and Soil Contamination Remediation: Toward Compatible Science, Policy and Perception. Colloquium no. 5. Washington, DC: National Academy Press, 1990. 261p. ISBN 0-309-04184-8.

This is a report on a colloquium sponsored by the Water Science and Technology Board held in Washington, DC, April 20–21, 1989. This colloquium discussed important issues in water science and technology to show how science influences groundwater and soil contamination cleanup policy. The participants were well versed in the cleanup of contaminated soil and groundwater and have experience with the public perception of this contamination. The participants are also familiar with the laws and statutes that control the responses of state and federal agencies to a contamination situation. The book begins with an overview and then proceeds to the issue papers. Biographical sketches of contributors and a list of people attending the colloquium are found in the appendices. Bibliographic references are provided.

Groundwater Contamination. Studies in Geophysics. Washington, DC: National Academy Press, 1984. 179p. ISBN 0-309-03441-8.

This study, by the Geophysics Study Committee for the Geophysics Research Forum, investigates the transport and control of groundwater contaminants and the magnitude and scientific understanding of contaminant transport. Some of the problems associated with waste disposal and the prevention of groundwater contamination are examined. Many figures, tables, and references are found in the book.

Groundwater Contamination from Hazardous Wastes. Princeton University Water Resources Program. Englewood Cliffs, NJ: Prentice-Hall, 1984. 163p. ISBN 0-13-366286-1.

Groundwater contamination from hazardous wastes has become a serious problem. This book was written for industrial engineers and managers, environmental scientists, and anyone responsible for the management of hazardous wastes and the prevention of groundwater contamination. Several case studies show how improper disposal or accidental spillage of hazardous materials can contaminate groundwater. An appendix gives parameters for siting land emplacement facilities in regard to their proximity to groundwater and surface water supplies in order to avoid pollution. Once an aquifer is polluted, it remains contaminated for a long time, whereas a polluted surface supply may be diluted and flushed out. Many figures are used throughout the book. Bibliographic references are found at the end of each chapter. Chapter 2 has three appendices.

Harrington, Winston, et al. *Economics and Episodic Disease: The Benefits of Preventing a Giardiasis Outbreak.* Baltimore, MD: Johns Hopkins University Press, 1991. 202p. ISBN 0-915707-59-4.

This book is based on a study of the Luzerne County, Pennsylvania, outbreak of this disease in 1983. A framework is developed for analyzing the valuation of the consequences of waterborne gastroenteric disease outbreaks for individuals, government agencies, and businesses.

King, Jonathan. *Troubled Water: The Poisoning of America's Drinking Water.* Emmaus, PA: Rodale, 1985. 256p. ISBN 0-87857-571-5.

This book explains how the government and industry allows drinking water to be poisoned, and what you can do to ensure a safe supply within the home. The book includes bibliographic references.

Lake, Elizabeth, William M. Hanneman, and Sharon M. Oster. *Who Pays for Clean Water? The Distribution of Water Pollution Control Costs.* An Urban Systems Research Report. Westview Replica Edition. Boulder, CO: Westview, 1979. 244p. ISBN 0-89158-586-9.

Because of amendments to the 1972 Water Pollution Control Act, Public Law 92-500, water treatment facilities are required to be updated. The cost of these updates will be financed by the American public through higher taxes, reduced public services, and increased prices. The distribution of the costs will affect various segments of the population differently, as shown in this text. Each chapter has many exhibits and footnotes.

Patrick, Ruth, Emily Ford, and John Quarles. *Groundwater Contamination in the United States.* 2d ed. Philadelphia: University of Pennsylvania Press, 1987. 513p. ISBN 0-8122-8079-2.

Since the publication of the first edition in 1983, much has been learned about how to mitigate the effects of and how to prevent groundwater contamination. It is the purpose of this edition to inform the public about changes in preventing and managing groundwater contamination. Federal and state laws pertaining to groundwater contamination are given in special chapters. The quality of groundwater and the effects of contamination on public health and the environment are also discussed. There are many figures, tables, and bibliographic references.

Postel, Sandra. *Defusing the Toxic Threat: Controlling Pesticides and Industrial Waste.* Worldwatch Paper no. 79. Washington, DC: Worldwatch Institute, 1987. 69p. ISBN 0-916468-80-1.

Pesticides from home and agricultural use and toxic chemicals from industry and abandoned waste sites contaminate the water used by people and cause considerable ecological damage. Water supplies can be protected from pesticide contamination by following the integrated pest management (IPM) philosophy, which consists of biological controls, cultural practices, genetic manipulations, and the judicious use of chemicals to minimize hazards to health and the environment.

Velz, Clarence J. *Applied Stream Sanitation.* 2d ed. A Wiley-Interscience Publication. New York: John Wiley, 1984. 800p. ISBN 0-471-86416-1.

This large volume complements *Stream Sanitation,* written by E. B. Phelps in 1944, which already contained much of the information we have today for dealing with stream sanitation. The present book offers useful tools for evaluating solutions to water pollution and helps to develop an appreciation of the complex problem of waste disposal and pollution in a technological society. The main concern in this book is liquid waste—the spent water of industries, communities, and households that ultimately reaches a watercourse. The three states of human waste products—solids, gases, and liquids—are interlinked. There are three appendices on statistical tools, bacterial enumeration, and computer adaptation. There are also three indexes—application and methods, authors, and rivers. There are many figures and bibliographic references.

Acid Precipitation

Acid Deposition Long-Term Trends. Washington, DC: National Academy Press, 1986. 506p. ISBN 0-309-03647-X.

This volume is the result of research by the Committee on Monitoring and Assessment of Trends in Acid Deposition, the Environmental Studies Board, the Commission on Physical Sciences, Mathematics, and Resources, and the National Research Council. The study was undertaken because the deposition of chemical pollutants from the atmosphere is one of the most important environmental issues of our time. The researchers investigated the role of industry in contributing to the pollution of the air and the effects on the environment. They concluded that acid deposition has adverse effects on human health, acidification of surface water and soil, and degradation of natural resources. The book includes five appendices, many tables and figures, and bibliographic references for each chapter.

Adams, Donald D., and Walter P. Page, eds. *Acid Deposition: Environmental, Economic and Policy Issues.* New York: Plenum, 1985. 560p. ISBN 0-306-42062-7.

This book shows that acid rain is one of the worst environmental crises of our time due to the damage done to lakes, soils, and forested ecosystems. The authors cover atmospheric transport, the chemical processes which produce acid rain, and ecological, economical, and environmental consequences of acid rain. Many of the chapters deal with topics presented at the Conference on the Environmental and Economic Effects of Acid Deposition held at the State University of New York at Plattsburgh in 1983. A lengthy chapter is devoted to the effects of acid rain on groundwater quality. The book includes figures, tables, and bibliographic references.

Adriano, D. C., and A. H. Johnson, eds. *Acidic Precipitation: Biological and Ecological Effects.* Vol. 2. Advances in Environmental Science. New York: Springer-Verlag, 1989. 368p.ISBN 0-387-97000-2.

This is the second volume in the Advances in Environmental Sciences series on acid precipitation. Many authors contributed to this volume which explains the environmental effects of acid precipitation on trees, soils, crops, and forest ecosystems. Several chapters are devoted to the effects on stream and lake ecosystems. Each chapter has bibliographic references as well as explanatory figures. Contributors are listed.

Dubenick, David V., ed. *Acid Rain Information Book.* 2d ed. Park Ridge, NJ: Noyes, 1984. 397p. ISBN 0-8155-0967-7.

Acid rain, caused by the emission of sulfur and nitrogen oxides to the atmosphere, is one of the most publicized environmental issues of the day. This book begins by giving sources of acid rain, and then discussing atmospheric transport, monitoring, regional transport, adverse and beneficial effects, current research, and legislation to control acid rain. Many bibliographic references are found at the end of each chapter. Book has many tables and figures pertaining to acid rain.

Gilleland, Diane Suitt, and James H. Swisher, eds. *Acid Rain Control: The Costs of Compliance.* Carbondale: Southern Illinois University Press, 1985. 177p. ISBN 0-8093-1205-0.

This volume contains the proceedings of a conference on acid rain sponsored by the Illinois Energy Resources Commission and the Coal Extraction and Utilization Research Center at Southern Illinois University at Carbondale, held in Carbondale on March 18, 1984. Participants took a look at the costs and techniques of complying with acid rain legislation. They studied the costs of acid rain control for the electric utilities and the coal industries and the use of technology in burning high-sulfur coal, low-sulfur coal, and other fuels. Acid deposition or acid rain is one of the most controversial environmental issues of the day. The book has some references, tables, and figures. Some papers include a question-and-answer discussion.

———. *Acid Rain Control: The Promise of New Technology.* Carbondale: Southern Illinois University Press, 1986. 208p. ISBN 0-8093-1292-1.

This volume gives papers and proceedings of a conference sponsored by the Illinois General Assembly and the Coal Extraction and Utilization Research Center, Southern Illinois University at Carbondale, and held in Carbondale on April 10, 1985. The conference centered on the environmental effects of acid rain.

Lefohn, Allen S., and Sagar V. Krupa, eds. *Acidic Precipitation: A Technical Amplification of NAPAP's Findings.* Pittsburgh, PA: Air Pollution Control Association, 1988. 239p. No ISBN.

This volume contains the proceedings of an Air Pollution Control Association (APCA) international conference held in Pittsburgh in 1988 after the National Acid Precipitation Assessment Program (NAPAP) published findings from its study on acid precipitation. Members of the APCA group did not agree with some of the results of the NAPAP group and held a meeting to discuss their differences. The proceedings include papers on the effects of acid rain on crops, forests, aquatic systems, and visibility. The volume has many figures and references.

Legge, A. H., and S. V. Krupa, eds. Acidic Deposition: Sulphur and Nitrogen Oxides. Chelsea, MI: Lewis, 1990. 659p. ISBN 0-87371-190-4.

The Alberta government, in conjunction with various segments of industry, established the Acid Deposition Research Program (ADRP) in 1983 to study the effects of acidic deposition on the environment. In addition, ADRP set out to provide environmental management and control with respect to acid-forming gases, and to disseminate this information to public and government bodies, to perform additional research in the area of acid rain, and to have public representation in the program. This large volume has many tables and bibliographic references. Members of ADRP and the Science and Public advisory boards are listed.

Mellanby, Kenneth, ed., *Air Pollution, Acid Rain and the Environment.* Watt Committee Report no 18. New York: Elsevier, 1988. 129p. ISBN 1-85166-222-7.

This volume was published on behalf of the Watt Committee on Energy, which became interested in acid rain caused by energy generation and use. Stricter standards for the cleaning of emissions have been adopted in Britain. Damage done by emissions depends on climatic and geological factors. Bibliographic references follow each section of the book. Objectives and history of the Watt Committee is given, together with a list of members. Many tables and figures are found in each section.

Park, Chris C. *Acid Rain: Rhetoric and Reality.* New York: Metheun, 1987. 272p. ISBN 0-416-92190-6.

The author investigates the problems, sources, and patterns of acid rain. He then discusses the scientific complexities and effects on surface water, soils, vegetation, buildings, and humans. The book also delves into cures and remedies to acid rain from a technological viewpoint. The book ends with the politics of acid rain—a matter of international concern both in the United States and Great Britain. A list of tables and figures can be found in the front of the book. Several pages of bibliographic references are included.

Roth, Philip, et al. *The American West's Acid Rain Test.* Research Report no 1. Washington, DC: World Resources Institute, 1985. 50p. ISBN 0-915825-07-4.

This project of the World Resources Institute was undertaken by the Energy and Resources Group at the University of California at Berkeley. The West is concerned mainly about the damage that acid precipitation can cause lakes, streams, and forests, but also about emissions from smelters and coal-fired power plants and how they affect the environment. People are affected by acidic deposition by eating food or drinking water containing toxic metals. These toxic metals also affect fish, birds, and animals. This small monograph has many notes and a good-sized bibliography.

Schmandt, Jurgen, Judith Clarkson, and Hilliard Roderick, eds. *Acid Rain and Friendly Neighbors: The Policy Dispute Between Canada and the United States.* rev. ed. Duke Press Policy Studies. Durham, NC: Duke University Press, 1988. 344p. ISBN 0-8223-0870-3.

This book presents the work of a team of 15 graduate students and two faculty members on U.S.-Canadian relations pertaining to acid rain. They had three questions to answer: (1) What is the extent of agreement between the two countries about acid rain and its environmental effects? (2) What steps are under way in the countries to control acid rain and what additional steps could help? (3) What joint measures have been taken by the two countries to resolve the problems? The project was funded by the Tom Slick Fund for World Peace. The book includes notes on each chapter and many figures and tables, along with two appendices. The results are given in the conclusion found on pages 253–263.

Marine Pollution

Bockholts, P., and I. Heidebrink, eds. *Chemical Spills and Emergency Management at Sea.* Boston, MA: Kluwer Academic Publishers, 1988. 488p. ISBN 0-7923-0052-1.

This volume presents the proceedings of the First International Conference on Chemical Spills and Emergency Management at Sea, held in

Amsterdam, The Netherlands, November 15–18, 1988. Papers dealt with such topics as prevention and preparedness for accidents, emergency response to chemical spills, ecological and environmental effects of chemical spills, and pollution of inland waterways. Specific oil spills are discussed. The book includes figures, tables, and references.

Clark, R. B. *Marine Pollution.* 2d ed. New York: Oxford University Press, 1989. 220p. ISBN 0-19-854263-1.

In the five years since the first edition of this book was published, the public has become much more concerned about marine pollution. The public feels that no wastes of any sort can be safely disposed of into the sea. Yet, there will always be wastes, and where will they be disposed of? If not in the sea, they will be disposed of on land, in freshwater, or in the atmosphere. Disposal or treatment of the large volume of wastes will have an environmental cost. The author begins by discussing pollution in general and then proceeds to focus on the various types of pollution. There are many figures in each chapter and a conclusion with a short list of references for further reading.

Haas, Peter M. *Saving the Mediterranean: The Politics of International Environmental Cooperation. The Political Economy of International Change Series.* New York: Columbia University Press, 1990. 303p. ISBN 0-231-07012-8.

This book describes evolving forms of international cooperation for treating pollution in the Mediterranean Sea, which is derived from industry, ships, offshore dredging and mining activities, and untreated human wastes. These pollutants disperse very slowly in semienclosed seas like the Mediterranean because of a lack of tides and weak coastal currents. The Mediterranean Action Plan (Med Plan) evolved as a collective effort to coordinate the marine pollution control practices of the 18 Mediterranean littoral countries. The book includes notes, an extensive bibliography, and lists of acronyms and abbreviations used. There is also an interesting chronology of major Mediterranean Action Plan dates.

Squires, Donald F. The Ocean Dumping Quandary: Waste Disposal in the New York Bight. Albany: State University of New York Press, 1983. 226p. ISBN 0-87395-689-3.

The author wrote this book to show the interaction between human activities and the ocean environment. The bight, bounded by the North Atlantic Ocean, the beaches of Long Island and New Jersey, and the Hudson-Raritan Bay estuary, is perhaps the most affected stretch of coastal ocean in the nation. The author discusses the resources and the industry of the bight and its use as a dumping ground for industrial and toxic wastes. He also examines the health effects of dumping in the bight and what the future holds for the area. Many illustrations and references are included.

Tippie, Virginia K., and Dana R. Lester, eds. *Impact of Marine Pollution on Society.* Center for Ocean Management Studies, University of Rhode Island. New York: Praeger, 1982. 313p. ISBN 0-03-059732-3.

This volume is a compilation of papers presented at the Conference on the Impact of Marine Pollution on Society sponsored by the Center for Ocean Management Studies at the University of Rhode Island in cooperation with the National Marine Pollution Program Office. The book begins by giving the status of marine pollution—ways of controlling it and our understanding of the human impact on the marine environment and international efforts to control marine pollution. It continues with discussions on specific examples of marine pollution, such as the *Amoco Cadiz* oil spill, disposal of contaminated dredged materials in Long Island Sound, and the discharge of sewage sludge off the coast of southern California. The book concludes with a section on future prospects and strategies. Most of the papers have figures, tables, and references.

Waters, W. G., II, T. D. Heaver, and T. Verrier. *Oil Pollution from Tanker Operations: Causes, Costs, Controls.* Centre for Transportation Studies. Vancouver: University of British Columbia, 1980. 216p. ISBN 0-919804-17-9.

The purpose of this book is to help people understand the facts and issues involved in controlling oil discharges by tanker operations. It discusses ways to reduce oil pollution of the sea and the economic principles of pollution control. A comparison of the benefits and costs of different methods of controlling operational discharges is examined. This book does not deal with spills caused by collisions or accidents, but only tanker discharges. The book contains many bibliographic references and three appendices.

Water Resources Development

Amoss, Harold L., ed. *Water: Measuring and Meeting Future Requirements.* Western Resources Papers no. 2. Boulder: University of Colorado Press, 1961. 261p. No ISBN.

This book is a record of the 1960 Western Resources Conference held in Boulder, Colorado on the subject of water. Papers dealt with problems of national scope and of regional planning. Of special interest was a paper on the water "industry" of California. Another paper discussed the engineering developments of the Bureau of Reclamation. Water used in recreation was also discussed. This conference provided the opportunity to discuss the new techniques and approaches to water use patterns and work being done in certain areas. Participants felt the incorporation of groundwater and surface water was necessary in calculating the present supply and future capabilities of water. The papers include few tables or illustrations and no references.

Humlum, Johannes. *Water Development and Water Planning in the Southwestern United States.* Aarhus, Denmark: Aarhus Universitet, Kulturgeografisk Institut, 1969. 240p. No ISBN.

This book reveals the growth in freshwater planning during the last 50 years in the southwestern United States and looks into plans for transferring freshwater from precipitous mountain areas to the semiarid regions of California. It shows the close relationship between where people live in California and Arizona and the freshwater supply distribution. The author spent several years in the 1960s as a visiting professor at the University of California, Riverside, during which time he made his studies on water development. The book has tables, maps, graphs, photographs,and bibliographic references.

Kneese, Allen V., and F. Lee Brown. *The Southwest Under Stress: National Resource Development Issues in a Regional Setting.* Baltimore, MD: Johns Hopkins University Press, 1981. 268p. ISBN 0-8018-2707-8.

This interesting volume begins by giving the background of the Southwest and the many problems such as water, Indians, and environment that settlers had to face and solve in order to develop the area. Part II focuses on the water problems, which have been increased by the movement of people and industries to the Sunbelt. The Upper Colorado River Basin is the source of renewable water supply and also has unexploited energy resources. The Colorado must also meet out-of-basin demands by diverting water between basins. The Yampa River and its coal resources are used to illustrate how water is essential for energy development. Water in the Southwest can be claimed by the Indian tribes, federal government, and water users such as farmers, mining companies, manufacturers, and municipalities. Proper management is necessary for such a scarce resource. The book includes many tables, figures, and a list of publications resulting from the Southwest Under Stress Project undertaken by Resources for the Future.

McCool, Daniel. *Command of the Waters: Iron Triangles, the Federal Water Development, and Indian Water.* Berkeley: University of California Press, 1987. 321p. ISBN 0-520-05846-1.

The author examines the water policy in the tripartite alliances known as the iron triangles. He discusses the Indian triangle and the non-Indian water development iron triangle and shows how the concept of iron triangles has changed over the years. People feel that the future federal water policy will have most impact on the West. Despite the expense, the government must provide water or compensate those who are injured by the lack of water. The appendix offers maps of the eastern, central, and western United States, showing locations of Indian reservations and the major water development projects. The book includes several pages of references.

McDonald, Adrian, and David Kay. *Water Resources: Issues and Strategies.* Themes in Resource Management. Essex, UK: Longman Scientific and Technical; copublished in New York with John Wiley, 1988. 284p. ISBN 0-470-21150-4.

The authors review the issues and opportunities associated with water supply, flooding and erosion control, power generation, and water quality. Water is the most essential item provided by nature to sustain life for humans, animals, and plants. Chapters cover such topics as water supply, the human resource, flooding control and management, water quality, water as a power resource, river basin management, and lessons for the future. A lengthy list of references is included.

Petersen, Margaret. *Water Resource Planning and Development.* Englewood Cliffs, NJ: Prentice-Hall, 1984. 316p. ISBN 0-13-945908-1.

The author pulls together material on water resource planning and development that appeared in various literature, manuals, and not easily available government publications, particularly documents from the U.S. Army Corps of Engineers and the United Nations. This volume was prepared as a teaching aid for a course the author gave in the Department of Civil Engineering and Mechanics at the University of Arizona. The book gives an overall view of principles, procedures, and problems in water resource planning. Each chapter has a list of references for further reading, some tables and figures, and seven appendices.

Ross, Lester. *Environmental Policy in China.* Bloomington: Indiana University Press, 1988. 240p. ISBN 0-253-31837-8.

This volume discusses the Chinese environmental policy and assesses the efficiency of alternative modes of policy implementation. Several related policy sectors such as forestry, water resources management, pollution control and natural hazards management are examined. The chapters on water policy and natural hazards show that China would have enough water were it not for the effects of regional and cyclical variation and population pressure. Management and distribution affect water resources. This book has bibliographical references and several tables and figures.

Management

Ali, Mohammed, George E. Radosevich, and Akbar Ali Khan, eds. *Water Resources Policy for Asia.* Rotterdam, The Netherlands: A. A. Balkema, 1987. 627p. ISBN 90-6191-684-4.

This large volume contains the Proceedings of the Regional Symposium on Water Resources Policy in Agro-Socio-Economic Development, held in Dhaka, Bangladesh, 4–8 August 1985. Papers dealt with such topics as water resources in various countries, national and transnational legal systems concerning water, and planning and implementation of national and transnational water resources policy. They discussed the need for comprehensive long-range planning on water issues because of it scarcity. Book has some tables and figures. An appendix gives information about the editors and authors.

Anderson, Terry L. ed. *Water Rights: Scarce Resource Allocation, Bureaucracy, and the Environment.* Pacific Studies in Public Policy. San Francisco, CA: Pacific Institute for Public Policy Research. Cambridge, MA: Ballinger Publishing Co., 1983. 348p. ISBN 0-88410-389-7.

The allocation of agricultural and urban water supplies in the United States is examined in this book. Water development in the United States is a political situation. As cities grow, a "water shortage" appears, and often new sources of water prove to be useless. This book discusses how markets and private ownership will provide a more efficient and equitable allocation of water. Many authors contributed to this volume, which discusses such subjects as the historical background for water policy, the role played by the Bureau of Reclamation, inefficient water use, restructuring institutions governing water allocation, and privatizing groundwater basins. The book addresses two of the most difficult water rights problems: pollution and instream use. Book has many figures, tables, and a selected bibliography.

Batie, Sandra S. and J. Paxton Marshall. *Emerging Issues in Water Management and Policy.* SNREC Publication, no. 17, SEPAC Publication, no. 4. Chicago, IL: Farm Foundation and the Southern Rural Development Center, 1983. 83p. No ISBN.

These are the proceedings of a workshop held in Blacksburg, Virginia, June 7–8, 1982. The main issues examined at the conference were water rights and laws in the South, water use and availability in the South, federal, state, and local governments' roles in water management, the potentials and problems of alternative water management, and the Virginia Polytechnic Institute and State University's role in water research and education. There are bibliographic references.

Black, Peter E. *Conservation of Water and Related Land Resources.* New York: Praeger Publishers, 1982. 209p. ISBN 0-03-060419-2.

The author begins by discussing the water law which focuses on the use of surface waters. He then offers a discussion of some regional, national, state, and local organizations. Two interesting chapters deal with policy, planning, and pollution in relation to water development projects and programs. The book concludes with a chapter on conservation. A list of cases, statutes and citations, and several appendices are found in the book. Book also has many figures, tables, a list of abbreviations, and a lengthy bibliography.

Brown, Carl, Joseph G. Monks and James R. Park. *Decision-Making in Water Resource Allocation.* Boston, MA: Lexington Books, 1973. 110p. ISBN 0-669-86736-1.

This book is the result of a study that addressed the allocation of water. The project used for the study was the Lower Amazon and Junction City Project in the Willamette Valley, Oregon. The purpose of the study was to identify the sources of demand for water resource development, to analyze the political/economic aspect of water resource development, to determine the decision-makers' idea of the value of water and legal constraints, to study the impact of decision-making on public and private demands, and to protect the rights of persons who owned the resource. This particular project was chosen because all levels of government were involved. Some conclusions of the study were: there was little participation by members in decision-making, there were inadequate channels of communication, the decision-making criteria changed at local, state, and federal levels, there was a feeling that people within the water control district benefited the most from the project, and there was also a feeling that the project, began for flood control and drainage, had become an irrigation project. There are several appendices and bibliographic references.

Cairns, John, Jr., and Ruth Patrick, eds. *Managing Water Resources.* Environmental Regeneration Series. New York: Praeger. 1986. ISBN 0-275-92200-6.

This book discusses the management, uses, and recycling of water. This book stresses human health and water quality, management of water for recreation, industrial, agricultural, and wildlife needs. Each of the seven chapters contains a bibliography.

Carroll, John E., ed. *International Environment Diplomacy: The Management and Resolution of Transfrontier Environment Problems.* New York: Cambridge University Press, 1988. 291p. ISBN 0-521-33437-3.

A transboundary environmental problem arises when a pollution-emitting activity benefits one nation (industry), and the environmental damage of that activity affects a bordering nation. The location of the border between two nations is the important factor when such a problem arises. This book begins by giving the framework for international diplomacy in industrialized nation-states. A few chapters deal with legal and diplomatic resolutions in international environmental diplomacy. A section of special interest deals with the case of acid rain, perhaps the greatest transboundary environmental issue. The book ends with a section on the problems of marine pollution. Each chapter was written by an expert in the subject area and has bibliographic references.

Dzurik, Andrew A. *Water Resources Planning.* Savage, MD: Rowman & Littlefield, 1990. 340p. ISBN 0-8476-7391-X.

This book gives a comprehensive survey of water resources planning and management, examining the concepts of hydrology. It covers federal, state, and local laws; water quality, supply, and demand; methods to evaluate water resource plans in relation to environmental impact; and cost-benefit analyses. Future trends in water resources planning and management are also discussed. This book has references, appendices, and a glossary.

Easter, K. William, John A. Dixon, and Maynard M. Hufschmidt, eds. *Watershed Resources Management: An Integrated Framework with Studies from Asia and the Pacific.* Studies in Water Policy and Management, no. 10. Boulder, CO: Westview, 1986. 236p. ISBN 0-8133-7300-X.

This book is the outcome of work completed on water resources and watershed management in developing countries at the East-West Center and the University of Minnesota. Four workshops were held during 1983–1985 dealing with watershed mismanagement of soil erosion and channel and reservoir sedimentation, problems on steep slopes in moist temperate forests, river and reservoir sedimentation associated with watershed management; and the final workshop dealt with integrated watershed management in developing countries. Six case studies were used in the watershed research: Hawaii, Hindu Kush-Himalayan Region, Northern Thailand, Nepal, Java, and the Philippines. This book has several tables and figures along with references for each chapter and a list of contributors.

El-Ashry, Mohamed T., and Diana C. Gibbons. *Troubled Waters: New Policies for Managing Water in the American West.* Study 6. Washington, DC: World Resources Institute, 1986. 89p. ISBN 0-915825-15-5.

Water is essential for economic growth in the arid and semiarid lands of the West. With the expansion of western cities, increased industry, and the growing demand for water some previously reliable water sources have become polluted and depleted. *Troubled Waters* is the first publication of the Arid Lands Project, undertaken by the World Resources Institute, dealing with land and water management. The Institute held a workshop in Tucson, Arizona, February 20–21, 1986, to study the agricultural cases of the Upper Colorado River Basin; Central Valley, California; and High Plains, Texas. Urban case studies were made for Denver, Tucson, and southern California. Participants are listed in an appendix. Bibliographical references are included.

Gibbons, Diana C. *The Economic Value of Water.* Baltimore, MD: Johns Hopkins University Press for Resources for the Future, 1986. 101p. ISBN 0-915707-23-3.

The author gives a framework for understanding water value and begins this book by discussing the price ranges of municipal water demand: if the supply decreases then the marginal value increases.

The use of water for irrigation and industry, and the associated costs, are then addressed. The author also studies the value of water for water-based recreation, navigation, and its use in hydropower. Many notes are found for each chapter along with figures and tables.

Grigg, Neil S. *Water Resources Planning.* New York: McGraw-Hill, 1985. 328p. ISBN 0-07-024771-4.

Water problems occur because of population increase and urbanization. This book has two parts: principles and techniques of planning, and case studies to illustrate the principles in action. Water management is a critical international issue, as illustrated by the desertification of Africa, losses from floods and droughts, the lack of safe drinking water, and the hunger problem in certain nations. Much of the technology to manage water is known, but the application of that technology has been tediously slow. The author contends that an understanding of politics is needed by water planners, as water planning is an exercise in public administration. Some of the case studies include regional water supplies in Tidewater, Virginia; Cache la Poudre Basin, Colorado; and the San Francisco Wastewater Plan. The book also addresses Flood Control Planning, Groundwater Planning, and River Basin Planning. References and conclusions are found at the end of each chapter.

Howe, Charles W., and K. William Easter. *Interbasin Tranfers of Water: Economic Issues and Impacts.* Baltimore, MD: Johns Hopkins University Press for Resources for the Future, 1971. 196p. ISBN 0-8018-1206-2.

To alleviate drought conditions, water has been transferred from one river basin to another, from one region of the country to another, and even from one part of the North American continent to another. This book examines the cost of interbasin transfers and suggests economic measures to be used in these projects. Emphasis is placed on the West because of the water shortage there. The book concludes by showing the need for more inclusive evaluations in water resources planning. Several pages of bibliographic references are given, and 65 tables are found throughout the book.

Keating, Michael. *To The Last Drop: Canada and the World's Water Crisis.* Toronto, Ontario: Macmillan, 1986. 265p. ISBN 0-7715-9704-5.

The author stresses the fact that threats to water supply and water quality make it necessary to adopt sound water management policies in North America and elsewhere. Water is a resource vital to human development and well-being, so it must be under careful management. This book has 15 chapters, four appendices, several maps, photos, and a short list of references.

Kneese, Allen V., and Stephen C. Smith, eds. *Water Research.* Baltimore, MD: Johns Hopkins University Press, 1966. 526p. No ISBN.

This book is a compilation of papers presented at the seminars in Water Research, sponsored by Resources for the Future and the seventh Western Resources Conference, held at Colorado State University, July 1965. Papers covered such topics as economic analysis, water management, evaluation problems, water reallocation, political and administrative problems, and hydrology and engineering research program needs. Each chapter has bibliographic references.

Koch, Stuart G. *Water Resources Planning in New England.* Hanover, NH: University Press of New England, 1980. 185p. ISBN 0-87451-176-3.

This book is concerned with the need for public participation in solving the water resources planning problem. The public participated in three regional planning studies conducted by the New England River Basins Commission, an agency which coordinates federal, state, and interstate plans for the management of water resources in the region. Public participation helps strengthen community bonds and often facilitates the implementation programs. The works cited in the book are listed along with notes.

Kosinski, Leszek A., W. R. Derrick Sewell, and Wu Chuanjun, eds. *Land and Water Management: Chinese and Canadian Perspectives.* Edmonton, Alberta: Department of Geography, University of Alberta, 1987. 214p. ISBN 0-88864-857-X.

This volume consists of selected papers from a Sino-Canadian conference on Territorial Development and Management held in Beijing in 1986. The conference provided personal contacts where ideas could be exchanged on research being done in the area of land and water management. Papers included in this book are by Canadian and Chinese geographers. This book has a list of tables and figures. References are found at the end of each paper.

Lee, Terence R. *Water Resources Management in Latin America and the Caribbean.* Studies in Water Policy and Management, no. 16, Boulder, CO: Westview, 1990. 208p. ISBN 0-8133-7999-7.

The people of Latin America and the Caribbean regions give special attention to the harnessing of water resources in order to improve their standards of living. Consequently, these geographical areas have great volumes of water stored in reservoirs and an increased amount of land under irrigation. Growth has also occurred in the generation of hydroelectricity and in the supply of water for consumption and industry. Investments in control and regulation of rivers in order to improve the economy of the region have increased the roles of the public sector and the government. This presentation gives case studies to show characteristics of water management in Latin America and the Caribbean. Some of the case studies cover Mendoza, Argentina; Bogota, Colombia; Lambayeque, Peru; and the Limari-Paloma water system in Chile. Conclusions are given as to the current state of water resource management and ways for improvement. Bibliographic notes are given for each chapter. This book has many tables and figures.

Mitchell, Bruce, ed. *Integrated Water Management: International Experience and Perspectives.* New York: Belhaven, 1990. 225p. ISBN 1-85293-026-8.

Authors from the United States, the United Kingdom, Canada, Poland, New Zealand, Japan, and Nigeria contributed to this book. The book focuses on the experiences in many countries to examine the various ways in which integrated water management was being approached. Successful approaches in one country could not necessarily be transferred to another country. People living and working in the countries were chosen to give their experiences. This book includes figures, tables, and a list of contributors. Each chapter has bibliographic references.

Moore, James W. *Balancing the Needs of Water Use.* Springer Series on Environmental Management. New York: Springer-Verlag, 1989. 280p. ISBN 0-387-96709-5.

This book studies the main components of the water-use cycle and the natural hydrologic cycle, examining the requirements of such water users as fish, wildlife, municipalities, agriculture, forestry, transportation, manufacturing, energy and storage and the impact each has on the environment. The book has many illustrations and appendices. References are found at the end of each chapter.

Postel, Sandra. *Water: Rethinking Management in an Age of Scarcity.* Worldwatch Paper, no. 62. Washington, DC: Worldwatch Institute, 1984. 65p. ISBN 0-916468-62-3.

The author paints a bleak picture of a water crisis in which no one knows how much water is used where, when, or by whom. Yet water is absolutely essential for life on the planet. Water shortages are greatest where population grows fastest. Much of the water has become unsafe for use because of pollution by disposal of chemicals and heavy metals. Deforestation diminishes runoffs while dams and reservoirs are built to collect the water. The author concludes that industry should pay for using water in production. This book has several tables and figures.

Reisner, Marc. *Cadillac Desert: The American West and Its Disappearing Water.* New York: Viking Penguin, 1986. 582p. ISBN 0-670-19927-3.

This is the story of the West's attempt to control and allocate water. It begins with the story of early settlers, continuing to the present day efforts to control water. The origin of the Bureau of Reclamation is discussed, as well as its role in the construction of dams, such as the Shasta, Hoover, and Grand Coulee, in order to generate cheap hydroelectricity for cities and towns. Any dams that the Bureau of Reclamation would not sponsor were built by the U.S. Army Corps of Engineers, which showed the competition between the two. *Cadillac Desert* paints a picture for the future of water in the West. This is a very readable book. It has several pages of notes and a bibliography.

Riebsame, William E., Stanley A. Changnon, Jr., and Thomas R. Karl. *Drought and Natural Resources Management in the United States: Impacts and Implications of the 1987–89 Drought.* Westview Special Studies in Natural Resources and Energy Management. Boulder, CO: Westview, 1991. 174p. ISBN 0-8133-8026-X.

Much of the United States experienced severe drought by the mid-summer of 1988. This book examines the 1987–1989 drought and studies the possibility of water management to reduce the future impacts of droughts. Four case studies are included: water and transportation management in the Mississippi River system, dryland farming in the Northern Great Plains, metropolitan water supply provision in Atlanta, and wildlife management in Yellowstone National Park. Problems, solutions, and responses to drought in each of these cases were evaluated to help improve drought management. This book has many figures and references are found with each chapter.

Sheaffer, John R., and Leonard A. Stevens. *Future Water: An Exciting Solution to America's Most Serious Resource Crisis.* New York: Morrow, 1983. 269p. ISBN 0-688-01575-1.

The authors discuss the problem of a serious water crisis by the year 2000 which they feel is even more serious than the energy crisis. They believe that the real solution is not finding new sources, but efficient management of existing supplies in closed, circular systems. The Federal Water Pollution Control Act Amendments of 1972 passed by Congress made it possible to clean up America's water, but still there are many chemicals causing contamination that have to be coped with. Water can be purified, where soil and plants act as the living filter, through the process of evaporation, condensation, rain/snow and the melting of snow. This book has many pictures and references for each chapter.

Viessman, Warren, Jr., and Claire Welty. *Water Management: Technology and Institutions.* New York: Harper and Row, 1985. 618p. ISBN 0-06-046818-1.

In a United Nations study its was found that two-thirds of the population of developing countries did not have safe and adequate drinking water. Even technically advanced nations are faced with impure water because of chemicals and waste disposal practices. The purpose of this book is to suggest ways to avoid water shortages in the future through technological advancements. Parts of this book are very technical, while other sections are nontechnical and easy reading. The book is to serve as a reference for professionals interested in water resource decision-making processes. Many figures and tables are found throughout the book and references are included for each chapter. Four appendices are included.

Economics

Saliba, Bonnie C., and David B. Bush. *Water Markets in Theory and Practice: Market Transfers, Water Values and Public Policy.* Studies in Water Policy and Management, no. 12. Boulder, CO: Westview, 1987. 273p. ISBN 0-8133-7465-0.

The authors describe what a water market is, why water markets develop, who is buying, selling, and leasing water, how trades are made, the prices paid, and the laws controlling market transactions. Water markets have been used for years in the Southwest and will continue to be used in coming years. Each chapter has bibliographic references. Some chapters have tables and figures.

Smith, Rodney T. *Troubled Waters: Financing Water in the West.* Washington, DC: Council of State Planning Agencies, 1984. 201p. ISBN 0-93482-33-7.

This book was written to help public officials properly manage the water supply. A chapter discusses the current state of water investment in the West. The use of the municipal bond is also discussed, relating the costs to municipalities, states, and the financial condition of the western states. Some of the other topics that are examined are: water use fees as a source of revenues, fiscal and legal issues, financial and administrative relations between state and local governments, and how the private sector may provide a means for financing water projects. Most of the policies and programs emphasize the West, but could apply to any region of the country. This book has many figures and tables and several pages of bibliographic references.

Storper, Michael, and Richard A. Walker. *The Price of Water: Surplus and Subsidy in the California State Water Project.* Institute of Government Studies. Berkeley: University of California, 1984. 30p. ISBN 0-87772-293-5.

This small monograph discusses the price of water and who pays for it in California. The appendix looks into transfers of surplus water between contractors along with several explanatory tables.

Water Rights

Andrews, Barbara T., and Marie Sansone. *Who Runs the Rivers? Dams and Decisions in the New West.* Stanford, CA: Stanford Environmental Law Society, 1983. 452p. No ISBN.

The authors examine water resource allocation west of the Mississippi, across the semiarid plains and mountainous regions where water has always been a major issue. The book presents the decision-making process for western water projects using the New Melones Dam on the Stanislaus River as a case study. The purpose of this book is to stimulate thought on the water allocation processes and problems in the West and what the role of state and federal governments should be in the development of western water. Also considered are the roles of the legislative branch, the agencies, and the courts in shaping decisions in water development. There is a selected bibliography and a list of interviewees.

Dunbar, Robert G. *Forging New Rights in Western Waters.* Lincoln: University of Nebraska Press, 1983. 278p. ISBN 0-8032-1663-7.

The author tells the story of irrigation and how it evolved in the western states, replacing the Doctrine of Riparian Rights with the Doctrine of Prior Appropriation which emphasized the equality of use. The book has several pages of notes for the sixteen chapters.

Eaton, David J., and John M. Andersen. *The State of the Rio, Grande/Rio Bravo: A Study of Water Resource Issues Along the Texas/Mexico Border.* Tucson: University of Arizona Press, 1987. 331p. ISBN 0-8165-0990-5.

This excellent volume on the Rio Grande discusses the availability and storage of surface water and the quality of the water. Another part discusses groundwater availability and quality as used in the United States and Mexico. An interesting chapter deals with the use of water along the Rio Grande, such as industrial, municipal, irrigation, livestock and manufacturing. There is concern about a shortage with the increased population along the river, thus both countries should reach bilateral agreements on water conservation. The section on potable water systems examines laws and regulations controlling drinking water, the existing water treatment facilities and gives some conclusions regarding the potable water supply in the area. The book finally ends with a discussion of diseases found in the area due to water and future projections on water uses. Book has many tables and figures, reference materials, glossary and bibliographic references.

Meyer, Michael C. *Water in the Hispanic Southwest: A Social and Legal History, 1550–1850.* Tucson: University of Arizona Press, 1984, 189p. ISBN 0-8165-0825-9.

Water has always been a concern of mankind—the availability or scarcity has always been a problem. The author traces the influence of water on the development of the land showing the inter- and intra-racial conflict over water. He also discusses conflict resolution and adaptation of Spanish and Mexican jurisprudence and various judicial systems to solving controversies produced by water. He really shows the role of water in the historical processes of the Southwest. The book has illustrations and bibliographic references.

Policy and Legislation

Anderson, Terry L. *Water Crisis: Ending the Policy Drought.* Baltimore, MD: Johns Hopkins University Press, 1983, 121p. ISBN 0-8018-3087-7.

The book is about water in the West and gives a provocative analysis of the problems of water shortage and property rights to water. The author documents obstacles that arise when water use is shifted from one area of use to another or from one kind of use to another. Such obstacles occur for shifts to environmental, recreational uses, and commercial water use. The author gives arguments for a private market in water rights and gives the advantages for such a market. The federal water development programs have become very expensive and unproductive. The author feels the water shortages in the West have been created by the people. The book has footnotes and illustrations.

Getches, David H. *Water Law in a Nutshell,* 2d ed. Boulder, CO: Westview, 1990. 459p. ISBN 0-314-73779-0.

Book cites generally applicable statutory and case law, with selected references to particular cases and states in the United States.

Goldfarb, William. Water Law. An Ann Arbor Science Book. Boston, MA: Butterworth, 1984. 233p. ISBN 0-250-40627-6.

Author wrote the book in a nonlegal style to make it readable for any audience. He begins by explaining what is meant by the term "law," "water law," and "water rights." The book is then divided into four main parts. Part 1 explains the law of water diversion and distribution by discussing the riparian system, prior appropriation, federal and state diversion law, interstate and intrastate transfers and the drainage law. Part 2 shows the part played by the federal government in the development and protection of water resources. Part 3 discusses the nontransformational uses such as use of rivers and streams, lakes, floodplain and wetland protection, acid precipitation prevention, and conservation and reuse. Part 4 deals with the Clean Water Act and how it applies to water treatment and land use. Bibliographic references are found at the end of each part.

Gottlieb, Robert. *A Life of Its Own: The Politics and Power of Water.* New York: Harcourt Brace Jovanovich, 1988. 332p. ISBN 0-15-195190-X.

The author wrote this book in a easy, readable style showing the many ways water affects us, such as the water we drink, the food we eat, irrigation, urban growth, and industrial hazards from toxicity. The destiny of the land and the people is controlled by both private interests and public agencies' control over the water industry. The book was based on the author's personal experience in the water industry. There are several pages of notes and a rather lengthy index as well as acknowledgements.

Ingram, Helen M. *Patterns of Politics in Water Resource Development: A Case Study of New Mexico's Role in the Colorado River Basin Bill.* Publications of the Division of Government Research, no. 79. Albuquerque: Institute for Social Research and Development, University of New Mexico, 1969. 95p. No ISBN.

The role of New Mexico in relation to the Colorado River Basin Bill shows clearly how important a part politics plays in determining water policy. Water development plans for New Mexico were included in the basin legislation on the basis of political calculation. The two New Mexico projects—Animas-La Plata Project and the Hooker Dam Project on the Gila River—were added to the Central Arizona Project for political reasons to help the passage of the Colorado River Basin Bill. Each project is discussed fully in the book.

———. *Water Politics: Continuity and Change.* Albuquerque: University of New Mexico Press, 1990, 148p. ISBN 0-8263-1189-X.

This is a revised edition of *Patterns of Politics in Water Resource Development,* published in 1969. Ingram shows the changes that have taken place in water policy and models from the 1960s to the 1990s. The Colorado River Basin Act is discussed as a water development model. The Animas-La Plata Project is used as a model for reclamation. Negotiations between New Mexico and Arizona regarding the Hooker Dam to increase water supply is discussed. The environmental challenge to the Hooker Dam and the way the challenge was handled politically is examined. There are several pages of bibliographic notes.

Kenski, Henry C. *Saving the Hidden Treasure: The Evolution of Ground Water Policy.* Contemporary Issues in Natural Resources and Environmental Policy, no. 2. Claremont, CA: Regina Books, 1990. 159p. ISBN 0-941690-26-1.

This is a timely book for the public that is becoming more conscious of groundwater contamination since it is the source of drinking water for most of the people. The detection of contaminants is a slow process, and often they cannot be traced back to their origin. Landfills and hazardous waste sites may not affect the groundwater for decades. Solution to the problem is also slow because of the cost. One of the promising developments for detecting and abating the situation is the discovery of bacteria that feed upon and detoxify hazardous wastes. He discusses the role science and technology must play in protecting groundwater in the future. General references are given for each chapter.

Morandi, Larry. *Enhancing Water Values: Proposed Legislation for Western Water Use.* Denver, CO: National Conference of State Legislatures, 1988. 15p. No ISBN.

This small publication supplements the National Conference of State Legislatures' 1988 publication *Reallocating Western Water: Equity Efficiency and the Role of Legislation.* The report has been modified so it can be used as a guide for incorporation into state water law. The objective is given for each piece of legislation. The report includes statutory language for such issues as authorization of water transfers, transfer of conserved water, temporary transfers, water exchanges, and conveyance of water through public facilities.

———. *Reallocating Western Water: Equity, Efficiency and the Role of Legislation.* Denver, CO: National Conference of State Legislatures, 1988. 35p. ISBN 1-55516-451-X.

This report deals with reallocating water to achieve greater water efficiency with the least amount of problems for users. It explains how water markets operate and legal obstacles liable to be encountered. The main emphasis is on legislative options to remove obstacles and encourage water use efficiency. Case studies for Colorado, Arizona, and California are used to show water market activity and legislative relationships.

National Research Council. Committee on Ground Water Quality Protection. *Ground Water Quality Protection: State and Local Strategies.* Washington, DC: National Academy Press, 1986. 309p. ISBN 0-309-03685-2.

This book identifies technical and institutional features of programs used in protecting groundwater.

Nikolaieff, George A., ed. *The Water Crisis.* The Reference Shelf, Volume 38, no. 6. New York: H.W. Wilson, 1967. 192p. No ISBN.

Water is precious but we are wasteful and careless in its use. The first part of the book deals with how much water we have and future needs. The second part treats the supply while the third deals with pollution. The next two parts treat ways to have clean water and who really pays and the last section examines ways to improve on nature. A bibliography including books, pamphlets, and documents is found at the end of the book.

Schmandt, Jurgen, Ernest T. Smerdon, and Judith Clarkson. *State Water Policies, a Study of Six States.* New York: Praeger, 1988. 205p. ISBN 0-275-93132-3.

The water management policies were examined for Arizona, California, Florida, North Carolina, Texas, and Wisconsin. Some of the conclusions made were: states are becoming more aware of the valuable resource that water is and must be managed very carefully, states now must assume responsibility for establishing sound management for water research, and each state has different problems to face in solving its water management situation. The book has many figures and tables. Notes are found at the end of each chapter.

Smith, Zachary A., ed. *Water and the Future of the Southwest.* Public Policy Series. Albuquerque: University of New Mexico Press, 1989. 278p. ISBN 0-8263-1156-3.

Many scholars contributed to this book. It is composed of three parts. Part one examines the legal aspects that govern water distribution and use along with the social and political ways in which water decisions are made in the Southwest. The second part examines the role the government plays in formulating water policies. The third part deals with water allocation and management examining who gets water and why. Each chapter has a summary and notes. Several tables can be found in the book. A short biography of each contributor is also in the book.

Articles and Government Documents

Water—General

General

Argent, Gala. "Earthquakes: How Water Purveyors Prepare." *Western Water* (November-December 1988): 6–11.

Balleau, W. P. "Water Appropriation and Transfer in a General Hydrogeologic System." *Natural Resources Journal* 28 (Spring 1988): 269–291.

Biswas, Asit K. "Water for the Third World." *Foreign Affairs* 60 (Fall 1981): 148–166.

Davis, Tony. "Managing to Keep Rivers Wild: Built To Tame Rivers and Generate Power, America's Grand Dams Are Hurting Fish and Plant Life Downstream; Dam Operators Are Adopting New Technologies To Reduce the Damage." *Technology Review* 89 (May–June 1986): 26–33.

Douglas, David. "Water Is Life: The International Drinking Water Supply and Sanitation Decade." *Amicus Journal* 7 (Spring 1986): 34–37.

Eichelberger, Charles H., Jr. "Drilling Water Wells in Pennsylvania." In *Water Resources in Pennsylvania: Availability, Quality and Management* edited by Shyamal Majumdar, E. Willard Miller, and Richard R. Parizek, 113–120. Easton, PA: The Pennsylvania Academy of Science, 1990.

"Focus: Memorandum of Agreement of Mitigation." *National Wetlands Newsletter* 12 (March–April 1990): 2–8.

Fritton, D. D. "Characterizing Water Flow in Unsaturated Soil." In *Water Resources in Pennsylvania: Availability, Quality and Management* edited by Shyamal Majumdar, E. Willard Miller, and Richard R. Parizek, 60–70. Easton, PA: The Pennsylvania Academy of Science, 1990.

Gast, William A. "Water Supply and Demand." In *Water Resources in Pennsylvania: Availability, Quality and Management* edited by Shyamal

Majumdar, E. Willard Miller, and Richard R. Parizek, 22–29. Easton, PA: The Pennsylvania Academy of Science, 1990.

Grisham, A., and W. M. Fleming. "Long-Term Options for Municipal Water Conservation" *American Water Works Association Journal* 81 (March 1989): 34–42.

Guruswany, Laksham. "Integrating Thoughtways: Re-opening of the Environmental Minds?" *Wisconsin Law Review* 3 (1989): 463–537.

Harrison, David C. "Institutional Barriers to National Water Policy." *Water Spectrum (Corps of Engineers)* 14 (Spring 1982): 1–7.

Hartshorn, J. K. " Tap Water Alternatives." *Western Water* (January–February 1987): 4–11.

Heath, Milton S., Jr. "Protection of Instream Flow and Lake Levels." *Popular Government* 50 (Spring 1985): 6–16.

Juhasz, Ferenc. "Water: Is There a Crisis?" *OECD Observer* (October–November 1989): 4–9.

Keller, Reiner. " The World's Fresh Water: Yesterday-Today-Tomorrow [Surface Water and Groundwater Found on Land Areas]." *Applied Geography and Development* 24 (1984): 7–23.

Kemper, Bob. "Water Resources and Utilization in Centre County, Pennsylvania." In *Water Resources in Pennsylvania: Availability, Quality and Management* edited by Shyamal Majumdar, E. Willard Miller, and Richard R. Parizek, 12–21. Easton, PA: The Pennsylvania Academy of Science, 1990.

Kinnersley, David. "Water: Tapping a World Resource; Four Years into the UN Water Decade, the Results in Some Parts of the Third World Are Encouraging." *South* (April 1985): 39+. 13 page section.

Miller, E. Willard, "Water Quantity, Quality and the Future." In *Water Resources in Pennsylvania: Availability, Quality and Management* edited by Shyamal Majumdar, E. Willard Miller, and Richard R. Parizek, 1–5. Easton, PA: The Pennsylvania Academy of Science, 1990.

New Jersey. General Assembly. Committee on Conservation, Natural Resources and Energy. *Public Hearing: Testimony on New Jersey's Water*

Supply Infrastructure: Trenton, New Jersey, August 31, 1988. Trenton, NJ: 1988. 116 p.

O'Meara, J. W. "A National Water Crisis: Can It Be Avoided?" *Reclamation Era* (April 1983): 4–11.

Ormsbee, L. E., et al. "Methodology for Improving Pump Operation Efficiency." *Journal of Water Resource Planning and Management* 115 (March 1989): 148–164.

Postel Sandra. "Needed: A New Water Policy: This Country Faces a Severe Water Shortage, and New Supplies Won't Fill It: Agriculture, Industry, and Urban Consumers Must Learn to Save Water, Allocate It More Wisely—and Soon." *Challenge* 28 (January–February 1986): 43–49.

"Power Station Reservoirs: Pros and Cons: Scientists, Officials Weigh Gains in Power Production, Irrigation against Floodplain Losses, Damage to Soil, Fish and Ecology [Soviet Union]." *Current Digest of the Soviet Press* 36 (December 12, 1984): 1–5.

Smerdon, Ernest T., ed. "Water Resources." *National Forum* 69 (Winter 1989): 2–35.

Starr, Joyce R. "Water Politics in the Middle East." *Middle East Insight* 7, no. 2–3 (1990): 64–70.

———. "Water Wars." *Foreign Policy* (Spring 1991): 17–36.

Stroud, Hubert B. "Water Resources at Cape Coral, Florida: Problems Created by Poor Planning and Development," *Land Use Policy* 8 (April 1991): 143–157.

"Symposium: Water Resources and Public Policy." *Policy Studies Review* 5 (November 1985): 349–450.

United Nations. Department of Technology Cooperation for Development. *Ground Water in Eastern and Northern Europe.* New York: U.N. Agent, 1990. 278 p. ISBN 92-1-104318-2.

———. *The Use of Non-Conventional Water Resources in Developing Countries.* Natural Resources/Water Series no. 14. New York: UN Agent, 1985. 278 p.

United Nations. Economic Commission for Europe, Committee on Water Problems. *Impact of Non-Conventional Sources of Energy on Water.* New York: UN Agent, 1985. 88 p.

"Water in America: Colourless, Odourless, Tasteless—Priceless." *Economist* (London) 301 (October 4, 1986): 35–38.

"Water Resources." *Banque Marocaine du Commerce Exterieur Information Review* (September 1986): 11–20.

"Weather and Water." *Western Water* (November–December 1982): 4–12.

"Where Your Water Comes From." *Western Water* (March–April 1989): 3–17. (Special Issue).

Woodard, Gary C. "Changing Weather, Climate Threatens Water Conservation Efforts." *Arizona's Economy* (September 1988): 1–4+.

Yagerman, Katherine S. "Underground Storage Tanks: The Federal Program Matures." *Environmental Law Reporter* 21 (March 1991): 10136–10147.

Availability and Demand

Abu Rizaiza, O. S., and M. N. Allam. "Water Requirements Versus Water Availability in Saudi Arabia." *Journal of Water Resource Planning and Management* 115 (January 1989): 564–574.

Balleau, W. P., "Water Appropriation and Transfer in a General Hydrogeologic System." *Natural Resources Journal* 28 (Spring 1988): 269–291.

Billings, R. Bruce, and W. Mark Day. "Elasticity of Demand for Residential Water: Policy Implications for Southern Arizona." *Arizona Review* 31 (Third Quarter 1983): 1–11.

Brown, Marion R., and William C. Thiesenhusen. "Access to Land and Water." *Land Reform* (FAO) nos. 1–2 (1983): 1–14.

Cooley, John K. "The War Over Water [Competition for Water Rights and Supplies in the Near East]." *Foreign Policy* (Spring 1984): 3–26.

Davis, Tony. "Trouble in a Thirsty City: Tucson's Water Budget Is Badly Out of Balance; Technological Advances and Cultural Changes Are in the Works—But Will They Be Enough?" *Technology Review* 87 (August–September 1984): 66–71.

Frey, Frederick W., and Thomas Naff. "Water: An Emerging Issue in the Middle East." *Annals of the American Academy of Political and Social Science.* 482 (November 1985): 65–84.

Gajewski, Gregory R., and Douglas Duncan. "1988 Drought Did Not Dry Up Credit." *Rural Development Perspectives* 6 (October 1989): 25–30.

Gleick, Peter H. "The Effects of Future Climatic Changes on International Water Resources: The Colorado River, the United States, and Mexico." *Policy Sciences* (Amsterdam) 21:1 (1988): 23–39.

Grigg, N. S., "Regionalization in Water Supply Industry: Status and Needs." *Journal of Water Resource Planning and Management* 115 (May 1989): 367–378.

Gunnell, Barbara. "Water Supply: An African Business Survey [Extent to Which the Targets of the United Nations Water Decade Are Being Met.]" *African Business* (September 1984): 69+. 9 page section.

Hartshorn, J. K. "Contract Controversies." *Western Water* (September–October 1989): 3–11.

Hollman, Kenneth W., et al. "A Profile of Water Systems in Mississippi." *Mississippi Business Review* 43 (February 1982): 3–9.

Jones, David H., ed. *Water Supply and Wastewater Disposal.* Poverty and Basic Needs Series. Washington, DC: World Bank, 1980. 46 p.

Krohe, James, Jr. "Illinois: The OPEC of Water [Problems and Prospects Surrounding Illinois' Water Resources]." *Illinois Issues* 8 (June 1982): 6–11.

Kromm, David E., and Stephen E. White. "Public Preferences for Recommendations Made by the High Plains-Ogallala Aquifer Study." *Social Science Quarterly* 67 (December 1986): 841–854.

Miraldi, Robert. "Draining Upstate: A Thirsty Big Apple Seeks More Water." *Empire State Report* 12 (April 1986): 11–14+.

Mohanty, R. P. "Strategic Planning for Water: A Case History [of a Multipurpose Reservoir Project in Orissa, India]." *Long Range Planning* 17 (October 1984): 79–87.

New Jersey. General Assembly. Committee on Conservation, Natural Resources and Energy. *Public Hearing: Testimony on New Jersey's Water Supply Infrastructure; Trenton, New Jersey, August 31, 1988.* Trenton, NJ: 1988. 116 p.

Parker, Richard. "Water Supply for Urban Southern California: An Historical and Legal Perspective." *Glendale Law Review* 8, nos. 1–2 (1988): 1-66.

Postel, Sandra. "Trouble on Tap: Valuing Water As the Precious Commodity It Really Is Would Go a Long Way Toward Staving Off Shortages in the Next Century." *World Watch* 2 (September–October 1989): 12–20.

Sheets, Kenneth R. "War over Water: Crisis of the 80s; Farmers, Cities, States, Business, Everyone Is Scrambling for Shrinking Supplies." *U.S. News* 95 (October 31, 1983): 57–60+.

Smith, J. A., "Water Supply Yield Analysis for Washington Metropolitan Area." *Journal of Water Resource Planning and Management* 115 (March 1989): 230–242.

"Symposium on Anticipating Transboundary Resource Needs and Issues in the U.S.-Mexico Border Region to the Year 2000." *Natural Resources Journal* 22 (October 1982): 729–1174.

Tameim, O., et al. "Water Supply Systems in Blue Nile Health Project." *Journal of Environmental Engineering* 113 (December 1987): 1219–1233. Discussion 115 (August 1989): 874–879.

Tucker, William. "Billions Down the Drain: Why Californians Are Squandering Their Most Precious Resource [Emphasis on the Proposed Peripheral Canal, Intended to Bring Water from Northern to Southern California]." *Reason* 14 (June 1982): 24–32.

U.S. Congress. House. Committee on Public Works and Transportation. Subcommittee on Water Resources. *Water Resources Needs: Hearing, January 28, 1982 (Mid-Winter Meeting of the United States Conference of Mayors, Washington, DC)* 97th Cong., 2nd sess., Washington, DC: GPO. 1982. 106 p.

———. *Water Resources Problems Affecting the Northeast: The Drought, and Present and Future Water Supply Problems: Hearings, March 19–30, 1981.* 97th Cong., 1st sess. Washington, DC: GPO, 1981. 1032 p.

U.S. Congress. Senate. Committee on Energy and Natural Resources. Subcommittee on Water and Power. *Water Supply Issues in the Middle Atlantic States: Hearings, July 22–August 7, 1985.* 99th Cong., 1st sess., Washington, DC: GPO, 1986. 605 p.

"Water for Arizona: The Central Arizona Project." *Western Water* (May–June 1985): 4–11.

Planning

Battaglia, A. Mark, Daniel R. Jones, and Richard R. Parizek. "Regional Environmental Planning for Water." In *Water Resources in Pennsylvania: Availability, Quality and Management* edited by Shyamal Majumdar, E. Willard Miller, and Richard R. Parizek, 518–537. Easton, PA: The Pennsylvania Academy of Science, 1990.

Carroll, John E. "Water Resources Management as an Issue in Environmental Diplomacy." *National Resources Journal* 26 (Spring 1986): 207–220.

"Environmental Planning and Management: Focus on Inland Waters." *Regional Development Dialogue* 8 (Autumn 1987): 1–238.

Feldman, David L. "Comparative Models of Civil-Military Relations and the U.S. Army Corps of Engineers." *Journal of Political and Military Sociology* 15 (Fall 1987): 229–244.

Hrezo, Margaret, and William E. Hrezo. "From Antagonistic to Cooperative Federalism on Water Resources Development: A Model for Reconciling Federal, State, and Local Programs, Policies and Planning." *American Journal of Economics and Sociology* 44 (April 1985): 199–214.

Jayal, N. D. "Research Priorities for Planning Water Resource Development." *Canadian Journal of Development Studies* 7, no. 1 (1986): 37–46 .

Lord, William B. "Water Resources Planning: Conflict Management." *Water Spectrum (Corps of Engineers)* 12 (Summer 1980): 1–11.

Mohanty, R. P. "Strategic Planning for Water: A Case History [of a Multipurpose Reservoir Project in Orissa, India]." *Long Range Planning* 17 (October 1984): 79–87.

Penning-Rowsell, Edmund C., and Dennis J. Parker. "The Changing Economic and Political Character of Water Planning in Britain." In *Progress in Resource Management and Environmental Planning,* edited by Timothy O'Riordan and R. Kerry Turner, 169–199. New York: John Wiley & Sons, 1983.

Robinson, Paul, et al. "In the Hands of the People: Establishing Planning Power for a Community." *Workbook* (Southwest Research and Info Center) 15 (Winter 1990): 144–157.

Schoolmaster, F. Andrew. "A Cartographic Analysis of Water Development Referenda in Texas, 1957–85." *Growth and Change* 18 (Fall 1987): 20–43.

Steinberg, Bory. "Planning Versus Implementation: The Story of the Seventies [Water Resources Projects of the U.S. Army Corps of Engineers]." *Water Spectrum (Corps of Engineers)* 14 (Winter 1981–1982): 28–38.

United Nations. Economic and Social Commission for Asia and the Pacific. *Water Use Statistics in the Long-Term Planning of Water Resources Development.* New York: UN Agent, 1989. 114 p.

U.S. Congress. Senate. Committee on Energy and Natural Resources. Subcommittee on Public Lands, National Parks and Forests. *Federal Cave Resources Protection Act and Restriction of Dams in Parks and Monuments: Hearing on June 16, 1988.* 100th Cong., 2d sess., Washington, DC: GPO, 1988. 126 p.

U.S. Water Resources Council. *Economic and Environmental Principles and Guidelines for Water and Related Land Resources Implementation Studies,* Washington, DC: GPO, 1983. 137 p.

Yarnal, Brent. " Past and Future Climatic Trends—Consequences on Water Resources Planning and Decision Making." In *Water Resources in Pennsylvania: Availability, Quality and Management,* edited by

Shyamal Majumdar, E. Willard Miller, and Richard R. Parizek, 30–41. Easton, PA: The Pennsylvania Academy of Science, 1990.

Economics

Hanke, Steve H., and James B. Anwyll "On the Discount Rate Controversy [How To Determine the Discount Rate in a Cost-Benefit Analysis of Water Projects]." *Public Policy* 28 (Spring 1980): 171– 183.

Muller, R. Andrew. "Some Economics of the GRAND Canal." *Canadian Public Policy (Guelph)* 14 (June 1988): 162–174.

Olson, Kent W. "Economics of Transferring Water to High Plains [Cost-Benefit Analysis of the Northern Component of the Oklahoma Statewide Water Conveyance System or Northern Interbasin Transfer System]." *Quarterly Journal of Business and Economics* 22 (Autumn 1983): 63–80.

U.S. Congress. House. Committee on Public Works and Transportation. Subcommittee on Water Resources. *Water Resources Development: Cost-Sharing Aspects of President's Water Policy Initiatives: Hearings March 14–April 10, 1979.* 96th Cong., 1st sess., Washington, DC: GPO, 1979. 1463 p.

Laws and Regulations

Bond, Marvin T., et al. "Recent Experiences with Water Legislation in Mississippi." *Public Administration Survey* 35 (Autumn 1987–Winter 1988): 1–5.

Caponera, Dante A. "Pattern of Cooperation in International Water Law: Principles and Institutions." *Natural Resources Journal* 25 (July 1985): 563–587.

Carruthers, Ian, and Roy Stoner "A Legal Framework in the Public Interest [Equitable Allocation of Irrigation Water from Large Groundwater Projects]." *Ceres* 15 (September–October 1982): 15–20.

Chan, Arthur H. "The Structure of Federal Water Resources Policy Making." *American Journal of Economics and Sociology* 40 (April 1981): 115–127.

Pages, Fernando, and John B. Herbich. "Review of Events Leading to the Passage of Public Law 99–662 (HR6), the Water Development Act of 1986." *Dock and Harbor Authority* 68 (November 1987): 153–159.

Sander, William. "Federal Water Resources Policy and Decision-Making: Their Formulation Is Essentially a Political Process Conditioned by Government Structure and Needs [Role of Cost-Benefit Analysis]." *American Journal of Economics and Sociology* 42 (January 1983): 1–12.

Teclaff, Ludwik A., ed. "Symposium on International Resources Law." *Natural Resources Journal* 25 (October 1985): 863–1024.

United Nations. Department of International Economics and Social Affairs. *Efficiency and Distributional Equity in the Use and Treatment of Water: Guidelines for Pricing and Regulations.* New York: U.N. Agent, 1980. 175 p.

United Nations. Department of Technology. Cooperation for Development. *Legal and Institutional Factors Affecting the Implementation of the International Drinking Water Supply and Sanitation Decade.* Natural Resources/Water Series no. 23. New York: U.N. Agent, 1989. 121 p.

U.S. Congress. House. Committee on Public Works and Transportation. Subcommittee on Water Resources. *To Amend the Water Resources Planning Act: Hearings, June 5–12, 1979 on H.R. 2610.* 96th Cong., 1st sess., Washington, DC: GPO, 1980. 375p.

U.S. Congress. Senate. *Colorado Ute Water Settlement Bill: Joint Hearing, December 3, 1947, before the Select Committee on Indian Affairs and the Committee on Energy and Natural Resources; on S. 1415, To Facilitate and Implement the Settlement of Colorado Ute Indian Reserved Water Rights Claims in Southwest Colorado.* 100th Cong., 1st sess., Washington, DC: GPO, 1988. 580 p.

U.S. Congress. Senate. Committee on Environment and Public. Works Subcommittee on Water Resources. *Water Resources Policy Issues: Hearings: Pts. 1–3, April 21–August 19, 1981.* 97th Cong., 1st sess., Washington, DC: GPO, 1981–1982. 3 parts.

U.S. General Accounting Office. *Water Supply Should Not Be an Obstacle To Meeting Energy Development Goals: Report to the Congress by the Comptroller General of the United States.* Washington, DC: GPO, 1980. 80 p.

Water Supply

Agriculture

"Agricultural Water Use: A Conversation with Marc Reisner." *Western Water* (January–February 1991): 3–11.

Baker, Dale E., and Donald M. Crider. "The Environmental Consequences of Agriculture in Pennsylvania." In *Water Resources in Pennsylvania: Availability, Quality and Management* edited by Shyamal Majumdar, E. Willard Miller, and Richard R. Parizek, 334–353. Easton, PA: Pennsylvania Academy of Science, 1990.

Bastian, Robert K., and Jan Beneforado. "Waste Treatment: Doing What Comes Naturally [Feasibility of Applying Both Wastewater and Sludge to Agricultural Land]." *Technology Review* 86 (February–March 1983): 58–66.

California Department of Water Resources. *Outlook for Water Consumption by California's Feed and Forage Industry Through 2010.* Sacramento, CA: 1982. 102 p.

Corbett, Patrick E. "The Overlooked Farm Crisis: Our Rapidly Depleting Water Supply." *Notre Dame Law Review* 61, no. 3 (1986): 454–477.

Day, D. G. "Water Allocation To Agriculture in the Hunter Valley [New South Wales]: Policy Change and Implementation." *Land Use Policy* 2 (January 1985): 187–204.

Ferrett, Robert L., and Robert M. Ward. "Agricultural Land Use Planning and Groundwater Quality." *Growth and Change* 14 (January 1983): 32–39.

Postel, Sandra L. "Needed: A New Water Policy: This Country Faces a Severe Water Shortage, and New Supplies Won't Fill It: Agriculture, Industry, and Urban Consumers Must Learn to Save Water, Allocate It More Wisely—and Soon." *Challenge* 28 (January–February 1986): 43–49.

Unwin, Tim. "Agriculture and Water Resources in the United Arab Emirates [Effects of Large-Scale Agricultural Production on Water Supplies]." *Arab Gulf Journal* 3 (April 1983): 75–85.

U.S. Congress. House. Committee on Agriculture. *Review of Soil and Water Conservation Programs Mandated under the Food Security Act of 1985; and Current and Developing Technologies and Research Being Employed in Today's Poultry and Egg Industries: Joint Hearing, May 23, 1988, Before the Subcommittee on Conservation, Credit and Rural Development and the Subcommittee on Livestock, Dairy, and Poultry.* 100th Cong., 2nd sess., Washington, DC: GPO, 1988. 111 p.

U.S. Congress. House. Committee on Agriculture. Subcommittee on Conservation, Credit, and Rural Development. *Effects of Interbasin Water Transfers on Agriculture: Hearing, June 15, 1983 on H.R. 1749.* 98th Cong., 1st sess., Washington, DC: GPO, 1983. 211 p.

———. *General Farm Bill of 1981: Hearings Pt. 6, March 12–16, on H.R. 1113 [and other bills].* 97th Cong., 1st sess., Washington DC: GPO, 1981. 508 p.

U.S. Congress. House. Committee on Science and Technology. Subcommittee on Natural Resources, Agriculture Research, and Environment. *Agriculture and Water Use: Present and Future Issues: Hearing, October 25, 1983.* 98th Cong., 1st sess., Washington, DC: GPO, 1984. 92 p.

U.S. Congress. House. *The National Ground Water Research Act of 1989: Joint Hearing, August 2 and October 3, 1989, Before the Subcommittee on Natural Resources, Agriculture Research and Environment of the Committee on Science, Space, and Technology, and the Subcommittee on Department Operations, Research, and Foreign Agriculture of the Committee on Agriculture.* 101st Cong., 1st sess., Washington, DC: GPO, 1990. 669 p.

U.S. Congress. Office of Technology Assessment. *Beneath the Bottom Line: Agricultural Approaches To Reduce Agrichemical Contamination of Groundwater.* Washington, DC: GPO, 1990. 337 p.

Wiggins, Steve. "The Planning and Management of Integrated Rural Development in Drylands; Early Lessons from Kenya's Arid and Semi–Arid Lands Programmes." *Public Administration and Development* 5 (April–June 1985): 91–108.

Irrigation

Barnett, Tony. "Small–Scale Irrigation in Sub–Saharan Africa: Sparse Lessons, Big Problems, Any Solutions?" *Public Administration and Development* 4 (March 1984): 21–47.

Billy, Bahe, and Philip Reno. " The Greening of the Navajo Indian Irrigation Project [New Mexico and Arizona]." *New Mexico Business* 33 (March 1980): 3–7.

Biswas, Asit K. "Irrigation in Africa." *Land Use Policy* 3 (October 1986): 269–285.

———. "Monitoring and Evaluation of Irrigated Agriculture: A Case Study of Bhima Project, India." *Food Policy* 12 (February 1987): 47–61.

Brisset, Claire. "Irrigation: The Absolute Imperative." *Ceres* 19 (September–October 1986): 20–24.

Carruthers, Ian. "Protecting Irrigation Investment: The Drainage Factor." *Ceres* 18 (July–August 1985): 15–21.

"Diverting Siberian Rivers: Pro and Con [for the Purpose of Irrigating Land in Central Asia]." *Current Digest of the Soviet Press* 34 (April 7, 1982): 1–5.

Easter, K. William. *Economic Failure Plagues Developing Countries' Irrigation: An Assurance Problem.* Staff Paper P90–44. St. Paul: University of Minnesota, Institute of Agriculture Forestry and Home Economics Department of Agriculture and Applied Economics, 1990. 37 p.

Frederick, Kenneth D. "The Future of Western Irrigation [Likely Changes in the Development of Western Irrigation and the Policy Implications of the Changing Water Situation]." *Southwestern Review of Management and Economics* 1 (Spring 1981): 19–33.

Gardner, Richard L., and Robert A. Young. "The Effects of Electricity Rates and Rate Structure on Pump Irrigation: An Eastern Colorado Case Study." *Land Economics* 60 (November 1984): 352–359.

Glenn, Bruce. "Water Conservation Opportunities on Federal Irrigation Projects [Western States]." *Reclamation Era* 65, no. 2 (1979): 12–17.

Gollehon, Noel R., et al. "Impacts on Irrigated Agriculture from Energy Development in the Rocky Mountain Region [Extent to Which Water May Be Diverted from Agriculture to Power Generation and Energy Resources Development]." *Southwestern Review of Management and Economics* 1 (Spring 1981): 61–88.

Goodenough, Richard. "Implications of Recent Changes in Californian Irrigation Systems." *Land Use Policy* 2 (April 1985): 135–147.

Harris, Tom. "The Kesterson Syndrome: The Federal Irrigation Projects That Made the Desert Bloom Are Now Killing Wildlife Throughout the West." *Amicus Journal* 11 (Fall 1989): 4–9.

"The Importance of Irrigated Agriculture to Nebraska's Economy." *Business in Nebraska* 44 (December 1988): 1–4.

"Irrigation." *Courier* (November–December 1990): 64–95.

Johnson, Mark R. "An Economic Reevaluation of the O'Neill Unit Irrigation Project [Proposed Federal Project in the Missouri River Basin, Intended to Supply Water to 77,000 Acres in North–Central Nebraska]." *Nebraska Journal of Economics and Business* 19 (Spring 1980): 3–21.

Knisel, W. G., and R. A. Leonard. "Irrigation Impact on Groundwater: Model Study in Humid Region." *Journal of Irrigation and Drainage Engineering* 115 (October 1989): 823–838.

Moris, Jon. "Irrigation as a Privileged Solution in African Development." *Development Policy Review* 5 (June 1987): 99–123.

Mossbarger, W. A., Jr., and R.W. Yost. "Effects of Irrigated Agriculture on Groundwater Quality in Corn Belt and Lake States." *Journal of Irrigation and Drainage Engineering* 115 (October 1989): 773–790.

"The Ogallala Aquifer [Economic Consequences of the Depletion of the Aquifer That Supplied Irrigation Water to the High Plains States: Based on a Study Commissioned by the U.S. Congress]." *Southwestern Review of Management and Economics* 2 (Spring 1982): 3–174.

O'Mara, Gerald T., *Issues in the Efficient Use of Surface and Groundwater in Irrigation [Developing Countries].* World Bank Staff Working Papers no. 707. Washington, DC: International Bank for Reconstruction and Development, 1984. 93 p.

Shirmohammadi, A., and W. G. Knisel. "Irrigated Agriculture and Water Quality in South." *Journal of Irrigation and Drainage Engineering* 115 (October 1989): 791–806.

Skold, Melvin D., et al. "Irrigation Water Distribution Along Branch Canals in Egypt: Economic Effects." *Economic Development and Cultural Change* 32 (April 1984): 547–567.

Steinberg, David I. *Irrigation and AID's Experience: A Consideration Based on Evaluations.* AID. Program Evaluation Report no. 8. Washington, DC: U.S. Agency for International Development, 1983. 244 p.

Takase, Kunico. "Irrigation Development and Cereal Production in Asia [Emphasis on Rice]." *Asian Development Review* 2, no. 2 (1984): 80–91.

Tarlock, A. Dan. "Supplemental Groundwater Irrigation Law: From Capture to Sharing." *Kentucky Law Journal* 73, no. 3 (1985): 695–722.

Industrial Use

Burchi, Stefano. "Legislation on Domestic and Industrial Uses of Water: A Comparative Review." *Natural Resources Journal* 24 (January 1984): 143–159.

California Department of Water Resources. *Water Use by Manufacturing Industries in California, 1979.* Sacramento, CA: 1982. 113 p.

Krohe, James, Jr. "Industrial Use of Water: The Assumption of Abundance [Illinois]." *Illinois Issues* 8 (September 1982): 6–11.

U.S. Congress. House. Committee on Public Works and Transportation. Subcommittee on Water Resources. *Study of the Effect of the Industrial Cost Exclusion (ICE) on the Construction Grants Program.* 97th Cong., 1st sess., Washington, DC: GPO, 1981. 403 p.

U.S. Congress. Senate. Committee on Environment and Public Works. Subcommittee on Environmental Pollution *Industrial Cost Recovery: Hearings, March 18–31, 1980.* 96th Cong., 2nd sess., Washington, DC: GPO, 1980. 342 p.

U.S. General Accounting Office. *Colorado River Basin Water Problems: How to Reduce Their Impact; Report to the Congress by the Comptroller General of the United States, May 4, 1979.* Report B-133053. Washington, DC: 1979. 137 p.

Management

Allen, Carol J., and William B. Hall. "The Pennsylvania Public Utility Commission's Role in the State's Water Resources Management." In *Water Resources in Pennsylvania: Availability, Quality and Management* edited by Shyamal Majumdar, E. Willard Miller, and Richard R. Parizek, 141–157. Easton, PA: Pennsylvania Academy of Science, 1990.

Bedient, Philip B., and Peter G. Rowe, eds. "Urban Watershed Management: Flooding and Water Quality." *Rice University Studies* 65 (Winter 1979): 1–205.

Cole-Miseh, Sally. "Great Lakes Water Quality Management." In *Water Resources in Pennsylvania: Availability, Quality and Management* edited by Shyamal Majumdar, E. Willard Miller, and Richard R. Parizek, 254–266. Easton, PA: Pennsylvania Academy of Science, 1990.

Dávid, László. "Environmentally Sound Management of Freshwater Resources." *Resources Policy* 12 (December 1986): 307–316.

Degenhardt, Christopher, and Stephen Smith. "Water Management for Urban Landscapes." *Urban Land* 42 (July 1983): 10–14.

Dixon, John A. "Managing Watershed Resources." *Annals of Regional Science* 21 (November 1987): 111–123.

Easter, K. William. *Institutional Arrangements for Managing Water Conflicts in Minnesota.* Staff Paper Series p 90–92. St. Paul: University of Minnesota, Department of Agricultural and Applied Economics, Institute of Agriculture, 1990. 29p.

"Groundwater Protection: Local Success Stories." *Management Information Service Report* 22 (February 1990): 1–13.

Kundell, James E., and Kathryn J. Hatcher. "The Policy Agenda for Integrated Water Management." *State and Local Government Review* 17 (Winter 1985): 162–173.

Lynch, James A., and Edward S. Corbett. "Management of Source Areas for Water Quality and Quantity." In *Water Resources in Pennsylvania: Availability, Quality and Management* edited by Shyamal Majumdar, E.

"Managing Florida's Water Resources." *Business and Economic Dimensions* 16, no. 1 (1980): 1–28.

"Managing Our Fresh Water Resources." *Impact of Science on Society* no. 1 (1983): 3–24+

Maniatis, Melina. "Stormwater Management." *Management Information Service Report* 22 (November 1990): 1–15.

Mayo, Alan L. "A 300–Year Water Supply Requirement: One County's Approach." *American Planning Association Journal* 56 (Spring 1990): 197–208.

Miller, D. A., et al. "The Role of Geographic Information Systems in Water Resources Management." In *Water Resources in Pennsylvania: Availability, Quality and Management* edited by Shyamal Majumdar, E. Willard Miller, and Richard R. Parizek, 538–550. Easton, PA: Pennsylvania Academy of Science, 1990.

Miller, Willard, and Richard R. Parizek, 499–517. Easton, PA: Pennsylvania Academy of Science, 1990.

Mrazik, Brian R., et al. *Integrated Watershed Management: An Alternative for the Northeast.* Research Bulletin no. 664. Amherst, MA: Agriculture Experiment Station, 1980. 50 p.

New Jersey. General Assembly. *Joint Public Hearing to Take Testimony on Possible Legislative Initiatives to Preserve Watershed Lands and Other Natural Resources: Parsippany, New Jersey, November 30, 1988: Before Senate Energy and Environment Committee and Assembly Conservation, Natural Resources and Energy Committee.* Trenton, NJ: 1988. 149 p.

New Jersey. General Assembly. County Government and Regional Authorities Committee. *Public Hearing: Assembly Bill No. 4365 Concerns the Ownership of Water Supply Utilities and Would Specifically Prohibit Water Supply Utilities from Being Corporately Related to Companies Which Are in the Real Estate Business.* Trenton, NJ: 1989. 5 p.

Organization for Economic Cooperation and Development. *Water Resource Management: Integrated Policies.* Paris: OECD, 1989. 199 p. ISBN 92-64-13285-6.

Plouffe, William L. "Forty Years after First Iowa: A Call for Greater State Control of River Resources." *Cornell Law Review* 71 (May 1986): 833–849.

Sewell, W. R. Derrick, and Asit K. Biswas. "Implementing Environmentally Sound Management of Inland Waters." *Resources Policy* 12 (December 1986): 293–306.

U.S. Congress. House. Committee on Agriculture. Subcommittee on Conservation, Credit, and Rural Development. *Upper Locust Creek Watershed in Missouri and Iowa; and Howard Creek Watershed in West Virginia: Hearing, April 5, 1989.* 101st Cong., 1st sess., Washington, DC: GPO, 1989. 17 p.

U.S. Congress. House. Committee on Science and Technology. Subcommittee on Investigations and Oversight. *Forecasting and Technology for Water Management: Hearings, October 13–14, 1983.* 98th Cong., 1st sess., Washington, DC: GPO, 1984. 492 p.

U.S. General Accounting Office. *Contracts to Provide Space in Federal Reservoirs for Future Water Supplies Should Be More Flexible: Report by the U.S. General Accounting Office.* Report B-157984. Washington, DC: 1980. 13 p.

Conservation

Baumann, Duane D., and John J. Boland. "Urban Water Supply Planning [Emphasis on Conservation]." *Water Spectrum (Corps of Engineers)* 12 (Fall 1980): 33–41.

Brooks, Robert P. "Wetlands and Deepwater Habitats in Pennsylvania." In *Water Resources in Pennsylvania: Availability, Quality and Management* edited by Shyamal Majumdar, E. Willard Miller, and Richard R. Parizek, 71–79. Easton, PA: Pennsylvania Academy of Science, 1990.

"Can Farm Water Conservation Save Californians From Building More Water Projects?" *Western Water* (November–December 1984): 4–12.

Cohen, Harry. "The Relationship between Water Conservation and Mineral Development [Some Emphasis on Water Pollution Problems Generated By Coal, Oil and Gas Development]." *Alabama Law Review* 31 (Summer 1980): 547–609.

"Egypt: A Case of Ecological Vulnerability [Land and Water Supply Problems]." *Conservation Foundation Letter* (August 1983): 1–8.

Glenn, Bruce. "Water Conservation Opportunities on Federal Irrigation Projects." *Reclamation Era* 65, no. 2 (1979): 12–17.

Marshall, Patrick G. "California: Enough Water for the Future? California Is Now in the Fifth Year of the Worst Drought in Its History, and Urban and Environmental Interests Are Calling for an End to Farmers' Guarantee of Current Supplies of Cheap Irrigation Water from Federal and State Water Projects." *Editorial Research Reports* (April 19, 1991): 222–234.

Morris, John R. "Water Conservation Progress in Denver." *Contemporary Policy Issues* 9 (July 1991): 35–45.

U.S. Department of Agriculture. Soil Conservation Service. *Washington's Soil and Water Conditions and Trends, 1982–1987.* Spokane, WA: 1990. 12 p.

Desertification

Biswas, Asit K. "Water Development and Management for Desertification Control: A Review of the Past Decade." *Land Use Policy* 4 (October 1987): 401–411.

Davis, Douglas R. "The Only Way to Manage a Desert: Utah's Liability Immunity for Flood Control." *Journal of Energy Law and Policy* 8, no. 1 (1987): 95–118.

"The Desert's Challenge and the Human Response." *Ceres* 19 (March–April 1986): 17–30.

Glantz, Michael H. "Man, State, and the Environment: An Inquiry Into Whether Solutions to Desertification in West African Sahel Are Known But Not Applied." *Canadian Journal of Development Studies* 1, no. 1 (1980): 75–97.

Helldén, Ulf. "Desertification Monitoring: Is the Desert Encroaching?" *Desertification Control Bulletin* no. 17 (1988): 8–12.

Hendry, Peter. "Where the Desert Stops: To Preserve Precious Agricultural Land the Chinese Are Mounting a Major Campaign Against Encroaching Sand." *Ceres* 17 (March–April 1984): 20–24.

Long, Bill L. "Desertification—in Perspective." *Population and Environment* 10 (Summer 1989): 237–244.

Mabbutt, J. A. "Implementation of the Plan of Action to Combat Desertification." *Land Use Policy* 4 (October 1987): 371–388.

Matheson, Alastair. "Stopping the Desert's Advance [Evaluates International Efforts to Combat the Growing Threat of Desertification in Africa]." *Africa Report* 29 (July–August 1984): 53–56.

Mensching, Horst G. "Natural Potential and Land Use in Drylands." *Applied Geography and Development* 32 (1988): 51–64.

Mortimore, Michael. "Shifting Sands and Human Sorrow: Social Response to Drought and Desertification." *Desertification Control Bulletin* no. 14 (1987): 1–14.

Postel, Sandra. "Land's End: Each Year Desertification Renders an Area of Land the Size of West Virginia Unusable for Any Purpose: Small-Scale Successes at Keeping Land Alive Show Promise for a Needed International Rescue Effort." *World Watch* 2 (May–June 1989): 12–20.

Sheridan, David. *Desertification of the United States [Western Regions].* Washington, DC: U.S. Council on Environmental Quality, 1981. 142 p.

Stiles, Daniel, and Ross Brennan. "The Food Crisis and Environmental Conservation in Africa." *Food Policy* 11 (November 1986): 298–310.

Tolba, Mostafa Kamal. "Desertification in Africa." *Land Use Policy* 3 (October 1986): 260–268.

U.S. Congress. Senate. Committee on Energy and Natural Resources. Subcommittee on Public Lands, National Parks and Forests. *California Desert Protection Act of 1987: Hearings, July 21 and 23, 1987, on S. 7, To Provide for the Protection of the Public Lands in the California Desert.* 100th Cong., 1st sess., Washington, DC: GPO, 1987. 1472 p.

Drought

"Africa's Future: Hostage to the Drought." *African Report* (July–August 1984): Entire issue.

Ahmad, Moid U. "Ground Water Resources: The Key To Combating Drought in Africa." *Desertification Control Bulletin* no. 16 (1988): 2–6.

Brown, Lester R. "The Growing Grain Gap: Drought in the United States and China May Drop World Grain Reserves to Their Lowest Level in 40 Years." *World Watch* 1 (September–October 1988): 10–18.

Burck, Charles G. "The Blight on Harvest and Herd [Aftereffects of the 1980 Drought]." *Fortune* 102 (September 22, 1980): 118–123.

Caldwell, John C., et al. "Periodic High Risk as a Cause of Fertility Decline in a Changing Rural Environment: Survival Strategies in the 1980–1983 South Indian Drought [Nine Villages in Karnataka]." *Economic Development and Culture Change* 34 (July 1986): 677–701.

Cronin, Mark X. "Lessons from a Drought Crisis: City Managers Can Learn from New York City's Crisis Planning." *Governance* (Summer–Fall 1986): 43–46.

Derrick, Jonathan. "West Africa's Worst Year of Famine [1984]." *African Affairs* 83 (July 1984): 281–299.

Dietz, Ton. "Migration to and from Dry Areas in Kenya." *Tijdschrift voor Economische en Sociale Geografie (Leiden)* 77, no. 1 (1986): 18–26.

Drabenstott, Mark, and Alan Barkema. "U.S. Agriculture Shrugs Off the Drought." *Federal Reserve Bank of Kansas City Economic Review* 73 (December 1988): 32–48.

"Drought—or Flood? How California Prepares." *Western Water* (January–February 1986): 4–10.

Dyer, J. A., et al. "A Scheme for Defining Drought Areas [Prairie Provinces]." *Canadian Farm Economics* 16 (October 1981): 1–8.

Epstein, Jack. "Tapping the Source: Scarcity and Pollution Have Made Water a Boom Industry, Particularly in Drought-Weary California." *California Business* 24 (August 1989): 20–23+.

Hansler, Gerald M. "Drought Management in the Delaware River Basin." In *Water Resources in Pennsylvania: Availability, Quality and Management* edited by Shyamal Majumdar, E. Willard Miller, and

Richard R. Parizek, 245–253. Easton, PA: Pennsylvania Academy of Science, 1990.

Haynes, Jos, et al. "Rural Indebtedness and the Need for Post-Drought Credit." *Quarterly Review of the Rural Economy* 5 (August 1983): 258–263.

Holm, John D., and Mark S. Cohen. "Enhancing Equity in the Midst of Drought: The Botswana Approach." *Ceres* 19 (November–December 1986): 20–24.

Hrezo, Margaret S., et al. "Integrating Drought Planning into Water Resources Management." *Natural Resources Journal* 26 (Winter 1986): 141–167.

Jackson, Henry F. "The African Crisis: Drought and Debt." *Foreign Affairs* (Summer 1985): 1081–1094.

Klaus, Dieter, "Geographical-Climatological Aspects of Drought-Based Food Crises." *Zeitschrift für Auslandische Landwirtschaft* 22 (January–March 1983): 8–26.

Krenz, Maria E. "Drought and Hunger in Africa: Denying Famine a Future." *Africa Today* (Fourth Quarter 1985): 55–61.

New Jersey. General Assembly. Agriculture and Environment Committee. *Briefing On (Water Supplies and Drought Conditions): Held: Trenton, New Jersey, July 31, 1985.* Trenton, NJ: 1985. 88 p.

Palmer, R. N., and K. J. Holmes. "Operational Guidance During Droughts: Expert System Approach." *Journal of Water Resource Planning and Management* 114 (November 1988): 647–666.

Rosine, John, and Nicholas Walraven. "Drought, Agriculture, and the Economy." *Federal Reserve Bulletin* 75 (January 1989): 1–12.

Smith, Hilary H. "Drought 1988: Farmers and the Macroeconomy." *Federal Reserve Bank of Dallas Economic Review* (September 1988): 15–22.

Sullivan, Gene D. "The Impact of Drought [How Continuation of 1980's Nationwide Drought Into 1981 Has Affected Agricultural Production in the Southeast; Benefits and Consequences of Increased

Irrigation].” *Federal Reserve Bank of Atlanta Economic Review* 66 (September 1981): 26–31.

Toulmin, Camilla. “Drought and the Farming Sector: Loss of Farm Animals and Post-Drought Rehabilitation.” *Development Policy Review* 5 (June 1987): 125–148.

U.S. Bureau of Reclamation. *The Drought of 1990 in the Western States and Outlook for 1991.* Washington, DC: 1991. 5 p.

Desalination

Brickson, Betty, et al. “Sea Water Desalination: An Ancient Concept Is Coming to the Aid of California’s Drought-Stricken Coastal Cities, But How Much Can It Contribute to Solving the State’s Water Problems?” *Western Water* (July–August 1991): 4–8.

Burns, O. K. “Desalting Practices in the United States.” *American Water Works Association Journal* 81 (November 1989): 38–42.

Gardner, Richard L., and Robert A. Young. “Assessing Salinity–Control Programs on the Colorado River.” *Resources* (Spring 1985): 10–13.

McCoy, William H. “Desalting: An Offer You May Not Refuse.” *Water Spectrum (Corps of Engineers)* 12 (Spring 1980): 48–53.

U.S. Congress. House. Committee on Agriculture. Subcommittee on Department Operations, Research and Foreign Agriculture. *Salinity Control in Colorado River Basin: Hearing, June 10, 1981.* 97th Cong., 1st sess., Washington, DC: GPO, 1981. 50 p.

U.S. Congress. Office of Technology Assessment. *Using Desalination Technologies for Water Treatment: Background Paper.* Washington, DC: GPO, 1988. 66 p.

Bottled Water

Ballentine, Carol L., and Michael L. Herndon. “The Water That Goes Into Bottles [How It Differs from That Comes Out of the Tap].” *FDA (Food and Drug Administration) Consumer* 17 (May 1983): 4–7.

Blair, Ian C. “Small Fish in a Big Pond? Bottled Water Sales Swim for Major Segment Status.” *Beverage World* 105 (December 1986): 42–44+.

Dingwall, James. "Battle of the Bubbles: A Small Montreal Bottler, Backed by Nestlé Has Grabbed a Share of the Booming Mineral Water Market: Perrier's Looking Less Effervescent." *Canadian Business* 53 (August 1980): 52–54+.

Fitzell, Phil. "The Emerging Water Industry [Increasing Number of Small Firms Selling Bottled Water]." *Venture* 3 (August 1981): 72–76.

Hartshorn, J. K. "Tap Water Alternatives." *Western Water* (January–February 1987): 4–11.

Sewell, Bradford H. "The Dark Cloud over Bottled Water." *Business and Society Review* (Fall 1986): 45–50.

U.S. Congress. House. Select Committee on Aging. Subcommittee on Health and Long-term Care. *Catalyst Altered Water: A Briefing, July 7, 1980.* 96th Cong., 2d sess., Washington, DC: 1980. 45 p.

"Water Quality Question [Whether California Has an Acid Rain Problem; Whether Bottled Water Is Safer Than Tap Water; Proposed Changes in U.S. Drinking Water Regulations]." *Western Water* (March–April 1984): 4–11.

Distribution Systems

Swamee, P. and A. K. Sharma. "Decomposition of Large Water–Distribution Systems." *Journal of Environmental Engineering* 116 (March–April 1990): 269–283.

Zessler, U., and U. Shamer. "Optimal Operation of Water Distribution Systems." *Journal of Water Resources Planning and Management* 115 (November 1989): 735–752.

Water Reuse

Anderson, Jim, and Jim Orput. "Wastewater Treatment for Your Community." *Minnesota Cities* 70 (March 1985): 5–6.

Bruvold, William H. "Public Participation in Environmental Decisions: Water Reuse [California]." *Pubic Affairs Report* 22 (February 1981): 1–6.

Dilger, Robert Jay. "Grantsmanship, Formulamanship, and Other Allocational Principles: Wastewater Treatment Grants." *Journal of Urban Affairs* 5 (Fall 1983): 269–286.

Harleman, Donald R. F. "Cutting the Waste in Wastewater Cleanups: Rather Than Mandate Technological Fixes to Water Pollution Problems: Congress Should Let Coastal Communities Use Innovative Approaches to Solve Them." *Technology Review* 93 (April 1990): 60–68.

Matthews, John E. *Industrial Reuse and Recycle of Wastewaters: Literature Review [1967–1978].* Ada, OK: Environmental Protection Agency, Robert S. Kerr Environmental Research Laboratory. 1980. 204 p.

Nichols, A. B. "Water Reuse Closes Water-Wastewater Loop." *Journal of Water Pollution Control Federation* 60 (November 1988): 1930–1937.

O'Toole, Laurence J., Jr. "Goal Multiplicity in the Implementation Setting: Subtle Impacts and the Case of Wastewater Treatment Privatization." *Policy Studies Journal* 18 (Fall 1989): 1–20.

U.S. Congress. House. Committee on Public Works and Transportation. Subcommittee on Investigations and Oversight. *Implementation of the Federal Water Pollution Control Act Concerning the Performance of the Municipal Wastewater Treatment Construction Grants Program: Report, October 1981.* 97th Cong., 1st sess., Washington, DC: GPO, 1981. 75 p.

U.S. Congress. Senate. Committee on Environment and Public Works. Subcommittee on Environmental Pollution. *Municipal Wastewater Treatment Construction Grants Program: Hearings, Pt. 2 June 29 and August 10, 1981, on S.975, a Bill to Revise and Extend Certain Provisions of the Federal Water Pollution Control Act, as Amended, for One year, and for Other Purposes and S.1274, a Bill to Amend Title II of the Clean Water Act, and for Other Purposes.* 97th Cong., 1st sess., Washington, DC: GPO, 1981. 269 p.

"Wastewater Management: Alternative Small Scale Treatment Systems." *Management Information Service Report* 17 (April 1985): 1–23.

Wetzel, Paul R. "The Use of Wetlands for Wastewater Treatment: An Introduction." *National Wetlands Newsletter* 6 (September–October 1984): 6–8.

Economics

Argent, Gala. "Water Marketing: Driven by Low Supplies." *Western Water* (May–June 1989): 4–11.

Belal, Rashida. "Environmental Taxes, 1981–1985." *Statistics of Income Bulletin* 6 (Spring 1987): 51–58.

Brice, R. L., and E. R. Unangst. "Long–Range Financial Planning for Water Utilities." *American Water Works Association Journal* 81 (May 1989): 48–52.

Christensen, Douglas A., et al. *The Potential Effect of Increased Water Prices on U.S. Agriculture.* Ames, IA: Center for Agricultural and Rural Development, Iowa State University, 1981. 121 p.

Ciminello, Paul. *State Strategies in Financing Water Treatment Facilities.* Growth and Environmental Management Series. Research Triangle Park, NC: Southern Growth Policies Board, 1986. 23 p.

Colby, Bonnie G. "Economic Impacts of Water Law: State Law and Water Market Development in the Southwest." *Natural Resources Journal* 28 (October 1988): 721–749.

Correll, Donald L., et al. "Financing a Major Water Supply Facility: A Case Study in Cooperation." *Public Utilities Fortnightly* 117 (June 26, 1986): 21–25.

Hartshorn, J. K., and Rita S. Sudman. "Water Marketing: A New Option? " *Western Water* (March–April 1986): 4–10.

Hope, Barney, and Michael Sheehan. "The Political Economy of Centralized Water Supply in California." *Social Science Journal* (Fort Collins) 20 (April 1983): 29–39.

Jonish, James E., and Charles E. Butler. "Municipal Water Pricing Practices in West Texas." *Texas Business Review* 57 (March–April 1983): 88–94.

McDonald, Brian, et al. "Evolving Urban Water Pricing Policies in Selected New Mexico Cities." *New Mexico Business* 33 (May 1980): 3–20.

Moore, W. John. "Mandates Without Money: The Rules of Federalism Are Changing: The Federal Government No Longer Links Direct Mandates to State and Local Governments with the Carrot of Federal Aid." *National Journal* 18 (October 4, 1986): 2366–2370.

Olson, Kent W. "Economics of Transferring Water to the High Plains." *Quarterly Journal of Business and Economics* 22 (Autumn 1983): 63–80.

Palmquist, Robert D. *Developing a Fair and Equitable Water Rate Structure.* Salt Lake City: University of Utah, Center for Public Affairs and Information, 1983. 49 p.

Rogers, Peter. "Water: Not as Cheap as You Think; We Must Change Our Water Policies So That the Cost of This Resource Reflects Its Value." *Technology Review* 89 (November–December 1986): 30–43.

Saliba, Bonnie C., et al. "Do Water Market Prices Appropriately Measure Water Values?" *Natural Resources Journal* 27 (Summer 1987): 615–651.

Stauffer, Thomas R. "The Price of Peace: The Spoils of War [Economic 'Imperatives' Behind the Lack of Progress Toward Peace; Focuses on the Question of Water Supply and on Israel's Interest in the Waters of the Litani River]." *American–Arab Affairs* (Summer 1982): 43–54.

Summer, Fern. "The Byzantine Art of Water Utility Ratemaking." *Environmental Forum* 4 (May 1985): 32–38.

United Nations. Economic Commission for Africa. *Economic Aspects of Drinking Water Supply and Sanitation in Africa with Particular Reference to Rural Areas.* New York: UN Agent, 1989. 63 p.

Wishart, David. "An Economic Approach To Understanding Jordan Valley Water Disputes." *Middle East Review* 21 (Summer 1989): 45–53.

Zamora, Jennifer, et al. "Pricing Urban Water: Theory and Practice in Three Southwestern Cities [Tucson, Ariz., Santa Fe, N. Mex., and Denver, Colo.]." *Southwestern Review of Management and Economics* 1 (Spring 1981): 89–113.

Laws and Regulations

Brieger, Heidi E. "LUST and the Common Law: A Marriage of Necessity." *Boston College Environmental Affairs Law Review* 13, no. 4 (1986): 521–551.

Davis, Peter N. "Protecting Waste Assimilation Streamflows by the Law of Water Allocation, Nuisance, and Public Trust, and by Environmental Statutes." *Natural Resources Journal* 28 (Spring 1988): 357–391.

DuMars, Charles T., ed. "New Challenges to Western Water Law." *Natural Resources Journal* 29 (Spring 1989): 331–592.

New Jersey. *Joint Public Hearing to Take Testimony On Possible Legislative Initiatives to Preserve Watershed Lands and Other Natural Resources: Parsippany, New Jersey, November 30, 1988; Before Senate Energy and Environment Committee and Assembly Conservation, Natural Resources and Energy Committee.* Trenton, NJ: 1988. 149 p.

Reinumagi, Irma U. "Diverting Water from the Great Lakes: Pulling the Plug On Canada." *Valparaiso University Law Review* 20 (Winter 1986): 299–338.

U.S. Congress. House. Committee on Agriculture. Subcommittee on Conservation, Credit, and Rural Development. *Upper Locust Creek Watershed in Missouri and Iowa; and Howard Creek Watershed in West Virginia: Hearing, April 5, 1989.* 101st Cong., 1st sess., Washington, DC: GPO, 1989. 17 p.

U.S. Congress. House. Committee on Interior and Insular Affairs. Subcommittee on Water and Power Resources. *Reclamation States Ground Water Protection and Management Act: Hearing, July 23, 1987, on H.R. 2320, to Direct the Secretary of the Interior to Improve Management of Ground Water in the Reclamation States.* 100th Cong., 1st sess., Washington, DC: GPO, 1988. 193 p.

U.S. Congress. House. Committee on Public Works and Transportation. Subcommittee on Water Resources. *Water Supply Policy of the Federal Government: Hearing, April 26, 1989.* 101st Cong., 1st sess., Washington, DC: GPO, 1989. 151 p.

U.S. Congress. Senate. Committee on Energy and Natural Resources. *Implementation of the Reclamation Act of 1982: Hearing, October 2, 1987.* 100th Cong., 1st sess., Washington, DC: GPO, 1988. 263 p.

"Water Law of the People's Republic of China." *Chinese Geography and Environment.* 20 (Fall 1989): 30–46.

Water Resources Development

General

Creighton, James L. *Public Involvement Manual: Involving the Public in Water and Power Resources Decisions.* Washington, DC: U.S. Department of the Interior, Water and Power Resources Service, 1980. 333 p.

Delli Priscoli, Jerry. "People and Water: Social Impact Assessment Research." *Water Spectrum* (Corps of Engineers) 13 (Summer 1981): 8–17.

Jacobson, Charles, et al. "Water, Electricity, and Cable Television: A Study of Contrasting Historical Patterns of Ownership and Regulation." *Urban Resources* 3 (Fall 1985): 9–18.

Miller, E. Willard. "Water Transportation in Pennsylvania." In *Water Resources in Pennsylvania: Availability, Quality and Management* edited by Shyamal Majumdar, E. Willard Miller, and Richard R. Parizek, 206–218. Easton, PA: Pennsylvania Academy of Science, 1990.

Parizek, Richard R., and Emmanuel K. Mundi. "Artificial Ground Water Recharge: An Under Used Water Resource Management Option." In *Water Resources in Pennsylvania: Availability, Quality and Management* edited by Shyamal Majumdar, E. Willard Miller, and Richard R. Parizek, 267–283. Easton, PA: Pennsylvania Academy of Science, 1990.

Rhodes, John J. "Developing a National Water Policy: Problems and Perspectives on Reform [United States]." *Journal of Legislation* 8 (Winter 1981): 1–15.

Rumbaugh, James O., III, et al. "Ground Water Modeling as a Management Tool for Pennsylvania's Water Resources, Parts I and II." In *Water Resources in Pennsylvania: Availability, Quality and Management* edited by Shyamal Majumdar, E. Willard Miller, and Richard R. Parizek, 284–300. Easton, PA: Pennsylvania Academy of Science, 1990.

"Saving the Nation's Great Water Bodies." *EPA (Environmental Protection Agency) Journal* 16 (November–December 1990): 2–64.

Siddiqui, Shams H. "Ground Water Protected Areas: Exploration, Development and Sustained Yields." In *Water Resources in Pennsylvania:*

Availability, Quality and Management edited by Shyamal Majumdar, E. Willard Miller, and Richard R. Parizek, 301–317. Easton, PA: Pennsylvania Academy of Science, 1990.

"Symposium: Water Resources and Public Policy." *Policy Studies Review* 5 (November 1985): 349–450.

Teclaff, Ludwik A., ed. "Symposium on International Resources Law." *Natural Resources Journal* 25 (October 1985): 863–1024.

United Nations. Economic Commission for Latin America and the Caribbean. *The Water Resources of Latin America and the Caribbean: Planning, Hazards and Pollution.* New York: UN Agent, 1990. 252 p.

U.S. Bureau of Reclamation. Guam Study Office. *Potential Water Resources Development: Guam Special Report and Environmental Assessment.* Agana, Guam: 1985. 87 p.

Groundwater

Deines, P., L. C. Hull, and Schuyler S. Stowe. "Study of Soil and Ground-Water Movement Using Stable Hydrogen and Oxygen Isotopes." In *Water Resources in Pennsylvania: Availability, Quality and Management* edited by Shyamal Majumdar, E. Willard Miller, and Richard R. Parizek, 121–140. Easton, PA: Pennsylvania Academy of Science, 1990.

Dycus, J. Stephen. "Development of a Groundwater Protection Policy." *Boston College Environmental Affairs Law Review* 11 (January 1984): 211–271.

Gburek, W. J., and H. B. Pionke. "Surface-Ground Water Interactions: Water Resources Implications for a Rural Upland Watershed." In *Water Resources in Pennsylvania: Availability, Quality and Management* edited by Shyamal Majumdar, E. Willard Miller, and Richard R. Parizek, 354–371. Easton, PA: Pennsylvania Academy of Science, 1990.

Giddings, Todd. "Ground-Water Heat Pumps." In *Water Resources in Pennsylvania: Availability, Quality and Management* edited by Shyamal Majumdar, E. Willard Miller, and Richard R. Parizek, 192–205. Easton, PA: Pennsylvania Academy of Science, 1990.

Frey, Robert F. "Overview of Surface and Ground-Water Quality in Pennsylvania." In *Water Resources in Pennsylvania: Availability, Quality and Management* edited by Shyamal Majumdar, E. Willard Miller, and Richard R. Parizek, 466–481. Easton, PA: Pennsylvania Academy of Science, 1990.

Whipple, W., and D. J. Van Abs. "Principles of a Ground-Water Strategy." *Journal of Water Resources Planning and Management* 116 (July–August 1990): 503–516.

Sources

Beck, Eckardt C. "Aboveground Worries About Underground Water: There's Plenty of Water in the Ground, But a Lot of It's In the Wrong Place and Some of It Is Getting Dirty." *Business and Society* (Summer 1981): 65–69.

Booth, Colin J. "Hydrogeological Significance of Subsurface Coal Mining." In *Water Resources in Pennsylvania: Availability, Quality and Management* edited by Shyamal Majumdar, E. Willard Miller, and Richard R. Parizek, 318–333. Easton, PA: Pennsylvania Academy of Science, 1990.

Chan, Arthur H. "Market Allocation of Interstate Groundwater: Evaluating Sporhase and the Commerce Chase." *Evaluation and Program Planning* 11, no. 4 (1988): 353–355.

Corbett, Patrick E. "The Overlooked Farm Crisis: Our Rapidly Depleting Water Supply." *Notre Dame Law Review* 61, no. 3 (1986): 454–477.

Forbes, Gregory S. "Precipitation in Pennsylvania." In *Water Resources in Pennsylvania: Availability, Quality and Management* edited by Shyamal Majumdar, E. Willard Miller, and Richard R. Parizek, 41–59. Easton, PA: Pennsylvania Academy of Science, 1990.

Lehr, Jay H. "The Importance of Ground Water in the National Water Resources Picture and the National Water Well Association." In *Water Resources in Pennsylvania: Availability, Quality and Management* edited by Shyamal Majumdar, E. Willard Miller, and Richard R. Parizek, 6–11. Easton, PA: Pennsylvania Academy of Science, 1990.

Parizek, Richard R. "Scientific Methods of Ground-Water Exploration." In *Water Resources in Pennsylvania: Availability, Quality and Management* edited by Shyamal Majumdar, E. Willard Miller, and Richard R. Parizek, 96–112. Easton, PA: Pennsylvania Academy of Science, 1990.

Parizek, Richard R., and E. Scott Bair. "Ground-Water Exploration Using Shallow Geothermal Techniques." In *Water Resources in Pennsylvania: Availability, Quality and Management* edited by Shyamal Majumdar, E. Willard Miller, and Richard R. Parizek, 80–95. Easton, PA: Pennsylvania Academy of Science, 1990.

Rajagopalan, S. P., and N. B. N. Prasad. "Subsurface Water in River Beds as Source of Rural Water Supply Schemes." *Journal of Water Resource Planning and Management* 115 (March 1989): 186–194.

United Nations. Department of Technical Cooperation for Development. *Ground Water in the Eastern Mediterranean and Western Asia.* National Resource–Water Series no. 9. New York: U.N. Agent, 1982. 230 p.

U.S. Department of the Interior. Office of Water Policy. *Directory of Groundwater Programs and Activities of the Federal Government.* Springfield, VA: National Technical Information Service, 1983. 79 p.

Yandle, Bruce. "Resource Economics: A Property Rights Perspective [Proposes a Model To Describe the Transfer of Natural Resources from Public to Private Ownership: Uses Groundwater Resources in the U.S. as an Example]." *Journal of Energy Law and Policy* 5, no. 1 (1983): 1–19.

Contamination

Brenner, Fred J., et al. "Non-Point Source Pollution: A Model Watershed Approach To Improve Water Quality." In *Water Resources in Pennsylvania: Availability, Quality and Management* edited by Shyamal Majumdar, E. Willard Miller, and Richard R. Parizek, 482–498. Easton, PA: Pennsylvania Academy of Science, 1990.

Burmaster, David E., and Robert H. Harris. "Groundwater Contamination: An Emerging Threat; Synthetic Organic Chemicals Are Contaminating Groundwater in Many Parts of the Nation, Possibly Posing Unacceptable Risks to Human Health." *Technology Review* 85 (July 1982): 50–58+.

Carpenter, Betsy. "Is Your Water Safe? Turning on the Faucet May Be Fraught with Health Risks." *U.S. News* 111 (July 12, 1991): 48–55.

Crosson, Pierre. "Implementation Policies and Strategies for Agricultural Non-Point Pollution [Controlling Water Pollution Caused By Runoff from Farm Fields]." *Southwestern Review of Management and Economics* 4 (Spring 1985): 27–36.

Culver, Alicia, and Rose Marie Audette. "Danger's in the Well." *Environmental Action* 16 (March–April 1985): 15–19.

Dworkin, Judith M. "Private Parties Rights To Recover Losses from Groundwater Contamination in Arizona." *Arizona State Law Journal* 3 (1985): 727–761.

Ferrett, Robert L., and Robert M. Ward. "Agricultural Land Use Planning and Groundwater Quality [Effects of Various Chemicals on Water Quality; Includes a Summary of Federal and State Legislative Actions Directed Toward Protecting Groundwater Resources]." *Growth and Change* 14 (January 1983): 32–39.

Fitchen, Janet M. "Cultural Aspects of Environmental Problems: Individualism and Chemical Contamination of Groundwater." *Science, Technology, and Human Values* 12 (Spring 1987): 1–12.

"Ground Water: The Endangered Resource." *Western Water* (September–October 1984): 4–11.

"Groundwater Problems Rise to Political Surface." *Conservation Foundation Letter* (September–October 1984): 1–6.

"Groundwater Supplies: Are They Imperiled? Underground Water Problems Keep Surfacing Ominously—Yet There Are No Comprehensive Management Systems for Protecting the Nation's Groundwater Supplies from Depletion or Contamination." *Conservation Foundation Letter* (June 1981): 1–8.

Halstead, John M., et al. "Impacts of Uncertainty on Policy Costs of Managing Nonpoint Source Ground Water Contamination." *Journal of Sustainable Agriculture* 1, no. 4 (1991): 29–48.

New Jersey. General Assembly. Agriculture and Environment Committee. *Public Hearing on Rockaway Experience, Water Pollution, and Related Issues; Held: Rockaway Township, New Jersey, March 10, 1982.* Trenton, NJ: 1982. 53 p.

Panasewich, Carol. "Protecting Groundwater from Pesticides." *EPA (Environmental Protection Agency) Journal* 11 (September 1985): 18–20.

Raucher, Robert L. "The Benefits and Costs of Policies Related to Groundwater Contamination." *Land Economics* 62 (February 1986): 33–45.

"Recent Developments in Waste Management [Hazardous and Nuclear Water; Groundwater Pollution; Sewage Treatment; Municipal Sludge Disposal]." *Urban Law Annual* 22 (1981): 317–405.

Shuey, Chris, et al. "Ground Water Contamination in New Mexico: Seeing What It's All About." *Workbook (Southwest Research and Information Center)* 12 (April–June 1987): 44–53.

Stone, Timothy, comp. *Jersey Ground Water Pollution Index, September 1974–January 1983.* New Jersey Geology Survey. Open File Report, no. 83-1. Trenton, NJ: Department of Environmental Protection, 1983. 99 p.

Thompson, Roger. "Preventing Groundwater Contamination." *Editorial Research Reports* (July 12, 1985): 519–536.

Tucker, Robert K. *Groundwater Quality in New Jersey: An Investigation of Toxic Contaminants.* Trenton, NJ: Department of Environmental Protection. Office of Cancer and Toxic Substances Research, 1981. 60 p.

U.S. Congress. House. Committee on Government Operations. Environment, Energy, and Natural Resources Subcommittee. *Ground Water Contamination: Hearings, June 22–29, 1983.* 98th Cong., 1st sess., Washington, DC: GPO, 1984. 630 p.

———. *Review of Ground Water Contamination and Depletion Problems in the Northwest: Hearing, November 28, 1983.* 98th Cong., 1st sess., Washington, DC: GPO, 1985. 158 p.

U.S. Congress. House. Committee on Merchant Marine and Fisheries. *Contaminated Marine Sediments: Hearing, March 20, 1990, Before the Subcommittee on Oversight and Investigations and the Subcommittee on Oceanography and Great Lakes, on Contaminated Marine Sediments in Navigable Waters and Harbor Areas of the United States and Federal Efforts To Address this problem.* 101st Cong., 2d sess., Washington, DC: GPO, 1990. 235 p.

U.S. Congress. House. Committee on Public Works and Transportation. Subcommittee on Water Resources. *Water Resources Research Activities of the U.S. Geological Survey and the National Ground Water Contamination Information Act: Hearings, July 8 and 23, 1987, on H.R. 791.* 100th Cong., 1st sess., Washington, DC: GPO, 1987. 357 p.

U.S. Congress. House. Committee on Science and Technology. Subcommittee on Natural Resources, Agriculture Research, and Environment. *Environmental Effects of Sewage Sludge Disposal: Hearing, May 27, 1981.* 97th Cong., 1st sess., Washington, DC: GPO, 1981. 109 p.

U.S. Congress. House. Committee on Science, Space, and Technology. Subcommittee on Natural Resources, Agriculture Research, and Environment. *H.R. 2253 the Ground Water Research, Development and Demonstration Act: Hearing July 21, 1987 and H.R. 791, the National Ground Water Contamination Information Act of 1987.* 100th Cong., 1st sess., Washington, DC: GPO, 1988. 437 p.

U.S. Congress. House. *State Solid Waste Plans and Ground Water Contamination: Joint Hearing, April 16, 1982, Before the Subcommittee on Commerce, Transportation, and Tourism of the Committee on Energy and Commerce and the Subcommittees on Water Resources of the Committee on Public Works and Transportation.* 97th Cong., 2d sess., Washington, DC: GPO, 1982. 217 p.

U.S. Congress. Office of Technology Assessment. *Beneath the Bottom Line: Agricultural Approaches To Reduce Agrichemical Contamination of Groundwater; Summary.* Washington, DC: GPO, 1990. 79 p.

U.S. Congress. Senate. Committee on Environment and Public Works. *Ground Water Protection: Joint Hearings, February 23–May 17, 1988, on S.20, S.1105, and H.R.791 before the Subcommittees on Water Resources, Transportation, and Infrastructure, and Hazardous Wastes and Toxic Substances.* 100th Cong., 2d sess., Washington, DC: GPO, 1988. 868 p.

———. *Groundwater Contamination By Toxic Substances: A Digest of Reports; A Report, November 1983.* 98th Cog., 1st sess., Washington, DC: GPO, 1983. 75 p.

U.S. Congress. Senate. Committee on Environment and Public Works. Subcommittee on Superfund, Ocean, and Water Protection. *The Seriousness and Extent of Ground Water Contamination: Hearing, August 1, 1989.* 101st Cong., 1st sess., Washington, DC: GPO, 1989. 174 p.

U.S. Congress. Senate. Committee on Environment and Public Works. Subcommittee on Toxic Substances and Environmental Oversight. *Ground Water Contamination: Hearings, November 1983–March 9, 1984.* 98th Cong., 1st and 2d sess., Washington, DC: GPO, 1984. 452 p.

Management

Baker, Brian P. *Production Response to Groundwater Pollution Control.* A.E. Research 87-18. Ithaca, NY: New York State Agriculture Experiment Station, 1987. 41 p.

Blomquist, William. "Exploring State Differences in Groundwater Policy Adoptions, 1980–1989." *Publius* 21 (Spring 1991): 101–115.

Bowman, J. A. "Ground-Water Management Areas in United States." *Journal of Water Resources Planning and Management* 116 (July–August 1990): 484–502.

Chan, Arthur H. "Market Allocation of Interstate Groundwater: Evaluating Sporhase and the Commerce Clause." *Evaluation and Program Planning* 11, no.4 (1988): 353–355.

Connall, Desmond D., Jr. "A History of the Arizona Groundwater Management Act." *Arizona State Law Journal* 1982, no. 2 (1982): 313–344.

"Groundwater Supplies: Are They Imperiled? Underground Water Problems Keep Surfacing, Ominously—Yet There Are No Comprehensive Management Systems for Protecting the Nation's Groundwater Supplies from Depletion or Contamination." *Conservation Foundation Letter* (June 1981): 1–8.

Hartig, John H., et al. "Overcoming Obstacles in Great Lakes Remedial Action Plans." *International Environmental Affairs* 3 (Spring 1991): 91–107.

Higdon, Philip R., and Terence W. Thompson. "The 1980 Arizona Groundwater Management Code." *Arizona State Law Journal,* no. 3 (1980): 621–671.

Kelly, Michael J. "Management of Groundwater through Mandatory Conservation." *Denver Law Journal* 61, no. 1 (1983): 1–24.

"Managing Our Fresh Water Resources." *Impact of Science on Society* no. 1 (1983): 3–24+.

Maniatis, Melina. "Stormwater Management." *Management Information Service Report* 22 (November 1990): 1–15.

Metzger, Philip C., et al. "The Developer's Stake in Groundwater Protection." *Urban Land* 45 (April 1986): 19–23.

Shelley, Fred M. "Groundwater Supply Depletion in West Texas; The Farmer's Perspective." *Texas Business Review* 57 (November–December 1983): 279–283.

Smith, Zachary A. "Federal Intervention in the Management of Groundwater Resources: Past Efforts and Future Prospects." *Publius* 15 (Winter 1985): 145–159.

United Nations. Economic Commission for Europe. *Charter on Ground-Water Management as Adopted by the Economic Commission for Europe at Its Forty-Fourth Session (1989) By Decision (E44)*. New York: UN Agent, 1989. 20 p.

U.S. Congress. House. Committee on Government Operations. Environment, Energy, and Natural Resources Subcommittee. *Environmental and Economic Benefits of Low-Input Farming: Hearing April 28, 1988.* 100th Cong., 2d sess., Washington, DC: GPO, 1989. 298 p.

U.S. Congress. House. Committee on Interior and Insular Affairs. Subcommittee on General Oversight and Investigations. *Nation's Ground Water Protection: Oversight Hearing, May 26, 1988.* 100th Cong., 2d sess., Washington, DC: GPO, 1988. 201 p.

U.S. Congress. House. Committee on Interior and Insular Affairs. Subcommittee on Water and Power Resources. *Reclamation States Ground Water Protection and Management Act: Hearing, July 23, 1987, on H.R.2320, To Direct the Secretary of the Interior To Improve Management of Ground Water in the Reclamation States.* 100th Cong., 1st sess., Washington, DC: GPO, 1988. 193 p.

U.S. Congress. Senate. Committee on Environment and Public Works. Subcommittee on Environmental Protection. *Protection of Groundwater Resources in the State of Maine: Hearing July 2, 1987.* 100th Cong. 1st sess., Washington, DC: GPO, 1987. 89 p.

U.S. Congress. Senate. Committee on Governmental Affairs. Subcommittee on Government Efficiency, Federalism, and the District of Columbia. *Protecting Our Nation's Groundwater: The Need for Better Program Coordination: Hearings, April 22 and May 18, 1988, on S.1992 To Promote Intergovernmental and Interagency Cooperation on Groundwater Management and Protection, and To Facilitate the Dissemination of Practical Information on Groundwater Protection to the Public.* 100th Cong., 2d sess., Washington, DC: GPO, 1988. 521 p.

Viessman, W. "Water Management; Challenge and Opportunity." *Journal of Water Resources Planning and Management.* 116 (March–April 1990): 155–169.

Weschler, Louis, and Helen Ingram. "Arizona Groundwater Reform: The Forces of Change." *Southwestern Review of Management and Economics.* 2 (Summer 1982): 13–21.

Wood, Mary Christina. "Regulating Discharges Into Groundwater: The Crucial Link in Pollution Control under the Clean Water Act." *Harvard Environmental Law Review* 12 (2): 569–626 (1988).

International Aspects

Conrad, Jobst. "Nitrate Debate and Nitrate Policy in FR Germany." *Land Use Policy* 5 (April 1988): 207–218.

European Communities Commission. *Intensive Farming and the Impact on the Environment and the Rural Economy of Restrictions on the Use of Chemical and Animal Fertilizers.* Luxembourg: European Communities Official Publications Office, 1989. 212 p. ISBN 92-826-0123-4.

Hayton, Robert D., and Albert E. Utton. "Transboundary Groundwaters: The Bellagio Draft Treaty." *Natural Resources Journal* 29 (Summer 1989): 663–722.

Rodgers, Ann B., and Albert E. Utton. "The Ixtapa Draft Agreement Relating to the Use of Transboundary Groundwaters." *Natural Resources Journal* 25 (July 1985): 713–772.

United Nations Department of Technical Cooperation for Development. *Ground Water in North and West Africa.* New York: U.N. Agent, 1988. 405 p.

Utton, Albert E. "The Development of International Groundwater Law." *Natural Resources Journal* 22 (January 1982): 95–118.

Laws and Regulations

Brown, Donald A. "Murky Standards for Groundwater: Since 1984, the EPA and the States Have Floated Along With a Jury-Rigged Policy That Encourages Groundwater Degradation." *Environmental Forum* 7 (May–June 1990): 16–21.

Buresh, James C. "State and Federal Land Use Regulation: An Application to Groundwater and Nonpoint Source Pollution Control." *Yale Law Review* 95 (June 1986): 1433–1458.

Carriker, Roy R., and William G. Boggess. "Agricultural Non-Point Pollution: A Regulatory Dilemma." *Applied Research and Public Policy* 3 (Summer 1988): 63–70.

Connall, Desmond D., Jr. "A History of the Arizona Groundwater Management Act." *Arizona State Law Journal*, no. 2 (1982): 313–344.

Durant, Robert F., and Michell D. Holmes. "Thou Shalt Not Covet Thy Neighbor's Water: The Rio Grande Basin Regulatory Experience." *Public Administration Review* 45 (November–December 1985): 821–831.

Durenberger, David. "Groundwater Policy: A Need for Federal Participation." *Forum for Applied Research and Public Policy* 1 (Spring 1986): 79–86.

Gelpe, Marcia R. "Pollution Control Laws Against Public Facilities." *Harvard Environmental Law Review* 13, no. 1 (1989): 69–146.

Gottlieb, Gail. "New Mexico's Mine Dewatering Act: The Search for Rehoboth [Attempt To Regulate the Use of Ground Water by the Uranium Mining Industry To 'Promote Maximum Economic Development of Mineral Resources While Ensuring That Such Development Does Not Impair Existing Prior Water Rights']." *Natural Resources Journal* 20 (July 1980): 653–680.

"Groundwater Liability Waivers." *Environmental Forum* 4 (January 1986): 28–35.

"Groundwater Problems Rise to Political Surface." *Conservation Foundation Letter* (September–October 1984): 1–6.

Hodges-Copple, John. *Protecting Public Water Supplies in the South: A Primer on the State's Role.* Growth and Environmental Management. Research Triangle Park, NC: Southern Growth Policies Board, 1988. 34 p.

Kovács, György. "Are Groundwater Resources Sufficiently Protected?" *Impact of Science on Society* no. 1 (1983): 35–47.

Malone, Linda A. "The Necessary Interrelationship between Land Use and Preservation of Groundwater Resources." *UCLA (Univ Cal Los Angeles) Journal of Environmental Law and Policy* 9, no. 1 (1990): 1–72.

Metzger, Philip C., et al. "The Developer's Stake in Groundwater Protection." *Urban Land* 45 (April 1986): 19–23.

Miniter, Richard. "Muddy Waters: The Quagmire of Wetlands Regulation." *Policy Review* (Spring 1991): 70–77.

Murphy, James. "Ground Water Protection: A Planning Process for Local Government." *Connecticut Government* 38 (Winter 1987): 5–8.

Ng, Lawrence. "A Drastic Approach To Controlling Groundwater Pollution." *Yale Law Journal* 98 (February 1989): 773–791.

Nunn, Susan Christopher. "The Political Economy of Institutional Change: A Distribution Criterion for Acceptance of Groundwater Rules." *Natural Resources Journal* 25 (October 1985): 867–892.

"Protecting GroundWater: The Hidden Resource." *EPA (Environmental Protection Agency) Journal* 10 (July–August 1984): 2–33.

Regens, James L. and Margaret A. Reams. "State Strategies for Regulating Groundwater Quality." *Social Science Quarterly* 69 (March 1988): 53–69.

Roberts, Rebecca S. and L. Mathis Butler. "Information for State Groundwater Quality Policymaking [Developing a Groundwater Quality Information Base That Relies on Land and Water Use Data, in Contrast to Groundwater Quality Data: Implications for State Policy Formulation]." *Natural Resources Journal* 24 (October 1984): 1015–1041.

Sater, Rachel J. "EPA's Pesticides-in-Groundwater Strategy: Agency Action in the Face of Congressional Inaction." *Ecology Law Quarterly* 17, no. 1 (1990): 143–177.

Smith, Zachary A. "Centralized Decisionmaking in the Administration of Groundwater Rights: The Experience of Arizona, California, and New Mexico, and Suggestions for the Future." *Natural Resources Journal* 24 (July 1984): 641–688.

———. "Rewriting California Groundwater Law: Past Attempts and Prerequisites To Reform." *California Western Law Review* 20 (Winter 1984): 223–257.

Swenson, Robert W. "A Primer of Utah Water Law [with a Section on Application to Groundwater]." *Journal of Energy Law & Policy* 5, no. 2 (1984): 165–196; 6, no. 1 (1985): 1– 54.

Tarlock, A. Dan. "Supplemental Groundwater Irrigation Law: From Capture to Sharing." *Kentucky Law Journal* 73, no. 3 (1984–1985): 695–722.

United Nations. Economic Commission for Europe. Committee on Water Problems. *Ground-Water Legislation in the ECE Region.* New York: U.N. Agent, 1986. 39 p.

U.S. Congress. House. Committee on Energy and Commerce. Subcommittee on Oversight and Investigations. *Ground Water Monitoring: Hearing, April 29, 1985.* 99th Cong., 1st sess., Washington, DC: GPO, 1985. 194 p.

U.S. Congress. House. Committee on Government Operations. Environment, Energy, and Natural Resources Subcommittee. *Review of Ground Water Protection Strategy Recently Proposed By the Environmental Protection Agency: Hearings, April 11–12, 1984.* 98th Cong. 2d sess., Washington, DC: GPO, 1984. 611 p.

———. *To Assess Progress Toward the Development of a National Groundwater Protection Program: Hearing, December 3, 1985.*, 99th Cong., 1st sess., Washington, DC: GPO, 1987. 99 p.

U.S. Congress. House. Committee on Interior and Insular Affairs. Subcommittee on General Oversight and Investigations. *Nation's Ground Water Protection: Oversight Hearing, May 26, 1988.* 100th Cong., 2d sess., Washington, DC: GPO, 1988. 201 p.

U.S. Congress. House. Committee on Public Works and Transportation. Subcommittee on Oversight and Review. *Implementation of the Federal Water Pollution Control Act: Hearings, October 30–November 1, 1979 (the Municipal Construction Grants Program and the State Management Assistance Program.* 96th Cong. 1st sess., Washington, DC: GPO, 1980. 1458 p.

U.S. Congress. House. Committee on Public Works and Transportation. Subcommittee on Water Resources. *To Repeal the Industrial Cost Exclusion Provision from the Federal Water Pollution Control Act: Hearings, April 28–29, 1981, on H.R. 2957, To Amend the Federal Water Pollution Control Act To Delete the Limitation of the Use of Public Treatment Works Grants for Treating, Storing or Conveying the Flow of Industrial Users Into Treatment Works.* 97th Cong., 1st sess., Washington, DC: GPO, 1981. 370 p.

U.S. Congress. Senate. Committee on Energy and Natural Resources. Subcommittee on Water and Power. *Groundwater-Related Programs of the USGS and the EPA: Hearings, June 5, and 18, 1987.* 100th Cong., 1st sess., Washington, DC: GPO, 1987. 463 p.

U.S. Congress. Senate. Committee on Environment and Public Works. Subcommittee on Environmental Protection. *Protection of Ground Water in the State of Maine: Hearing, July 2, 1987.* 100th Cong., 1st sess., Washington, DC: GPO, 1987. 89 p.

U.S. Congress. Senate. Committee on Governmental Affairs. Subcommittee on Government Efficiency, Federalism, and the District of Columbia. *Protecting Our Nation's Groundwater: The Need for Better Program Coordination: Hearings, April 22 and May 18, 1988, on S.1992, To Promote Intergovernmental and Interagency Cooperation on Groundwater Management and Protection and To Facilitate the Dissemination of Practical Information on Groundwater Protection to the Public.* 100th Cong., 2d sess., Washington, DC: GPO, 1988. 521 p.

Weston, R. Timothy, and Joel R. Burcat. "Legal Aspects of Pennsylvania Water Management." In *Water Resources in Pennsylvania: Availability, Quality and Management* edited by Shyamal Majumdar, E. Willard Miller, and Richard R. Parizek, 219–244. Easton, PA: The Pennsylvania Academy of Science, 1990.

Wood, Mary Christina. "Regulating Discharges into Groundwater; The Crucial Link in Pollution Control under the Clean Water Act." *Harvard Environmental Law Review* 12, no. 2 (1988): 569–626.

Woodard, Gary C. "Salvaging Arizona's Groundwater Code." *Arizona Review* (1990): 34–40.

Water Quality

General

Barnaby, Wendy. "The Purity of the Water Supply." *Science and Public Affairs* 5, pt. 1 (1990): 23–29.

Clark R. M. "Identifying Vulnerable Surface Water Utilities." *American Water Works Association Journal* 81 (February 1989): 60–67.

Davis, Peter N. "Federal and State Water Quality Regulation and Law in Missouri." *Missouri Law Review* 55 (Spring 1990): 411–507.

Duncan, Jeanne. "Drinking Water: How Much of the Fear Is Real?" *Western Water* (July–August 1987): 4–11.

Goldman, George, et al. "California's Water: Quality and Quantity, and the [Sacramento–San Joaquin] Delta." *Public Affairs Report* 23 (April 1982): 1–12.

Gough, William R., and Burt A. Waite. "Oil and Gas Exploration and Water Quality Considerations." In *Water Resources in Pennsylvania: Availability, Quality and Management* edited by Shyamal Majumdar, E. Willard Miller, and Richard R. Parizek, 384–398. Easton, PA: Pennsylvania Academy of Science, 1990.

Hartshorn, J. K. "Tap Water Alternatives." *Western Water* (January–February 1987): 4–11.

Levine, Elyse. "Is Your Water Safe To Drink? The Answer Is Yes for the Majority of the Homes Served by Large Municipal Systems, But It's Still a Question Everyone Should Consider." *World and I* 5 (April 1990): 238–245.

Pedersen, William F., Jr. "Turning the Tide on Water Quality." *Ecology Law Quarterly* 15, no. 1 (1988): 69–102.

Redwine, James C. "Water Quality Concerns from the Electric Utility Industry Perspectives." In *Water Resources in Pennsylvania: Availability, Quality and Management* edited by Shyamal Majumdar, E. Willard

Miller, and Richard R. Parizek, 551–569. Easton, PA: Pennsylvania Academy of Science, 1990.

Ribaudo, Marc O. "Targeting the Conservation Reserve Program to Maximize Water Quality Benefits." *Land Economics* 65 (November 1989): 320–332.

Roberts, Rebecca, and L. Mathis Butler. "Information for State Groundwater Quality Policymaking [Developing a Groundwater Quality Information Base that Relies on Land and Water Use Data, in Contrast to Groundwater Quality Data]." *Natural Resources Journal* 24 (October 1984): 1015–1041.

Robinson, Keith, et al. *New Jersey 1988 State Water Quality Inventory Report.* Trenton, NJ: Department of Environmental Protection, Division of Water Resources, 1988. 5 p.

Santschi, P. H., et al. "The Self-Cleaning Capacity of Surface Waters After Radioactive Fallout: Evidence From European Waters After Chernobyl, 1986–1988." *Environmental Science & Technology* 24 (April 1990): 519–527.

Stanfield, Rochelle L. "Enough and Clean Enough? Across the Country, Fear Is Mounting About Whether There Will Be Enough Clean Water; Environmentalists and Resource Specialists Predict Troubled Times Ahead." *National Journal* 17 (August 17, 1985): 1876–1887.

U.S. Environmental Protection Agency. Office of Water. *National Water Quality Inventory: 1988 Report to Congress.* Washington, DC: 1990. 226 p.

———. *The Quality of Our Nation's Water; A Summary of the 1988 National Water Quality Inventory.* Washington, DC: 1990. 24 p.

Zeldin, Marvin. *Our Drinking Water: A Threatened Resource.* New York: Public Affairs Committee, 1983. 24 p.

Management

Lee, Terrence Richard. "Managing Water Resources in Latin America." *Natural Resources Journal* 30 (Summer 1990): 581–607.

Leschine, Thomas M. "Setting the Agenda for Estuarine Water Quality Management: Lessons from Puget Sound." *Ocean and Shoreline Management* 13, no. 3–4 (1990): 295–313.

Rassi, Clifford. "Improving Rural New York's Water Systems." *Rural Development Perspectives* 3 (February 1987): 21–25.

U.S. Congress. House. Committee on Energy and Commerce. Subcommittee on Oversight and Investigations. *Groundwater Monitoring at RCRA Land Disposal Facilities: Hearing, April 27, 1989.* 101st Cong., 1st sess., Washington, DC: GPO, 1989. 357 p.

U.S. Congress. Senate. Committee on Governmental Affairs. Subcommittee on Oversight of Government Management. *Oversight of EPA and the Great Lakes Water Quality Agreement: Hearings, April 28 and May 23, 1989.* 101st Cong., 1st sess., Washington, DC: GPO, 1989. 307 p.

Drinking Water

Dingwall, James. "Battle of the Bubblies: A Small Montreal Bottler, Backed By Nestlé, Has Grabbed a Share of the Booming Mineral Water Market: Perrier's Looking Less Effervescent." *Canadian Business* 53 (August 1980): 52–54+.

"Drinking Water." *Courier* (March–April 1986): 62–96.

Duncan, Jeanne. "Drinking Water: How Much of the Fear Is Real?" *Western Water* (July–August 1987): 4–11.

Meier, J. R. and F. B. Daniel. "The Role of Short-Term Tests in Evaluating Health Effects Associated with Drinking Water." *American Works Association Journal* 82 (October 1990): 48–56.

Safe Drinking Water Policy Research Project. Options for Community Response to the Safe Drinking Water Act. Policy Research Project Report no. 35. Austin: University of Texas, Lyndon B. Johnson School of Public Affairs, 1979. 117 p.

Sorkin, Alan L. "The Impact of Water and Sanitation on Health and Development." In *Research in Human Capital and Development, 1988,* edited by Ismail Sirageldin and Alan Sorkin, 77–98. Greenwich, CT: JAI Press, 1988.

United Nations. Department of Technical Cooperation for Development. *Legal and Institutional Factors Affecting the Implementation of*

International Drinking Water Supply and Sanitation Decade. New York: U.N. Agent, 1989. 121 p.

U.S. Congress. *The Safe Drinking Water Act: As Amended by Public Law 96-63, September 6, 1979 and Public Law 96-502, December 5, 1980.* 96th Cong., 2d sess., Washington, DC: GPO, 1980. 42 p.

U.S. Congress. House. Committee on Interstate and Foreign Commerce. *Hazardous Waste and Drinking Water: Joint Hearing, August 22, 1980, Before the Subcommittee on Health and the Environment and the Subcommittee on Transportation and Commerce.* 96th Cong, 2d sess., Washington, DC: GPO, 1981. 195 p.

U.S. Congress. House. Committee on Interstate and Foreign Commerce. Subcommittee on Health and the Environment. *Quality of Drinking Water—1980: Hearings, June 6–August 18, 1980, on Gathering Information To Help the Subcommittee Get a Better Grasp of Our Drinking Water Problems Across the Nation.* 96th Cong., 2d sess., Washington, DC: GPO, 1980. 714 p.

U.S. Congress. Senate. Committee on Environment and Public Works. *A Legislative History of the Safe Drinking Water Act; Together with a Section-by-Section Index.* 97th Cong., 2d sess., Washington, DC: GPO, 1982. 1103 p.

U.S. Congress. Senate. Committee on Environment and Public Works. Subcommittee on Toxic Substances and Environmental Oversight. *Safe Drinking Water: Hearings, March 29–July 28, 1982, on S.1866, a Bill To Assure Safe Drinking Water, and S.2131, a Bill To Amend the Safe Drinking Water Act, To Provide for the Protection of Certain Recharge Areas Overlying Sole Source Underground Water Supplies.* 97th Cong., 2d sess., Washington, DC: GPO, 1982. 762 p.

Wehr, Elizabeth. "Congress Facing Pressures To Loosen, Tighten Federal Safe Drinking Water Statute: Contamination Increasing." *Congressional Quarterly Weekly Report* 40 (May 1, 1982): 973–976.

"Where Your Water Comes From." *Western Water* (March–April 1989): 3–17. Special Issue.

Zeldin, Marvin. *Our Drinking Water; A Threatened Resource.* Public Affairs Pamphlet, no. 613, New York: Public Affairs Committee, 1983. 24 p.

Contamination

Conrad, Jobst. "Nitrate Debate and Nitrate Policy in FR Germany." *Land Use Policy* 5 (April 1988): 207–218.

Douglas, David. "Water Is Life: The International Drinking Water Supply and Sanitation Decade." *Amicus Journal* 7 (Spring 1986): 34–37.

"Drinking Water." *Courier* (March–April 1986): 62–96.

Eaton, Scott F. "Legal Consequences in New Hampshire from Asbestos Contamination of Drinking Water by Asbestos Cement Pipe." *IDEA* 21, no. 2 (1980): 115–139.

Hornberger, R. J., et al. "Acid Mine Drainage from Active and Abandoned Coal Mines in Pennsylvania." In *Water Resources in Pennsylvania: Availability, Quality and Management* edited by Shyamal Majumdar, E. Willard Miller, and Richard R. Parizek, 434–451. Easton, PA: Pennsylvania Academy of Science, 1990.

Krohe, James, Jr. "Mr. Clean's Impossible Sewer Dream: Making U.S. Waters Safe for Oysters and Clams, as Well as Oyster and Clam Eaters, Will Not Be Cheap." *Across the Board* 20 (September 1983): 30–37.

Marinelli, Janet. "It Came From Beneath Long Island: Vinyl Chloride, Benzene, Aldicarb and Trichloroethane Are But a Few of the Contaminants That Have Leached Into the Drinking Water of Three Million Residents." *Environmental Action* 14 (May 1983): 8–12.

Mosher, Lawrence. "A Host of Pollutants Threaten Drinking Water from Underground: The Federal Government Is Only Beginning To Identify the Scope of the Problem and To Take Steps To Protect the Drinking Water of Half the Population." *National Journal* 12 (August 16, 1980): 1353–1356.

New Jersey. General Assembly. Agriculture and Engineering Committee. *Public Hearing on Toms River Experience—Water Pollution and Related Issues: Held, Toms River, New Jersey, March 25, 1982.* Trenton, NJ: 1982. 67 p.

Olson, Erik D. "Nation's Drinking Water at Risk: Groundwater Contamination." *State Government News* 29 (June 1986): 4–7.

"Protecting Our Drinking Water." *EPA (Environmental Protection Agency) Journal* 12 (September 1986): 2–28.

Roe, David. "An Incentive–Conscious Approach To Toxic Chemical Controls." *Economic Development Quarterly* 3 (August 1989): 179–187.

Rose, Arthur W., and P. Evan Dresel. "Deep Brines in Pennsylvania." In *Water Resources in Pennsylvania: Availability, Quality and Management* edited by Shyamal Majumdar, E. Willard Miller, and Richard R. Parizek, 420–431. Easton, PA: Pennsylvania Academy of Science, 1990.

Russell, Christine. "California Gets Tough on Toxics." *Business and Society Review* (Summer 1989): 47–54.

Stanfield, Rochelle L. "Enough and Clean Enough? Across the Country, Fear is Mounting About Whether There Will Be Enough Clean Water; Environmentalists and Resource Specialists Predict Troubled Times Ahead." *National Journal* 17 (August 17, 1985): 1876–1887.

Teclaff, Ludwick A., and Eileen Teclaff. "International Control of Cross-Media Pollution: An Ecosystem Approach." *Natural Resources Journal* 27 (Winter 1987): 21–53.

U.S. Congress. House. Committee on Energy and Commerce. Subcommittee on Health and the Environment. *Lead Contamination: Hearings, December 10, 1987 and July 13, 1988.* 100th Cong., 1st and 2d sess., Washington, DC: GPO, 1988. 677 p.

U.S. Congress. House. Committee on Government Operations. Environment, Energy, and Natural Resources Subcommittee. *Ground Water Contamination; Hearings, June 22– 29, 1983.* 98th Cong., 1st sess., Washington, DC: GPO, 1984. 630 p.

"Water and Wastewater Treatment: A Thriving Business." *Chemical Engineering (U.S.)* 96 (December 1989): 85–103.

Disposal

Gansell, Stuart I. "Status of Domestic Wastewater Disposal in Pennsylvania." In *Water Resources in Pennsylvania: Availability, Quality and Management* edited by Shyamal Majumdar, E. Willard Miller, and Richard R. Parizek, 372–383. Easton, PA: The Pennsylvania Academy of Science, 1990.

Jones, David Howell, ed. *Water Supply and Waste Disposal.* Poverty and Basic Needs Series. Washington, DC: World Bank, 1980. 46 p.

U.S. Congress. House. Committee on Public Works and Transportation. Subcommittee on Water Resources. *Inadequate Water Supply and Sewage Disposal Facilities Associated with "Colonias" Along the United States and Mexican Border: Hearings, March 11–12, 1988.* 100th Cong., 2d sess., Washington, DC: GPO, 1988. 551 p.

"Water and Sanitation: The Pure and the Impure [Developing Countries]." *UNICEF News* no. 1 (1980): 3–31.

Acid Precipitation

"Acid Rain: A Major Threat to the Ecosystem." *Conservation Foundation Letter* (December 1982): 1–8.

Adams, R. J., et al. "Pollution, Agriculture and Social Welfare: The Case of Acid Deposition." *Canadian Journal of Agricultural Economics* 34 (March 1986): 3–19.

Alm, Leslie R. "The United States—Canadian Acid Rain Debate: The Science-Politics Linkage." *American Review of Canadian Studies* 20 (Spring 1990): 59–79.

Audette, Rose Marie L. "Acid Rain Is Killing More Than Lakes and Trees." *Environmental Action* 18 (May–June 1987): 10–13.

Batterman, S. A. "Selection of Reception Sites for Optimized Acid Rain Control Strategies." *Journal of Environmental Engineering* 115 (October 1989): 1046–1058.

Boehmer-Christiansen, Soñja. "Black Mist and Acid Rain Science as Fig Leaf of Policy." *Political Quarterly* 59 (April–June 1988): 145–160.

Bricker, O. P., and K. C. Rice. "Acidic Deposition to Streams." *Environmental Science & Technology* 23 (April 1989): 379–385.

"The Clean Air Act: Pro & Con." *Congressional Digest* 68 (February 1989): 35–64.

Day, John, and David Hodgson. "Pollution Control: The Cases of Acid Rain and Lead in Petrol." *Economics (London)* 23 (Autumn 1985): 106–112.

Dudek, Daniel J. "Emissions Trading: Environmental Perestroika or Flimflam? The Author Challenges the American Utility Industry To Accept Acid Rain Controls and Use Creatively the Flexibility Offered by the Administration's Clean Air Act Bill." *Electricity Journal* 2 (November 1989): 32–43.

Eisenreich, Steven J. "Toxic Fallout in the Great Lakes: The Buildup of Organic Pollutants from Far-Away Places Shows That the Acid Rain Problem in Lakes Is Only the Beginning." *Issues in Science and Technology* 4 (Fall 1987): 71–75.

Fisk, David P. "Will Acid Rain on a Modest Proposal? [for a Compact of Midwestern and Northeastern States To Share the Cost of Controlling Acid Rain]." *Glendale Law Review* 6, no. 2 (1984): 108–124.

Gallogly, Margaret R. "Acid Precipitation: Can the Clean Air Act Handle It?" *Boston College Environmental Affairs Law Review* 9, no. 3 (1981): 687–744.

Gladwell, Malcolm. "Rain, Rain, Go Away: Canadians May Be Outraged by Acid Rain: But the Americans Believe It's All a Plot To Make Them Buy Canadian Electricity." *Saturday Night* 103 (April 1988): 48–54.

Gorham, Eville. "What To Do About Acid Rain: Industry and the EPA Say the Complex Problem of Acid Rain Is Still Too Little Understood To Warrant Undertaking Multibillion-Dollar Cleanup Programs; But the Evidence Calls for Action Now." *Technology Review* 85 (October 1982): 58–63.

Hamm, A., et al. "Documentation of Areas Potentially Inclined to Water Acidification in the F.R.G." *Water Research* 23 (January 1989): 1–5.

Krohe, James, Jr. "Can We Stop Acid Rain? And Who Should Pay the Bill?" *Across the Board* 21 (February 1984): 14–25.

Lind, Douglas. "Umbrella Equities: Use of Federal Common Law of Nuisance To Catch the Fall of Acid Rain." *Urban Law Annual* 21 (1981): 143–178.

Magnet, Myron. "How Acid Rain Might Dampen the Utilities." *Fortune* 108 (August 8, 1983): 58–60+.

Milstone, Nancy H. "A Common Law Solution To the Acid Rain Problem." *Valparaiso University Law Review* 20 (Winter 1986): 277–297.

Moller, Erik K. "The United States–Canadian Acid Rain Crisis: Proposal for an International Agreement." *UCLA (Univ Cal Los Angeles) Law Review* 36 (1989): 1207–1240.

New Jersey. Department of Environmental Protection. Division of Environmental Quality. *The Effects of Acid Rain in New Jersey: Report of Public Hearings, October 17, 1983.* Trenton, NJ: 1984. 25 p.

Olem, H., and P. H. Berthouex. "Acidic Deposition and Cistern Drinking Water Supplies." *Environmental Science & Technology* 23 (March 1989): 333–340.

Ray, Dixie Lee. "The Great Acid Rain Debate: No One in Washington (or Ottawa) Knows What He's Talking About." *American Spectator* 20 (January 1987): 21–25.

Regens, James L. "Acid Rain Policymaking and Environmental Federalism: Recent Developments, Future Prospects." *Publius* 19 (Summer 1989): 75–91.

Regens, James L., and Robert W. Rycroft. "Options for Financing Acid Rain Controls." *Natural Resources Journal* 26 (Summer 1986): 519–549.

Robinson, Raymond M. "Physical Dimensions and Solutions of the Acid Rain Problem [Approaches in Canada: Emphasis on U.S. Sources of Canadian Acid Rain]." *Canada–United States Law Journal* 5 (1982): 111–117.

Schindler, D.W. "Biological Impoverishment in Lakes of the Midwestern and Northeastern United States from Acid Rain." *Environmental Science & Technology* 23 (May 1989): 573–580.

"Senate Clean Air Bill: Title–by–Title Provisions of S.1630 as Passed On April 3, 1990." *Congressional Quarterly Weekly Report* 48 (May 12, 1990): 43 page supplement.

Sharpe, William E., and David R. Dewalle. "Acid Precipitation and Water Quality in Pennsylvania: Making the Connection." In *Water Resources in Pennsylvania: Availability, Quality and Management* edited by Shyamal Majumdar, E. Willard Miller, and Richard R. Parizek, 452–465. Easton, PA: The Pennsylvania Academy of Science, 1990.

Stanfield, Rochelle L. "Antiacid Remedy: With Another Debate on Acid Rain Coming Up in Congress, West Germany Offers Lessons in Pollution Control; But It Took Trouble in the Black Forest to Get Action." *National Journal* 19 (June 27, 1987): 1655–1659.

Steel, Brent S., and Dennis L. Soden. "Acid Rain Policy in Canada and the United States: Attitudes of Citizens, Environmental Activists, and Legislators." *Social Science Journal (Fort Worth).* 26, no. 1 (1989): 27–44.

Sweet, William. "Acid Rain." *Editorial Research Reports* (June 20, 1980): 447–464.

Tonnessen, Kathy, and John Harte. "Acid Rain and Ecological Damage: Implications for Sierra Nevada Lake Studies." *Public Affairs Report* 23 (December 1982): 1–9.

Trisko, Eugene M., and Robert E. Wayland. "Acid Rain Control and Public Utility Regulation [Problems Raised for Electric Utilities and Their Regulators by Prospective Legislation To Impose More Stringent Plant Emission Controls]." *Public Utilities Fortnightly* 114 (August 30, 1984): 15–22; 115 (April 15, 1985): 29–33.

U.S. Congress. House. Committee on Energy and Commerce. Subcommittee on Health and the Environment. *Acid Deposition Control Act of 1987: Hearings, July 9–10, 1987, on H.R. 2666, a Bill To Amend the Clean Air Act.* 100th Cong., 1st sess., Washington, DC: GPO, 1988. 806 p.

———. *Acid Deposition Control Act of 1986: Hearings, April 29–May 7, 1986, on H.R. 4567, a Bill To Amend the Clean Air Act To Reduce Acid Deposition and for Other Purposes.* 99th Cong., 2d sess., Washington, DC: GPO, 1986. 3 pts.

———. *Acid Rain Control: Hearings, Pts. 1–2, December 1, 1983–March 5, 1984, on H.R. 3400, a Bill To Amend the Clean Air Act To Control Certain Sources of Sulfur Dioxide and Nitrogen Oxides To Reduce Acid Deposition and for Other Purposes.* 98th Cong., 1st and 2d sess., Washington, DC: GPO, 1984. 2 pts.

———. *Acid Rain Control Proposals: Hearings, April 6, 1989, on H.R. 144 and H.R. 1470, Bills To Amend the Clean Air Act To Provide Further Controls of Certain Stationary Sources of Sulfur Dioxides and Nitrogen Oxides To Reduce and Control Acid Deposition, To Provide for the Commercialization of Clean Coal Technologies for Existing Stationary Sources, and for Other Purposes.* 101st Cong., 1st sess., Washington, DC: GPO, 1989. 375 p.

U.S. Congress. House. Committee on Interior and Insular Affairs. Subcommittee on Mining, Forest Management, and Bonneville Power Administration. *Effects of Air Pollution and Acid Rain on Forest Decline: Oversight Hearing, June 7, 1984.* 98th Cong., 2d sess., Washington, DC: GPO, 1984. 230 p.

U.S. Congress. House. Committee on Interstate and Foreign Commerce. Subcommittee on Oversight and Investigations. *Acid Rain: Hearings, February 26–27, 1980.* 96th Cong., 2d sess., Washington, DC: GPO, 1980. 784 p.

U.S. Congress. House. Committee on Science, Space and Technology. Subcommittee on Natural Resources, Agriculture Research, and Environment. *National Acid Precipitation Assessment Program: Hearing, April 27, 1988.* 100th Cong., 2d sess., Washington, DC: GPO, 1988. 247 p.

U.S. Congress. House. Committee on Small Business. Subcommittee on Environment and Labor. *Effects of Proposed Acid Rain Legislation on Workers and Small Business in the High–Sulfer Coal Industry: Hearing, November 6, 1989.* 101st Cong., 1st sess., Washington, DC: GPO, 1990. 96 p.

U.S. Congress. Senate. Committee on Commerce, Science, and Transportation. *National Ocean Policy Study. Impact of Acid Rain on Coastal Waters; Hearing, June 8, 1988, on the Impact of Acid Precipitation on Coastal Waters and Reauthorization of Title III of the Marine Protection, Research, and Sanctuaries Act.* 100th Cong., 2d sess., Washington, DC: GPO, 1988. 91 p.

U.S. Congress. Senate. Committee on Energy and Natural Resources. *Acid Precipitation and the Use of Fossil Fuels: Hearing, August 19, 1982, To Review the Issue of Acid Precipitation and Fossil Fuel Use in Our National Economy.* 97th Cong., 2d sess., Washington, DC: GPO, 1982. 1542 p.

———. *Clean Coal Technology Development and Strategies for Acid Rain Control: Hearings, June 9 and 10, 1986.* 99th Cong., 2d sess., Washington, DC: GPO, 1987. 935 p.

———. *Effects of Acid Rain: Hearing, Pt. 1, May 28, 1980, on the Phenomenon of Acid Rain and Its Implications for a National Energy Policy.* 96th Cong., 2d sess., Washington, DC: GPO, 1980. 752 p.

———. *Effects of Acid Rain: Hearings, Pt. 2, June 21, 1980, on the Nature, Source and Effects of Acid Rain and How the Potential Growth in Emissions from Coal Burning Powerplants Will Affect Acid Rain Problems.* 96th Cong., 2d sess. Washington DC: GPO, 1981. 121 p.

U.S. Congress. Senate. Committee on Environment and Public Works. *Acid Rain: Hearing, October 29, 1981, on S.1706 [and Other Bills].* 97th Cong., 1st sess. Washington, DC: GPO, 1982. 787 p.

———. *Acid Rain in the West: Hearing, August 12, 1985.* 99th Cong., 1st sess., Washington, DC: GPO, 1987. 285 p.

———. *The New Clean Air Act: Hearings, September 25–October 2, 1986, on S.2203, a Bill To Establish a Program To Reduce Acid Deposition, and for Other Purposes.* 99th Cong., 2d sess., Washington, DC: GPO, 1986. 671 p.

U.S. Congress. Senate. *Economic Impact of Acid Rain: Hearing, September 23, 1980, Before the Select Committee on Small Business and the Committee on Environment and Public Works.* 96th Cong., 2d sess., Washington, DC: GPO, 1980. 224 p.

"Water Quality Question [Whether California Has an Acid Rain Problem: Whether Bottled Water Is Safer Than Tap Water: Proposed Changes in U.S. Drinking Water Regulations]." *Western Water* (March–April 1984): 4–11.

Wilcher, Marshall E. "The Acid Rain Debate in North America: 'Where You Stand Depends on Where You Sit.' " *Environmentalist* 6 (Winter 1986): 289–298.

Control

Adam, O., and Y. Kott. "Evaluation of Water Quality as Measured by Bacterial Maximum Growth Rate." *Water Research* 23 (November 1989): 1407–1412.

Bruvold, W. H. "A Critical Review of Methods Used for the Sensory Evaluation of Water Quality." *Critical Reviews in Environmental Control* 19, no. 4 (1989): 291–308.

Cheremisinoff, P. N. "Treating Wastewater." *Pollution Engineering* 22 (September 1990): 60–65.

Garman, D. E. J. *The Effects of Land Use on Water Quality: An Australian Perspective.* Kingsford, N. S. W., Australia: Water Resource Foundation, 1981. 37 p. ISBN 0-85838-059-5.

Huber, C. V. "A Concerted Effort for Water Quality." *Journal of Water Pollution Control Federation* 61 (March 1989): 310–315.

New Jersey. Drinking Water Quality Institute. *Maximum Contaminant Level Recommendations for Hazardous Contaminants in Drinking Water.* Trenton, NJ: 1987. 52 p.

Preston, L. A. "A New Horizon for Water Quality in Japan." *Journal of Water Pollution Control Federation* 61 (May 1989): 578–581+.

Rajagopal, R., and G. Tobin. "Expert Opinion and Ground-Water Quality Protection: The Case of Nitrate in Drinking Water." *Ground Water* 27 (November–December 1989): 835–847.

Smith, D. G. "A Better Water Quality Indexing System for Rivers and Streams." *Water Research* 24 (October 1990): 1237–1244.

Stukel, T. A., et al. "A Longitudinal Study of Rainfall and Coliform Contamination in Small Community Drinking Water Supplies." *Environmental Science & Technology* 24 (April 1990): 571–575.

Thompson, Paul. "Poison Runoff! New Answers To a Pervasive Problem." *Environmental Forum* 6 (July–August 1989): 5–11.

U.S. Congress. House. Committee on Agriculture. Subcommittee on Department Operations, Research and Foreign Agriculture. *Review of Ground Water Quality Concerns: Hearings, October 12, 1987–May 24, 1988.* 100th Cong., 1st and 2d sess., Washington, DC: GPO, 1989. 615 p.

———. *Salinity Control in Colorado River Basin: Hearing, June 10, 1981.* 97th Cong., 1st sess., Washington, DC: GPO, 1981. 50 p.

U.S. Congress. House. Committee on Public Works and Transportation. Subcommittee on Investigations and Oversight. *Hazardous Waste Contamination of Water Resources (Access to EPA Superfund Records):*

Hearing, December 2, 1982. 97th Cong., 2d sess., Washington, DC: GPO, 1983. 100 p.

van der Wende, E., et al. "Biofilms and Bacterial Drinking Water Quality." *Water Research* 23 (October 1989): 1313–1322.

Technology

Adam, O., and Y. Kott. "Evaluation of Water Quality as Measured by Bacterial Maximum Growth Rate." *Water Research* 23 (November 1989): 1407–1412.

Ciminello, Paul. *State Strategies in Financing Water Treatment Facilities.* Growth and Environmental Management Series. Research Triangle Park, NC: Southern Growth Policies Board, 1986. 23 p.

Kawamura, S., et al. "More and Better Water for Thirsty São Paulo, Brazil." *American Water Works Association Journal* 81 (October 1989): 32–38.

Nelson, A. C. and K. J. Dueker. "Exurban Living Using Improved Water and Wastewater Technology." *Journal of Urban Planning and Development* 115 (December 1989): 101–113.

Pessen, D., et al. "Design of Dilution Junctions for Water Quality Control." *Journal of Water Resources Planning and Management* 115 (November 1989): 829–845.

Schwartz, Larry N., and Mary F. Smallwood. "Regulation of Wastewater Discharge To Florida Wetlands." *National Wetlands Newsletter* 8 (November–December 1986): 5–9.

Taylor, J. S., et al. "Cost and Performance of a Membrane Pilot Plant." *American Water Works Association Journal* 81 (November 1989): 52–60.

U.S. Congress. Office of Technology Assessment. *Using Desalination Technologies for Water Treatment: Background Paper.* Washington, DC: GPO, 1988. 66 p.

U.S. Environmental Protection Agency. Center for Environmental Research Information. *Environmental Pollution Control Alternatives: Drinking Water Treatment for Small Communities.* Cincinnati, OH: 1990. 82 p.

"Water and Wastewater Treatment: A Thriving Business." *Chemical Engineering* 96 (December 1989): 85–103.

Recreation

Davis, Joseph A. "Congress Tames Development To Keep Rivers Wild." *Congressional Quarterly Weekly Report* 44 (November 22, 1986): 2941–2943.

Fickes, Roger. "Pennsylvania's Science River Program." In *Water Resources in Pennsylvania: Availability, Quality and Management* edited by Shyamal Majumdar, E. Willard Miller, and Richard R. Parizek, 180–191. Easton, PA: Pennsylvania Academy of Science, 1990.

Graefe, Alan R. "Water-Based Recreation in Pennsylvania." In *Water Resources in Pennsylvania: Availability, Quality and Management* edited by Shyamal Majumdar, E. Willard Miller, and Richard R. Parizek, 158–168 . Easton, PA: Pennsylvania Academy of Science, 1990.

New Jersey. Department of Environmental Protection. Office of Science and Research. *A Study of Toxic Hazards to Urban Recreational Fishermen and Crabbers.* Trenton, NJ: 1985. 68 p.

New Jersey. Department of Health. Environmental Health Service. *A Study of the Relationship between Illnesses in Swimmers and Ocean Beach Water Quality: Progress Report.* Trenton, NJ: 1988. 167 p.

"Turmoil over the Tuolumne [Whether To Declare This California River Part of the Federal Wild and Scenic River System or Develop Its Hydroelectric Potential]." *Western Water* (July–August 1984): 4–11.

Weiss, Richard A. "Whitewater Recreation in Pennsylvania." In *Water Resources in Pennsylvania: Availability, Quality and Management* edited by Shyamal Majumdar, E. Willard Miller, and Richard R. Parizek, 169–179. Easton, PA: Pennsylvania Academy of Science, 1990.

Laws and Regulations

Attey, John W. and Drew R. Liebert. "Clean Water, Dirty Dams: Oxygen Depletion and the Clean Water Act [of 1982; Legislative and Regulatory Policies That Would Prevent Environmental Degradation Caused By Dams and Hydroelectric Plants]." *Ecology Law Quarterly* 11, no. 4 (1984): 703–729.

"Future Scope of the Clean Water Act: Pro & Con." *Congressional Digest* 64 (December 1985): 290–314.

Gray, Kenneth F. "The Safe Drinking Water Act Amendments of 1986: Now a Tougher Act To Follow." *Environmental Law Reporter* 16 (November 1986): 10338–10345.

Hanford, Priscilla L., and Alvin D. Sokolow. "Mandates as Both Hardship and Benefit: The Clean Water Program in Small Communities." *Publius* 17 (Fall 1987): 131–146.

Hawke, Neil, and Joan Himan. "Water Pollution Law: Plugging the Leaks." *Journal of Planning and Environmental Law* (October 1988): 670–673.

Heath, Milton S., Jr. "Ground Water Quality Law in North Carolina." *Popular Government* 52 (Winter 1987): 39–49.

Licata, Jane M., and Charles A. Licata. "Citizen Suits: Help or Hinderance in the Enforcement of Environmental Statutes? The Clean Water Act Experience." *Environmental Forum* 3 (March 1985): 20–25.

Liebesman, Lawrence R., and Elliott P. Laws. "The Water Quality Act of 1987: A Major Step in Assuring the Quality of the Nation's Waters." *Environmental Law Reporter* 17 (August 1987): 10311–10329.

New Jersey. Senate. Energy and Environment Committee. *Public Hearing: Senate Bill no. 2787 (The Clean Water Enforcement Act): Brick, New Jersey, October 12, 1988.* Trenton, NJ: 1988. 98 p.

Plumlee, John P., and Jay D. Starling. "Citizen Participation in Water Quality Planning: A Case Study of Perceived Failure [Based on Interviews with Participants in the EPA-Sponsored Projects in Texas]." *Administration and Society* 16 (February 1985): 455– 473.

Polebaum, Elliot E., and Matthew D. Slater. "Preclusion of Citizen Environmental Enforcement Litigation by Agency Action." *Environmental Law Reporter* 16 (January 1986): 10013–10018.

"Protecting Our Drinking Water." *EPA (Environmental Protection Agency) Journal* 12 (September 1986): 2–28.

Rastatter, Clem L. "Congress Braces for New Fight on Water Quality." *Conservation Foundation Letter* (September 1981): 1–8.

Rosenthal, Alon. "Going With the Flow; USDA's Dubious Commitment to Water Quality." *Environmental Forum* 5 (September–October 1988): 15–18.

Smith, Turner T., Jr., and Steven J. Koorse. "New Safe Drinking Water Act Liability for Corporate America." *Environmental Law Reporter* 18 (October 1988): 10422–10430.

United Nations. Department of Technical Cooperation for Development. *Legal and Institutional Factors Affecting the Implementation of the International Drinking Water Supply and Sanitation Decade.* New York: UN Agent, 1989. 121 p.

U.S. Congress. House. Committee on Agriculture. Subcommittee on Department Operations, Research, and Foreign Agriculture. *Ground Water Quality and Quantity Issues: Hearing, July 23, 1981.* 97th Cong., 1st sess., Washington, DC: GPO, 1981. 106 p.

———. *Ground Water Quality: Hearings, July 28 and September 16, 1987, on H.R. 791.* 100th Cong., 1st sess., Washington, DC: GPO, 1988. 328 p.

———. *Review of Ground Water Quality Concerns: Hearings, October 12, 1987–May 24, 1988.* 100th Cong, 1st and 2d sess., Washington, DC: GPO,. 1989. 615 p.

U.S. Congress. House. Committee on Government Operations. Environment, Energy, and Natural Resources Subcommittee. *Environmental Protection Agency: Private Meetings and Water Protection Programs: Hearings, October 21, and November 4, 1981.* 97th Cong., 1st sess., Washington, DC: GPO, 1982. 628 p.

U.S. Congress. House. Committee on Interstate and Foreign Commerce. Subcommittee on Health and the Environment. *Safe Drinking Water-Oversight: Hearing, November 5, 1979, on Has the Environmental Protection Agency Been Too Strict in Controlling the Level of Barium in Drinking Water.* 96th Cong., 1st sess., Washington, DC: GPO, 1980. 529 p.

U.S. Laws, Statues. *The Clean Water Act, As Amended by the Water Quality Act of 1987 Public Law 100–4.* 100th Cong., 2d sess., Washington, DC: GPO, 1988. 214 p.

———. *The Clean Water Act, As Amended through December 1981.* 97th Cong., 2d sess., Washington, DC: GPO, 1982. 132 p.

———. *The Safe Drinking Water Act as Amended by the Safe Drinking Water Act Amendments of 1986 (Public Law 99–939, June 19, 1986)*. 99th Cong., 2d sess., Washington, DC: GPO, 1986. 63 p.

U.S. Congress. Senate. Committee on Commerce, Science, and Transportation. *National Ocean Policy Study. Impact of Acid Rain on Coastal Waters: Hearing, June 8, 1988, on the Impact of Acid Precipitation on Coastal Waters and Reauthorization of Title III of the Marine Protection, Research, and Sanctuaries Act.* 100th Cong., 2d sess., Washington, DC: GPO, 1988. 91 p.

U.S. Congress. Senate. Committee on Environment and Public Works. *A Legislative History of the Safe Drinking Water Act: Together with a Section–by–Section Index.* 97th Cong., 2d sess., Washington, DC: GPO, 1982. 1103 p.

———. *A Legislative History of the Water Quality Act of 1987 (Public Law 100-4) Including Public Law 97-440 [and Other Bills], Together with a Section-by-Section Index.* 100th Cong., 2d sess., Washington, DC: GPO, 1988. 4 vols. 2768 p.

U.S. Congress. Senate. Committee on Environment and Public Works. Subcommittee on Environmental Pollution. *Clean Water Act Amendments of 1982: Hearings, July 21– 29, 1982, on S.777, A Bill To Amend the Federal Water Pollution Control Act To Restrict the Jurisdiction of the United States over the Discharge of Dredged or Fill Material to Those Discharges Which Are Into Navigable Waters, and for Other Purposes.* 97th Cong., 2d sess., Washington, DC: GPO, 1982. 1373 p.

Van Putten, Mark C., and Bradley D. Jackson. "The Dilution of the Clean Water Act." *University of Michigan Journal of Law Reform* 19 (Summer 1986): 863–901.

"Water Quality Question [Whether California Has an Acid Rain Problem; Whether Bottled Water Is Safer Than Tap Water; Proposed Changes in United States Drinking Water Regulations]." *Western Water* (March–April 1984): 4–11.

"What Are the Practical Implications of Proposition 65?" *Environmental Forum* 5 (May–June 1988): 16–20.

Water Rights

General

Arrandale, Tom. "Western Water." *Editorial Research Reports* (January 30, 1987): 42–51.

Bennett, Carla J. "Quantification of Indian Rights: Foresight or Folly?" *UCLA (Univ Cal Los Angeles) Environmental Law and Policy* 8:2 (1989): 267–285.

Brendecke, C. M., et al. "Network Models of Water Rights and System Operations." *Journal of Water Resources Planning and Management* 115 (September 1989): 684–696.

Dillman, Jeffrey D. "Water Rights in the Occupied Territories." *Journal of Palestine Studies* 19 (Autumn 1989): 47–71.

Gould, George A. "Wyoming Water Rights—a Primer." *Wyoming Issues* 3 (Summer 1980): 13–19.

Gray, Brian E. "A Reconsideration of Instream Appropriate Water Rights in California." *Ecology Law Quarterly* 16, no. 3 (1989): 667–717.

Ingram, Helen. "Water Rights in the Western States." *Proceedings of the Academy of Political Science* 34, no. 3 (1982): 134–143.

Moore, Michael R. "Native American Water Rights: Efficiency and Fairness." *Natural Resources Journal* 29 (Summer 1989): 763–791.

Ognibene, Peter J. "Indian Water Rights Clouding Plans for the West's Economic Development." *National Journal* 14 (October 30, 1982): 1841–1845.

Ramgolam, Roopchand, and Floyd L. Corty. *Water Use and Water Rights in Louisiana.* D.A.E. Research Report no. 593. Baton Rouge, LA: Agriculture Experiment Station, 1982. 36 p.

Samelson, Kirk S. "Water Rights for Expanded Uses on Federal Reservations [National Forests]." *Denver Law Journal* 61, no. 1 (1983): 67–76.

Sewell, W. R. Derrick, and Philip Dearden, eds. "Wilderness: Past, Present, and Future." *Natural Resources Journal* 29 (Winter 1989): 1–302.

Tristani, M. Gloria. "Interior Turns Off Tap for Wilderness Areas." *Natural Resources Journal* 29 (Summer 1989): 877–894.

U.S. Congress. Senate. Committee on Finance. *Soil and Water Conservation Tax Credits: Joint Hearing, March 5, 1984, Before the Subcommittee on Energy and Agricultural Taxation and Subcommittee on Oversight of the Internal Revenue Service, on S.152 and S.1280.* 98th Cong., 2d sess., Washington, DC: GPO, 1984. 58 p.

"Water Rights Symposium [Williamsburg, Virginia, March 1983; Eastern States]." *William and Mary Law Review* 24 (Summer 1983): 535–793.

Economics

Anderson, Terry L. "The Market Alternative for Hawaiian Water." *Natural Resources Journal* 25 (October 1985): 893–910.

Brajer, Victor, and Wade E. Martin. "Water Rights Markets: Social and Legal Considerations: Resource's 'Community' Value, Legal Inconsistencies and Vague Definition and Assignment of Rights Color Issues." *American Journal of Economics and Sociology* 49 (January 1990): 35–44.

Brookshire, David S., et al. "Economics and the Determination of Indians Reserved Water Rights." *Natural Resources Journal* 23 (October 1983): 749–765.

Chan, Arthur H. "To Market or Not To Market: Allocating Water Rights in New Mexico." *Natural Resources Journal* 29 (Summer 1989): 629–643.

Summers, Lyle C. "An Economic Framework for Valuing Transient Water Rights in the Arid West." *Appraisal Journal* 49 (January 1981): 9–14.

Laws and Regulations

Blumm, Michael C. "A Trilogy of Tribes v. FERC [Federal Energy Regulatory Commission]: Reforming the Federal Role in Hydropower Licensing." *Harvard Environmental Law Review* 10, no. 1 (1986): 1–59.

Burchi, Stefano, "Legislation on Domestic and Industrial Uses of Water: A Comparative Review." *Natural Resources Journal* 24 (January 1984): 143–159.

Burness, H. S., et al. "United States Reclamation Policy and Indian Water Rights." *Natural Resources Journal* 20 (October 1980): 807–826.

Burton, Lloyd. "The American Indian Water Rights Dilemma: Historical Perspective and Dispute—Settling Policy Recommendations." *UCLA (Univ Cal Los Angeles) Journal of Environmental Law and Policy* 7, no. 1 (1987): 1–66.

Caponera, Dante A. "Patterns of Cooperation in International Water Law: Principles and Institutions." *Natural Resources Journal* 25 (July 1985): 563–587.

Du Mars, Charles, and Helen Ingram. "Congressional Quantification of Indian Reserved Water Rights: A Definite Solution or a Mirage? [Discusses the History of the Navajo Indian Irrigation Project and the Current Controversy That Surrounds It]." *Natural Resources Journal* 20 (January 1980): 17–43.

Fisher, Todd A. "The Winters of Our Discontent: Federal Reserved Water Rights in the Western States [United States Supreme Court Decision in *Winters v. United States,* That in Reserving Land for a Federal Enclave Such as an Indian Reservation, National Forest, or Military Reservation, the Federal Government Also Implicitly Reserves a Sufficient Quantity of Water To Carry Out the Purpose of the Reserved Land]." *Cornell Law Review* 69 (June 1984): 1077–1093.

Folk-Williams, John A. "The Use of Negotiated Agreements To Resolve Water Disputes Involving Indian Rights." *Natural Resources Journal* 28 (Winter 1988): 63–103.

Gellis, Ann J. "Water Supply in the Northeast: A Study in Regulatory Failure." *Ecology Law Quarterly* 12, no. 3 (1985): 429–479.

Gottlieb, Gail. "New Mexico's Mine Dewatering Act: The Search for Rehoboth [Attempt To Regulate the Use of Ground Water by the Uranium Mining Industry To 'Promote Maximum Economic Development of Mineral Resources While Ensuring That Such Development Does Not Impair Existing Prior Water Rights']." *Natural Resources Journal* 20 (July 1980): 653–680.

Grant, Douglas L. "Public Interest Review of Water Rights Allocation and Transfer in the West: Recognition of Public Values." *Arizona State Law Journal* 19 (Winter 1987): 681–718.

Griffith, Gwendolyn. "Indian Claims to Groundwater: Reserved Rights or Beneficial Interests?" *Stanford Law Review* 33 (November 1980): 103–130.

Gross, Sharon P. "The Galloway Project and the Colorado River Compacts: Will the Compacts Bar Transbasin Water Diversions?" *Natural Resources Journal* 25 (October 1985): 935–960.

Hite, James. "Interbasin Water Transfers in Riparian Doctrine States: The Case for Interregional Compensation." *Growth and Change* 17 (October 1986): 10–24.

Kosloff, Laura H. "Water for Wilderness: Colorado Court Expands Federal Reserved Rights." *Environmental Law Reporter* 16 (January 1986): 10002–10007.

Landry, Stephanie. "The Galloway Proposal and Colorado Water Law: The Limits of the Doctrine of Prior Appropriation." *Natural Resources Journal* 25 (October 1985): 961–983.

Leahy, John D., and James Belanger. "Arizona Law Where Ground and Surface Water Meet." *Arizona State Law Journal* 20 (Fall 1988): 657–748.

Marks, Jason. "The Duty of Agencies To Assert Reserved Water Rights in Wilderness Areas." *Ecology Law Quarterly* 14, no. 4 (1987): 639–683.

New Jersey. Senate. Committee To Study Coastal and Ocean Pollution. *Public Hearing on Senate Bill no. 2787 "Clean Water Enforcement Act": Trenton, New Jersey, April 5, 1989.* Trenton, NJ: 1989. 149 p.

Pacheco, Thomas H. "How Big Is Big? The Scope of Water Rights Suits under the McCarran Amendment." *Ecology Law Quarterly* 15, no. 4 (1988): 627–669.

Planck, Ulrich. "Issues of Water in Agrarian Reform Legislation of the Near East." *Land Reform (FAO)* nos. 1–2 (1987): 58–82.

Ray, Jayanta Kumar. "The Farakka Agreement between India and Bangladesh [on Sharing Waters of the Ganga]." *International Studies (New Delhi)* 17 (April–June 1978), 235–246.

Romm, Jeff, and Sally K. Fairfax. "The Backwaters of Federalism. Receding Reserved Water Rights and the Management of National Forests." *Policy Studies Review* 5 (November 1985): 413–430.

Simms, Richard A. "National Water Policy in the Wake of United States v. New Mexico [Examines the Legal Basis of 'Non–Reserved Federal Water Rights']." *Natural Resources Journal* 29 (January 1980): 1–16.

Smith, Zachary A. "Centralized Decisionmaking in the Administration of Groundwater Rights; The Experience of Arizona, California, and New Mexico, and Suggestions for the Future." *Natural Resources Journal* 24 (July 1984): 641–688.

Stevens, Anastasia S. "Pueblo Water Rights in New Mexico." *Natural Resources Journal* 28 (Summer 1988): 535–583.

Swenson, Robert W. "A Primer of Utah Water Law [with a Section on Application to Groundwater]." *Journal of Energy Law and Policy* 5 (2): 165–196 (1984); 6 (1): 1–54 (1985).

U.S. Congress. House. Committee on Interior and Insular Affairs. Subcommittee on Water and Power Resources. *San Luis Rey Indian Water Rights Settlement Act: Hearing, September 9, 1986, on H.R. 4468, To Provide for the Settlement of Water Rights Claims of the La Jolla, Rincon, San Pasqual, Pauma, and Pala Bands of Mission Indians in San Diego County, CA, and for Other Purposes.* 99th Cong., 2d sess., Washington, DC: GPO, 1987. 144 p.

U.S. Congress. Senate. *Colorado Ute Water Settlement Bill: Joint Hearing, December 3, 1987, Before the Select Committee on Indian Affairs and the Committee on Energy and Natural Resources; on S. 1415, To Facilitate and Implement the Settlement of Colorado Ute Indian Reserved Water Rights Claims in Southwest Colorado.* 100th Cong., 1st sess., Washington, DC: GPO, 1988. 580 p.

U.S. Congress. Senate. Committee on Energy and Natural Resources. Subcommittee on Public Lands and Reserved Water. *Indian Reserved Water Rights.* 98th Cong., 2d sess., Washington, DC: GPO, 1985. 404 p.

U.S. Congress. Senate. Select Committee on Indian Affairs. *Indian Water Policy: Hearing, April 6, 1989, 101st Cong., 1st sess.* Washington, DC: GPO, 1989. 341 p.

———. *Seminole Water Claims Settlement Act: Hearing, November 5, 1987, on S.1684, To Settle Seminole Indian Land Claims Within the State of Florida.* 100th Cong., 1st sess., Washington, DC: GPO, 1988. 132 p.

Utton, Albert E. "In Search of an Integrating Principle for Interstate Water Law: Regulation Versus the Market Place." *Natural Resources Journal* 25 (October 1985): 985–1004.

Wescoat, James L., Jr. "On Water Conservation and Reform of the Prior Appropriation Doctrine in Colorado." *Economic Geography* 61 (January 1985): 3–24.

Williams, Stephen F. "The Law of Prior Appropriation: Possible Lessons for Hawaii." *Natural Resources Journal* 25 (October 1985): 911–934.

Zaleski, Alexander V. "A New Authority for Massachusetts: Best Solution for a Difficult Task." *National Civic Review* 74 (December 1985): 531–537.

Pollution

General

Brickson, Betty, and Rita S. Sudman. "A Briefing on California Water Issues." *Western Water* (September–October 1990): 3–11.

Gallob, Joel A. "Birth of North American Transboundary Environmental Plaintiff: Transboundary Pollution and the 1979 Draft Treaty for Equal Access and Remedy." *Harvard Environmental Law Review* 15, no. 1 (1991): 85–148.

Garrett, Theodore L. "NRDC v. EPA [Natural Resources Defense Council. Environmental Protection Agency]: The D.C. Circuit's Long-Awaited Decision in the NPDES [National Pollutant Discharge Elimination System] Permit Rules Litigation" *Environmental Law Reporter* 19 (May 1989): 10223–10229.

Lovich, Nicholas P., et al. "Water Pollution Control in Democratic Societies: A Cross-National Analysis of Sources of Public Benefits in Japan and the United States." *Policy Studies Review* 5 (November 1985): 431–450.

Malone, Linda A. "The Necessary Interrelationship between Land Use and Preservation of Groundwater Resources." *UCLA (Univ Cal Los Angeles) Journal of Environmental Law and Policy* 9 (1): 1–72 (1990).

Maniatis, Melina. "Stormwater Management." *Management Information Service Report* 22 (November 1990): 1–15.

New Jersey. General Assembly. Agriculture and Environment Committee. *Public Hearing on Rockaway Experience, Water Pollution and Related Issues: Held: Rockaway Township, N.J., March 10, 1982.* Trenton, NJ: 1982. 53 p.

"Pollution Control: Special Feature." *Dock and Harbour Authority (Gt. Brit)* 66 (June 1985): 27–38.

"The Problem of Water Pollution in the United States."*Environmental Action* 15 (May 1983): Entire Issue.

Protasel, Greg J. "Program Success and Program Failure in a Multi-Organizational Environment: The Formation of Nonpoint Source Water Pollution Control Policy in the U.S. [Pollution caused by Rain Run-off]." In *Public Policy Formation,* edited by Robert Eyestone, 263–290. Greenwich, CT: JAI Press, 1984.

Teclaff, Ludwik A., and Eileen Teclaff. "International Control of Cross-Media Pollution: An Ecosystem Approach." *Natural Resources Journal* 27 (Winter 1987): 21–53.

U.S. Congress. House. Committee on Government Operations. Government Activities and Transportation Subcommittee. *Coast Guard Capabilities for Oilspill Cleanup: Hearing, August 26, 1982.* 97th Cong., 2d sess., Washington, DC: GPO, 1983. 42 p.

U.S. Congress. House. Committee on Merchant Marine and Fisheries. *Contaminated Marine Sediments: Hearing, March 20, 1990, Before the Subcommittee on Oversight and Investigations and the Subcommittee on Oceanography and Great Lakes, on Contaminated Marine Sediments in Navigable Waters and Harbor Areas of the United States and Federal Efforts To Address This Problem.* 101st Cong., 2d sess., Washington, DC: GPO, 1990. 235 p.

U.S. Congress. House. Committee on Merchant Marine and Fisheries. Subcommittee on Coast Guard and Navigation. *Investigation Into*

Coastal Oil Spills: Hearing, June 21, 1990, To Examine the Rash of Recent Oil Spills Along U.S. Coasts, Emphasizing the Need for Oil Spill Legislation and Ratification of International Protocols. 101st Cong., 2d sess., Washington, DC: GPO, 1990. 83 p.

U.S. Congress. House. Committee on Merchant Marine and Fisheries. Subcommittee on Oversight and Investigations. *Floatable Pollution in the New Jersey–New York Harbor Complex: Hearing, May 22, 1989, on Oversight of Legislation Governing the Problem of Pollution in the New York–New Jersey Harbor Complex and a Look At Long-term Solutions to these Long-term Problems.* 101st Cong., 1st sess., Washington, DC: GPO, 1989. 302 p.

U.S. Environmental Protection Agency. Center for Environmental Research Information. *Environmental Pollution Control Alternatives: Drinking Water Treatment for Small Communities.* Cincinnati, OH: EPA, 1990. 82 p.

U.S. Environmental Protection Agency. Office of Marine and Estuarine Protection. *Saving Bays and Estuaries: A Primer for Establishing and Managing Estuary Projects.* Washington, DC: GPO, 1989. 5 p.

U.S. Environmental Protection Agency. Office of Water. *The Quality of Our Nation's Water; A Summary of the 1988 National Water Quality Inventory.* Washington, DC: GPO, 1990. 24 p.

———. *A Review of Sources of Ground-Water Contamination from Light Industry.* Washington, DC: GPO, 1990. 48 p.

U.S. Congress. Senate. Committee on Environment and Public Works. *Small Communities Environmental Assistance: Hearing, May 15, 1990, on S. 1296, S.1331, S. 1514, and S. 2184, Bills to Assist Small Communities To Finance Environmental Facilities.* 101st Cong., 2d sess., Washington, DC: GPO, 1990. 220 p.

Valencia, Mark J. "Sea of Japan: Transnational Marine Resource Issues and Possible Cooperative Responses." *Marine Policy* 14 (November 1990): 507–525.

Technology

Landau, Jack L. "Economic Dream or Environmental Nightmare? the Legality of the 'Bubble Concept' in Air and Water Pollution Control [an Alternative Emission Reduction Options Policy? Which Would Allow a Plant To Reduce Its Total Emission or Effluents to a Legal Level in Whatever Way Would Be Most Cost-Effective]." *Boston College Environmental Affairs Law Review* 8, no. 4 (1980): 741–781.

Sherwin, Emily L. "The Bubble Concept in Water Pollution Control [Makes Entire Industrial Processes or Plants, Rather Than Individual Pipes and Outlets Contained Within Them, Directly Subject to Environmental Protection Agency Regulations]." *Boston University Law Review* 60 (July 1980): 686–712.

Economics

Brown, Gardner M., Jr., and Ralph W. Johnson. "Pollution Control by Effluent Charges: It Works in the Federal Republic of Germany, Why Not in the U.S.? [Charges Levied on Direct Discharges for Specified Effluents Into Public Waters]." *Natural Resources Journal* 24 (October 1984): 929–966.

Cheatham, Leo R. "How Much Water Clean-Up Do We Want to Purchase?" *Mississippi Business Review* 43 (August 1982): 3–14.

Gianessi, Leonard P., and Henry M. Peskin. "The Distribution of the Costs of Federal Water Pollution Control Policy." *Land Economics* 56 (February 1980): 85–102.

Matthews, Deborah E. "Designing User Charges That Work: Well Designed User Charges Can Help Clean Up Boston's Harbor." *Governance* (Summer–Fall 1986): 39–42.

Pineles, Barry A. "Cost-Benefit Analysis and the Federal Water Pollution Control Act Amendments of 1972: A Proposal for Congressional Action." *Iowa Law Review* 67 (July 1982): 1057–1079.

Rothfelder, Marty. "Reducing the Cost of Water Pollution Control under the Clean Water Act." *Natural Resources Journal* 22 (April 1982): 407–421.

Pesticides

Foster, Douglas. "The Growing Battle over Pesticides in Drinking Water: Disturbing New Studies on Toxic 'Trickle-Down' [of Agricultural

Pesticides Into California Groundwater Supplies].” *California Journal* 14 (May 1983): 177–179.

Organization for Economic Cooperation and Development. *Water Pollution Fertilizers and Pesticides.* Paris, France: OECD, 1986. 144 p.

Segerson, Kathleen. “Liability for Groundwater Contamination from Pesticides.” *Journal of Environmental Economics and Management* 19 (November 1990): 227–243.

U.S. Congress. House. Committee on Energy and Commerce. Subcommittee on Health and the Environment. *Dioxin Contamination of Food and Water: Hearing, December 7, 1988.* 100th Cong., 2d sess., Washington, DC: GPO, 1989. 216 p.

U.S. Congress. Senate. Committee on Environment and Public Works. *Environmental Issues Related to the Use of Pesticides: Hearing, June 10, 1988.* 100th Cong., 2d sess., Washington, DC: GPO, 1988. 136 p.

Toxins

Eisenreich, Steven J. “Toxic Fallout in the Great Lakes: The Buildup of Organic Pollutants from Far Away Places Shows That the Acid Rain Problem in Lakes Is Only the Beginning.” *Issues in Science and Technology* 4 (Fall 1987): 71 –75.

Licht, Judy, and Jeff Johnson. “Without a Paddle: America’s Sewers Are Flooded with Toxics, Sludges and No One Knows Exactly What Else.” *Environmental Action* 17 (September–October 1985): 10–13.

Logan, Terry J., and Stephen M. Yaksich. “Lake Erie: A New Prognosis.” *Water Spectrum (Corps of Engineers)* 12 (Summer 1980): 26–34.

Rebuffoni, Dean. “The Mississippi River [Some Emphasis on Pollution Abatement Problems].” *EPA (Environmental Protection Agency) Journal* 6 (July–August 1980): 22–27.

Roe, David. “An Incentive-Conscious Approach to Toxic Chemical Controls.” *Economic Development Quarterly* 2 (August 1989): 179–187.

“Special Report: Troubled Waters.” *Amicus Journal* 11 (Summer 1989): 10–31.

"Tackling Nonpoint Water Pollution." *EPA (Environmental Protection Agency) Journal* 12 (May 1986): 2–23.

Teclaff, Ludwick A. and Eileen Teclaff. "Transboundary Toxic Pollution and the Drainage Basin Concept." *Natural Resources Journal* 25 (July 1985): 589–912.

"Toxics in Water: A Hidden Threat." *EPA (Environmental Protection Agency) Journal* 11 (September 1985): 2–23+.

U.S. Congress. House. Committee on Government Operations. Environment, Energy, and Natural Resources Subcommittee. *Toxic Chemical Contamination of Ground Water: EPA Oversight: Hearings, July 24–September 18, 1980.* 96th Cong., 2d sess., Washington, DC: GPO, 1981. 410 p.

U.S. Congress. House. Committee on Interstate and Foreign Commerce. Subcommittee on Transportation and Commerce. *Hazardous Waste Disposal: Our Number One Environmental Problem: Hearing, June 9, 1980.* 96th Cong., 2d sess., Washington, DC: GPO, 1980. 97 p.

U.S. Congress. House. Committee on Public Works and Transportation. Subcommittee on Investigations and Oversight. *Diffuse Toxic Pollutants in the Great Lakes Ecosystem: Hearing, April 14, 1988.* 100th Cong., 2d sess., Washington, DC: GPO, 1988. 108 p.

U.S. Congress. House. Committee on Public Works and Transportation. Subcommittee on Water Resources. *Toxic Pollution in the Great Lakes: Hearing, March 2, 1988.* 100th Cong., 2d sess., Washington, DC: GPO, 1988. 145 p.

U.S. Congress. House. Committee on Science and Technology. Subcommittee on Natural Resources and Environment. *Coordination of Federal Research and Monitoring Programs for Toxic and Hazardous Substances in the Great Lakes Region: Hearing, November 19, 1979.* 96th Cong., 1st sess., Washington, DC: GPO, 1980. 451 p.

U.S. Congress. Senate. Committee on Environment and Public Works. Subcommittee on Environmental Pollution. *Municipal Wastewater Treatment Construction Grants Program: Hearings, Pt. 2, June 29 and August 10, 1981, on S.975, a Bill To Revise and Extend Certain Provisions of the Federal Water Pollution Control Act, as Amended, for One Year, and for Other Purposes and S. 1274, a Bill To Amend Title II of the Clean Water*

Act, and for Other Purposes. 97th Cong., 1st sess., Washington, DC: GPO, 1981. 269 p.

Watson, William D., and Ronald G. Ridker. "Revising Water Pollution Standards in an Uncertain World." *Land Economics* 57 (November 1981): 485–506.

Laws and Regulations

Davis, Peter N. "Federal and State Water Quality Regulation and Law in Missouri." *Missouri Law Review* 55 (Spring 1990): 411–507.

———. "Protecting Waste Assimilation Streamflows By the Law of Water Allocation, Nuisance, and Public Trust, and By Environmental Statutes." *Natural Resources Journal* 28 (Spring 1988): 357–391.

Dycus, J. Stephen. "Development of a National Groundwater Protection Policy." *Boston College Environmental Affairs Law Review* 11 (January 1984): 211–271.

Garrett, Theodore L. "NRDC [Natural Resources Defense Council] v. EPA: the D.C. Circuits Long-Awaited Decision in the NPDES [National Pollutant Discharge Elimination System] Permit Rules Litigation." *Environmental Law Reporter* 19 (May 1989): 10223–10229.

Gelpe, Marcia R. "Pollution Control Laws Against Public Facilities." *Harvard Environmental Law Review* 13, no. 1 (1989): 69–146.

Haas, Peter M. "Do Regimes Matter? Epistemic Communities and Mediterranean Pollution Control." *International Organization* 43 (Summer 1989): 377–403.

Harrison, Peter and W. R. Derrick Sewell. "Water Pollution Control By Agreement: The French System of Contracts." *Natural Resources Journal* 20 (October 1980): 765–786.

Hodges-Copple, John. *Protecting Public Water Supplies in the South: A Primer on the State's Role.* Growth and Environmental Management. Research Triangle Park, NC: Southern Growth Policies Board, 1988. 34 p.

Mingst, Karen A. "The Functionalist and Regime Perspectives: The Case of Rhine River Cooperation [in the Context of Recent Pollution

Problems of the International River]." *Journal of Common Market Studies* 20 (December 1981): 161–173.

New Jersey. General Assembly. *Joint Public Hearing: Senate Bill no. 2188 (the Clean Water Amendment Act): Before Senate Environmental Quality Committee and Assembly Energy and Environment Committee: Old Bridge, New Jersey, February 13, 1990.* Trenton, NJ: 1990. 200 p.

New Jersey. Senate. Special Committee To Study Coastal and Ocean Pollution. *Public Hearing: The Pretreatment of Industrial Waste Waters Prior To Discharge Into Publicly Owned Treatment Works: Trenton, NJ: September 15, 1987.* Trenton, NJ: 1987. 138 p.

Ogalla, Bondi D. "Water Pollution Control in Africa: A Comparative Legal Survey." *Journal of African Law* 33 (Autumn 1989): 149–156.

Oppenheimer, Todd. "Humpty Dumpty." *Amicus Journal* 10 (Winter 1988): 14–23.

Sorenson, Jay B. "The Assurance of Reasonable Toxic Risk? [in the Context of the 1980–82 New Mexico Water Quality Control Commission Regulation on the Control of Organic Toxic Contaminants in Groundwater]." *Natural Resources Journal* 24 (July 1984): 549–569.

U.S. Congress. House. Committee on Merchant Marine and Fisheries. *Antifouling Points: Hearing, July 3, 1987, Before the Subcommittee on Fisheries and Wildlife Conservation and the Environment and the Subcommittee on Oceanography, on H.R. 1046 and H.R. 2210.* 100th Cong., 1st sess., Washington, DC: GPO, 1988. 234 p.

U.S. Congress. House. Committee on Public Works and Transportation. Subcommittee on Investigations and Oversight. *Hazardous Waste Contamination of Water Resources; Hearing, March 10–July 9, 1982.* 97th Cong., 2d sess., Washington, DC: GPO, 1983. 368 p.

———. *Implementation of the Federal Waste Pollution Control Act Concerning the Performance of the Municipal Wastewater Treatment Construction Grants Program; Report, October 1981.* 97th Cong., 1st sess. Washington, DC: GPO, 1981. 75 p.

———. *Implementation of the Federal Waste Pollution Control Act (Performance of Federally Assisted Wastewater Treatment Systems): Hearing, March 10, 1981.* 97th Cong., 1st sess., Washington, DC: GPO, 1982. 128 p.

U.S. Congress. House. Committee on Public Works and Transportation. Subcommittee on Oversight and Review. *Nonpoint Pollution and the Areawide Waste Treatment Management Program under the Federal Water Pollution Control Act: Summary of Hearings, July 11–18, 1979.* 96th Cong., 2d sess., Washington, DC: GPO, 1980. 52 p.

U.S. Congress. House. Committee on Public Works and Transportation. Subcommittee on Water Resources. *Industrial Cost Recovery: Hearing, March 12, 1980, on H.R. 6667, To Amend the Federal Water Pollution Control Act Relating to Authorization Extensions and Industrial Cost Recovery.* 96th Cong., 2d sess., Washington, DC: GPO, 1980. 447 p.

———. *The Need for Legislative Changes in the Construction Grant Program of the Federal Water Pollution Control Act.* 97th Cong., 1st sess., Washington, DC: GPO, 1982. 1776 p.

Wen, Boping. "On the Water Pollution Prevention Control Law." *Chinese Law and Government* 19 (Spring 1986): 50–63.

Yeager, Peter C. "Structural Bias in Regulatory Law Enforcement: The Case of the U.S. Environmental Protection Agency." *Social Problems (Soc Study Social Problems)* 34 (October 1987): 330–344.

Chemical Pollution

General

Belton, Thomas J., et al. *A Study of Dioxin (2,3,7,8-Tetrachlorodibenzo-P-Dioxin) Contamination in Select Finfish, Crustaceans and Sediments of New Jersey Waterways.* Department of Environmental Protection. Office of Science and Research. Trenton, NJ: 1985. 102 p.

Fitchen, Janet M. "Cultural Aspects of Environmental Problems: Individualism and Chemical Contamination of Groundwater." *Science, Technology, and Human Values* 12 (Spring 1987): 1–12.

Marinelli, Janet. "It Came From Beneath Long Island; Vinyl Chloride, Benzene, Aldicarb and Trichloroethane Are But a Few of the Contaminants That Have Leached Into the Drinking Water of Three Million Residents." *Environmental Action* 14 (May 1983): 8–12.

Trost, Cathy. "Hooker Chemical's Michigan Mess: Another Tale of Poisoned Wells, and Dying Lakes, and Regulators Who Looked the Other Way." *Business and Society Review* (Winter 1981–1982): 32–39.

U.S. Congress. House. Committee on Public Works and Transportation. Subcommittee on Water Resources. *Hazardous Chemicals under the Federal Water Pollution Control Act: Hearings, April 15–17, 1980.* 96th Cong., 2d sess., Washington, DC: GPO, 1980. 385 p.

———. *PBB (Polybrominated Biphenyls) Pollution Problem in Michigan: Hearing, November, 19, 1979.* 96th Cong., 1st sess., Washington, DC: GPO, 1980. 89 p.

U.S. Congress. Senate. *Toxic Chemical Pollution in Puget Sound: Joint Hearing, October 25, 1982, Before the Committee on Environmental and Public Works and Commerce, Science, and Transportation.* 97th Cong., 2d sess., Washington, DC: GPO, 1983. 153 p.

Yaniga, Paul M. "Organic Chemical (Hydrocarbon) Contamination of Aquifers: Assessment, Abatement and Cleanup." In *Water Resources in Pennsylvania: Availability, Quality and Management,* edited by Shyamal Majumdar, E. Willard Miller, and Richard R. Parizek, 399–419. Easton, PA: Pennsylvania Academy of Science, 1990.

Industrial Control

La Grega, M. D., et al. "Stabilization of Acidic Refinery Sludges." *Journal of Hazardous Materials* 24 (September 1990): 169–187.

Randall, William H. "The EPA's Proposed Regulations on Wastewater Discharges from Power Plants: Will They Help or Hinder? [Proposed EPA's Regulations for Steam-Electric Power Generating Plants]." *Public Utilities Fortnightly* 107 (January 15, 1981): 24–27.

Nitrates

Conrad, Jobst. "Nitrate Debate and Nitrate Policy in FR Germany." *Land Use Policy* 5 (April 1988): 207–218.

Great Britain. House of Lords. Select Committee on the European Communities. *Nitrate in Water: Report; 16th Report, Session 1988–89.* House of Lords Paper 73-I. Lanham, MD: UNIPUB, 1989. 51 p.

———. *Nitrate in Water: With Evidence; 16th Report, Session 1988–89.* House of Lords Paper 73. Lanham, MD: UNIPUB, 1989. 288 p.

Arsenic

Cullen, W. R., and K. J. Reimer. "Arsenic Speciation in the Environment." *Chemical Reviews* 89 (June 1989): 713–764.

Fox, K. R. "Field Experience with Point-of-Use Treatment Systems for Arsenic Removal." *American Water Works Association Journal* 81 (February 1989): 94–101.

Mok, W. M., and C. M. Wai. "Distribution and Mobilization of Arsenic Species in the Creeks Around the Blackbird Mining District, Idaho." *Water Research* 23 (January 1989): 7–13.

Welch, A. H., et al. "Arsenic in Ground Water of the Western United States." *Ground Water* 26 (May–June 1988): 333–347.

Chlorination

Clark, R. M., et al. "Analysis of Inactivation of Giardia Lamblia by Chlorine." *Journal of Environmental Engineering* 115 (February 1989): 80–90.

Krasner, S. W., et al. "The Occurrence of Disinfection By-products in U.S. Drinking Water." *American Water Works Association Journal* 81 (August 1989): 41–53.

Corrosives

Schock, M. R., and C. H. Neff. "Trace Metal Contamination from Brass Fittings." *American Water Works Association Journal* 80 (November 1988): 47–56.

Fluoridation

Easley, Michael W. "The Status of Community Water Fluoridation in the United States." *Public Health Reports* 105 (July–August 1990): 348–353.

"The Fluoridation Controversy: Understanding the Opposition and Effectively Meeting the Challenge." *Health Matrix* 2 (Summer 1984): 65–77.

Leo, Harald. "The Fluoridation Status of U.S. Public Water Supplies." *Public Health Reports* 101 (March–April 1986): 157–162.

McNeil, Donald R. "America's Longest War: The Fight over Fluoridation." *Wilson Quarterly* 9 (Summer 1985): 140–153.

Wild, Russell. "Fluoride Miracle Cure or Public Menace?" *Environmental Action* 16 (July–August 1984): 14–19.

Lead Poisoning

Cosgrove, E., et al. "Childhood Lead Poisoning: Case Study Traces Source to Drinking Water." *Journal of Environmental Health* 52 (July–August 1989): 346–349.

Lee, R. G., et al. "Lead at the Top: Sources and Control." *American Water Works Association Journal* 81 (July 1989): 52–62.

Schock, M. R. "Understanding Corrosion Control Strategies for Lead." *American Water Works Association Journal* 811 (July 1989): 88–100.

Phosphate

Dorioz, J. M., et al. "Phosphorus Dynamics in Watersheds: Role of Trapping Processes in Sediments." *Water Research* 23 (February 1989): 147–158.

Selenium

Deason, J. P. "Irrigation–Induced Contamination: How Real a Problem?" *Journal of Irrigation and Drainage Engineering* 115 (February 1989): 9–20.

Hall, S. K., et al. "Agricultural Drainage Water—How Should It Be Regulated in California?" *Journal of Irrigation and Drainage Engineering* 115 (February 1989): 3–8.

Moore, S. B. "Selenium in Agricultural Drainage: Essential Nutrient or Toxic Threat?" *Journal of Irrigation and Drainage Engineering* 115 (February 1989): 21–28.

Squires, R. C., et al. "Economics of Selenium Removal from Drainage Water." *Journal of Irrigation and Drainage Engineering* 115 (February 1989): 48–57.

Saltwater Intrusion

Savenije, H. H. G. "Influence of Rain and Evaporation on Salt Intrusion in Estuaries." *Journal of Hydraulic Engineering* 114 (December 1988): 1509–1524.

Urish, D. W., and M. M. Ozbilgin. "The Coastal Ground-Water Boundary." *Ground Water* 27 (May–June 1989): 310–315.

Wilkinson, D. L. "Avoidance of Seawater Intrusion Into Ports of Ocean Outfalls." *Journal of Hydraulic Engineering* 114 (February 1988): 218–228, Discussion, 115 (July 1989): 1015–1017.

Floods

General

Alper, Donald K., and Robert L. Manahan. "Regional Transboundary Negotiations Leading to the Skagit River Treaty: Analysis and Future Application." *Canadian Public Policy (Guelph)* 12 (March 1986): 163–174.

"Drought—or Flood? How California Prepares." *Western Water* (January–February 1986): 4–10.

Krohe, James, Jr. "Dams, Floods, Rainmaking and Droughts [Illinois]." *Illinois Issues* 8 (August 1982): 23–29.

Mercer, David. "Australia's Constitution, Federalism and the Tasmanian Dam Case." *Political Geography Quarterly* 4 (April 1985): 91–110.

Middleton, M. J., et al. "Structural Flood Mitigation Works and Estuarian Management in New South Wales: Case Study of the Macleay River." *Coastal Zone Management Journal* 13, no. 1 (1985): 1–23.

Miller, Christopher, and Geraldine Bachman. "Planning for Hurricanes and Other Coastal Disturbances." *Urban Land* 43 (January 1984): 18–23.

Tompkins, Mark E. "South Carolina's Diked Tidal Wetlands: The Persisting Dilemmas." *Coastal Management* 15, no. 2 (1987): 135–155.

U.S. Congress. House. Committee on Public Works and Transportation. Subcommittee on Oversight and Review. *Flooding of the Red River of the North and Its Tributaries: Hearing, July 2, 1979.* 96th Cong., 1st sess., Washington, DC: GPO, 1979. 306 p.

U.S. Congress. House. Committee on Public Works and Transportation. Subcommittee on Water Resources. *Flooding Problems: State of Arizona: Hearings, June 1 and 2, 1979.* 96th Cong., 1st sess., Washington, DC: GPO, 1979. 380 p.

Wiley, William N. *Elevated Housing: Flood Protection through Raising Existing Structures.* Research Report no. 186. Frankfort, KY: Legislative Research Commission, 1981. 62 p.

Control

Argent, Gala. "Keeping California Above Water: Flood Control." *Western Water* (January–February 1989): 4–11.

Laska, Shirley Broadway. "Involving Homeowners in Flood Mitigation." *American Planning Association Journal* 52 (Autumn 1986): 42–66.

Ross, Lester. "Flood Control Policy in China: The Policy Consequences of Natural Disasters." *Journal of Public Policy* 3 (May 1983): 209–231.

Effects

Dingle, Michael W. "The Flooding of an American Canaan: The Endangered Species Act and the Value of Wildlife [Focuses on the Dispute over the Construction of the Tellico Dam, on the Little Tennessee River]." *Urban Law Annual* 22 (1981): 161–198.

Dzurik, Andrew A. "Floodplain Management Trends [Role of the U.S. Federal Government in Flood Control]." *Water Spectrum (Corps of Engineers)* 12(Summer 1980): 35–42.

Ives, Jack D. "Deforestation in the Himalayas: The Cause of Increased Flooding in Bangladesh and Northern India?" *Land Use Policy* 6 (July 1989): 187–193.

Jansen, Robert B. *Dams and Public Safety.* Water Resources Technology Publication. Washington, DC: U.S. Bureau of Reclamation, 1983. 332 p.

Steffen, Constance C. "The Great Salt Lake: Major Economic Impacts of High Lake Levels." *Utah Economics and Business Review* 43 (September–October 1983): 1–7.

United Nations. Economic Commission for Europe. *Application of Environmental Impact Assessment: Highways and Dams.* New York: U.N. Agent, 1987. 210 p.

U.S. Congress. House. Committee on Foreign Affairs. Subcommittee on Asian and Pacific Affairs. *Disaster Relief for Bangladesh: Hearing and Markup, September 23, 1988, on H.R. 5389 and H. Con. Res. 303.* 100th Cong., 2d sess., Washington, DC: GPO, 1989. 92 p.

Management

Bedient, Philip B., and Peter G. Rowe, eds. "Urban Watershed Management: Flooding and Water Quality [Proceedings of a Symposium Held at Rice University, May 25–28, 1978]." *Rice University Studies* 65 (Winter 1979): 1–205.

California. Department of Water Resources. *California Flood Management: An Evolution of Flood Damage Prevention Programs.* Bulletin 199, Sacramento, CA: 1980. 277 p.

Cigler, Beverly A., et al. "Rural Community Responses to a National Mandate: An Assessment of Floodplain Land Use Management." *Publius* 17 (Fall 1987): 113–30.

Davis, Tony. "Managing to Keep Rivers Wild: Built To Tame Rivers and Generate Power, America's Grand Dams Are Hurting Fish and Plant Life Downstream: Dam Operators Are Adopting New Techniques To Reduce the Damage." *Technology Review* 89 (May–June 1986): 26–33.

Farr, Cheryl. "Land Development and the Environment: Decision Making for Flood-Prone Areas." *Public Management* 64 (February 1982): 5–13.

New Jersey. General Assembly. Independent and Regional Authorities Committee. *Public Hearing: A Record of Testimony on Assembly Bills 2047, 2048, and 2570 (Flood Control Bills): Wayne, New Jersey, May 26, 1987.* Trenton, NJ: 1987. 98 p.

U.S. Congress. House. Committee on Interior and Insular Affairs. *Colorado River Management: Oversight Hearings, September 7–8, 1983.* 98th Cong., 1st sess., Washington, DC: GPO, 1983. 798 p.

U.S. Congress. House. Committee on Science and Technology. Subcommittee on Investigations and Oversight. *Forecasting and Technology for Water Management: Hearings, October 13–14, 1983.* 98th Cong., 1st sess., Washington, DC: GPO, 1984. 492 p.

U.S. Federal Emergency Management Agency. *Preparing for Hurricanes and Coastal Flooding: A Handbook for Local Officials.* Washington, DC: U.S. Federal Emergency Management Agency, 1983. 136 p.

U.S. Water Resources Council. *Floodplain Management Handbook.* Washington, DC: GPO, 1981. 69 p.

Dams and Hydroelectric Development

Aiken, S. Robert, and Colin H. Leigh. "Hydroelectric Power and Wilderness Protection." *Impact of Science on Society* 36, no. 1 (1986): 85–96.

Attey, John W., and Drew R. Liebert. "Clean Water; Dirty Dams: Oxygen Depletion and the Clean Water Act [of 1982: Legislative and Regulatory Policies That Would Prevent Environmental Degradation Caused by Dams and Hydroelectric Power Plants]." *Ecology Law Quarterly* 11, no. 4 (1984): 703–729.

Boxer, Baruch. "China's Three Gorges Dam: Questions and Prospects." *China Quarterly* (March 1988): 94–108.

Dingle, Michael W. "The Flooding of an American Canaan: The Endangered Species Act and the Value of Wildlife [Focuses on the Dispute over the Construction of the Tellico Dam, on the Little Tennessee River]." *Urban Law Annual* 22 (1981): 161–198.

Fearnside, Philip M. "China's Three Gorges Dam: 'Fatal' Project or Step Toward Modernization?" *World Development* 16 (May 1988): 615–630.

Grieves, Robert. "China Focus: Chongquing & the Three Gorges." *Asian Business* 22 (September 1986): 56–59.

Jansen, Robert B. *Dams and Public Safety*. Water Resources Technology Publication. Washington, DC: U.S. Bureau of Reclamation. 1983. 332 p.

Marble, Anne D. "Identification and Assessment of Aquatic Resources for Power Plant Sites on Maryland's Eastern Shore [Siting Procedure Which Takes into Account the Protection and Preservation of Economically Valuable Species of Finfish and Shellfish]." *Coastal Zone Management Journal* 7, no. 1 (1980): 49–70.

Railsback, S. F., et al. "Aeration at Ohio River Basin Navigation Dams." *Journal of Environmental Engineering* 116 (March–April 1990): 361–375.

Reisner, Marc. "America's Newest Old Energy Source: Hydro Power: Wherever Water Flows and Drops an Appreciable Distance, Someone, Somewhere Is Contemplating a Dam." *Amicus Journal* 6 (Spring 1985): 42–52.

———. "Water Folly: Dam the Rivers and Damn the Taxpayers: How Power Politics Is Costing You Billions of Dollars." *Common Cause Magazine* 12 (November–December 1986): 12–17.

Robinson, Paul, et al. "In the Hands of the People: Establishing Planning Power for a Community." *Workbook (Southwest Research and Info Center)* 15 (Winter 1990): 144–157.

"The Three Gorges Project." *Chinese Geography and Environment* 1 (Fall 1988): 1–102; (Winter, 1988): 3–119.

United Nations. Economic Commission for Europe. *Application of Environmental Impact Assessment: Highways and Dams*. New York: UN Agent, 1987. 210 p.

U.S. Congress. House. Committee on Energy and Commerce. Subcommittee on Energy Conservation and Power. *Conduit Hydroelectric Act of 1983: Hearing, May 17, 1983, on H.R. 1618, a Bill To Amend the Federal Power Act To Encourage Conduit Hydroelectric Facilities*. 98th Cong., 1st sess., Washington, DC: GPO, 1983. 97 p.

U.S. Congress. House. Committee on Government Operations. Environment, Energy and Natural Resources Subcommittee. *The Columbia Dam: ATVA [Tennessee Valley Authority] Project*. 96th Cong., 2d sess., Washington, DC: GPO, 1980. 262 p.

U.S. Congress. Senate. *Environmental Effects of the Harry S. Truman Dam: Joint Hearing, July 29, 1980, Before the Subcommittee on Water Resources of the Committee on Environment and Public Works and the Subcommittee on Federal Spending Practices and Open Government of the Committee on Governmental Affairs.* 96th Cong., 2d sess., Washington, DC: GPO, 1980. 338 p.

U.S. Congress. Senate. Committee on Energy and Natural Resources. Subcommittee on Public Lands, National Parks and Forests. *Federal Cave Resources Protection Act and Restriction of Dams in Parks and Monuments; Hearing, June 16, 1988.* 100th Cong., 2d sess., Washington, DC: GPO, 1988. 126 p.

Wallop, Malcolm. "Energy Development and Water; Dilemma for the Western United States." *Journal of Energy and Development* 8 (Spring 1983): 203–209.

"Water and Energy." *State Government* 55, no. 4 (1982): 111–138.

Oil Pollution

General

Conrad, Jon M. "Oil Spills: Policies for Prevention, Recovery, and Compensation." *Public Policy* 28 (Spring 1980): 143–170.

Dempsey, Paul S. "Compliance and Enforcement in International Law: Oil Pollution of the Marine Environment by Ocean Vessels." *Northwestern Journal of International Law and Business* 6 (Summer 1984): 459–561.

Dempsey, Paul S., and Lisa L. Helling. "Oil Pollution by Ocean Vessels–an Environmental Tragedy: The Legal, Regime of Flags of Convenience, Multilateral Conventions and Coastal States." *Denver Journal of International Law and Policy* 10 (Fall 1980): 37–87.

Exxon Corporation Public Affairs Department. *Fate and Effects of Oil in the Sea.* New York: 1979. 11 p.

———. *Fate and Effects of Oil in the Sea.* Exxon Background Series. New York: 1985. 12 p.

O'Neill, Trevor. "Oil Spills Contingency Plans and Policies in Norway and the United Kingdom." *Coastal Zone Management Journal* 8, no. 4 (1980): 289–317.

Tomasek, Robert D. *United States–Mexico Relations: Blowout of the Mexican Oil Well Ixtoc I.* Reports 1981 no. 20. Hanover, NH: American Universities Field Staff, 1981. 10 p.

U.S. Coast Guard. *Polluting Incidents In and Around U.S. Waters, Calendar Year 1980 and 1981.* Washington, DC: GPO, 1982. 41 p.

U.S. Congress. House. Committee on Energy and Commerce. Subcommittee on Transportation, Tourism, and Hazardous Materials. *Regulation of Above-Ground Oils and Waste Containers: Hearing, January 26, 1988.* 100th Cong., 2d sess., Washington, DC: GPO, 1988. 181 p.

U.S. Congress. Office of Technology Assessment. *Coping With an Oiled Sea; An Analysis of Oil Spill Response Technologies: Background Paper.* 101st Cong., 1st sess., Washington, DC: GPO, 1990. 70 p.

U.S. Congress. Senate. *Campeche Oil Spill: Joint Hearing, December 5, 1979, Before the Committee on Commerce, Science, and Transportation and Committee on Energy and Natural Resources.* 96th Cong., 1st sess., Washington, DC: GPO, 1980. 237 p.

Economics

Bederman, David J. "High Stakes in the High Arctic: Jurisdiction and Compensation for Oil Pollution from Offshore Operations in the Beaufort Sea." *Alaska Law Review* 4 (June 1987): 37–69.

Bobba, A. G., and S. R. Joshi. "Groundwater Transport of Radium-226 and Uranium from Port Granby Waste Management Site to Lake Ontario." *Nuclear and Chemical Waste Management* 8, no. 4 (1988): 199–209.

Cech, I., et al. "Radon Distribution in Domestic Water of Texas." *Ground Water* 25 (September–October 1988): 561–569; *Discussion* 27 (May–June 1989): 403–407.

Cheek, Leslie. "Pollution: The Peril Around Us [Liability for Environmental Damage]." *Journal of Insurance* 42 (November–December 1981): 2–7.

Eaton, Judith R. "Oil Spill Liability and Compensation; Time To Clean Up the Law." *George Washington Journal of International Law and Economics* 19 (3): 787–827 (1985).

Gaskell, N. J. J. "The Amoco Cadiz: Liability Issues." *Journal of Energy and Natural Resources Law* 3, no. 3 (1985): 169–194; 3 no. 4 (1985): 225–242.

Hartje, Volkmar J. "Oil Pollution Caused by Tanker Accidents: Liability Versus Regulation [Whether Liability Costs Are More Likely To Stimulate Preventive Measures Than Direct Regulation of Tank Ships]." *Natural Resources Journal* 24 (January 1984): 41–60.

Jacobsen, Douglas A., and James D. Yellen. "Oil Pollution: The 1984 London Protocols and the Amoco Cadiz [Oil Tanker Spill, March 1978; International Pollution Liability Schemes]." *Journal of Maritime Law and Commerce* 15 (October 1984): 467–488.

"Oil Spills and Cleanup Bills: Federal Recovery of Oil Spill Cleanup Costs [Under Federal Water Pollution Control Act and By Other Means]." *Harvard Law Review* 93 (June 1980): 1761–1785.

Organization for Economic Cooperation and Development. *Combatting Oil Spills, Some Economics Aspects.* Paris: OECD, 1982. 140 p. ISBN 92-64-12341-5.

———. *The Cost of Oil Spills.* Paris: OECD, 1982. 252 p. ISBN 92-64-12339-3.

Randle, Russell V. "The Oil Pollution Act of 1990: Its Provisions, Intent, and Effects." *Environmental Law Reporter* 21 (March 1991): 10119–10135.

Smets, Henri, "The Oil Spill Risk: Economic Assessment and Compensation Limit." *Journal of Maritime Law and Commerce* 14 (January 1983): 23–43.

"Superfund's Petroleum Exclusion [Four Perspectives on Whether the Exemption of Petroleum and Petroleum Products from Coverage under the Cleanup Program Should Be Revoked in Light of the Emergence of Leaking Underground Storage Tanks]." *Environmental Forum* 3 (August 1984): 32–38.

U.S. Congress. House. Committee on Merchant Marine and Fisheries. Subcommittee on Coast Guard and Navigation. *Oil Pollution Liability: Hearing, April 20, 1983, on H.R. 2222 (H.R. 2115, H.R. 2368), a Bill To Provide a Comprehensive System of Liability and Compensation for Oilspill Damage and Removal Costs and for Other Purposes.* 98th Cong, 1st sess., Washington, DC: GPO, 1983. 414 p.

———. *Oil Pollution Liability: Hearing, March 27, 1985, on H.R. 1232, a Bill To Provide a Comprehensive System of Liability and Compensation for Oilspill Damage and Removal Costs, and for Other Purposes.* 99th Cong, 1st sess., Washington, DC: GPO, 1985. 290 p.

U.S. Congress. House. Committee on Ways and Means. *The Comprehensive Oil Pollution and Compensation Act, and the Hazardous Waste Containment Act: Hearing, June 2, 1980, on H.R. 85 and H.R. 7020, Bills Which Would Impose Certain Fees on Crude Oil and Certain Chemicals To Finance a 'Superfund' for the Cleanup of Oil Spills, Chemical Spills, or Hazardous Waste Sites.* 96th Cong, 2d sess., Washington, DC: GPO, 1980. 276 p.

U.S. National Oceanic and Atmospheric Administration. National Ocean Service. *Assessing the Social Costs of Oil Spills: The Amoco Cadiz Case Study.* Washington, DC: GPO, 1983. 144 p.

Treaties

Boyle, Alan E. "Marine Pollution under the Law of the Sea Convention." *American Journal of International Law* 79 (April 1985): 347–372.

International Maritime Organization. *Regulations for the Prevention of Pollution By Oil: Articles of the International Convention for the Prevention of Pollution from Ships 1973 and of the Protocol of 1978 Relating Thereto: Annex I of MARPOL 73/78 Including Proposed Amendments and Unified Interpretation of the Provision of Annex I.* London: IMO, 1982. 165 p.

Kale, Joseph J. "Water Pollution and Commercial Fishermen: Applying General Maritime Law to Claims for Damages to Fisheries in Ocean and Coastal Waters." *North Carolina Law Review* 61 (January 1983): 313–363.

Kindt, John Warren. "Vessel-Source Pollution and the Law of the Sea." *Vanderbilt Journal of Transnational Law* 17 (Spring 1984): 287–328.

Meese, Sally A. "When Jurisdictional Interests Collide: International, Domestic, and State Efforts To Prevent Vessel Source Oil Pollution." *Ocean Development and International Law* 12, no. 1-2 (1982): 71–139.

Popp, Alfred H. E. "Recent Developments in Tanker Control in International Law [Tanker Safety and Pollution Prevention]." In *Canadian Yearbook of International Law, 1980,* edited by C. B. Bourne, 3–30. Vancouver, British Columbia: University of British Columbia Press, 1984.

Wang, Cheng-Pang. "A Review of the Enforcement Regime for Vessel-Source Oil Pollution Control." *Ocean Development and International Law* 16, no. 4 (1986): 305–339.

Rivers and Harbors

Evans, Shelley M. "Control of Marine Pollution Generated by Offshore Oil and Gas Exploration and Exploitation: The Scotian Shelf." *Marine Policy* 10 (October 1986): 258–270.

Kiss, Alexandre. "The Protection of the Rhine Against Pollution." *Natural Resources Journal* 25 (July 1985): 613–637.

New Jersey. General Assembly. Agriculture and Environment Committee. *Pubic Hearing on Gasoline Storage Tanks as a Source of Water Pollution and Related Issues: Held, Trenton, New Jersey, March 4, 1982.* Trenton, NJ: 1982. 43 p.

New Jersey. Senate. Energy and Environment Committee. *Public Hearing: Oil Spill Prevention and Response Capability: Camden, New Jersey, April 19, 1989.* Trenton, NJ: 1989. 114 p.

Schwabach, Aaron. "The Sandoz Spill: The Failure of International Law To Protect the Rhine from Pollution." *Ecology Law Quarterly* 16, no. 2 (1989): 443–480.

U.S. Coast Guard. *Activities Relating to the Port and Tanker Safety Act of 1978: A Report to Congress.* Washington, DC: GPO, 1980. 445 p.

U.S. Congress. House. Committee on Merchant Marine and Fisheries. Subcommittee on Oceanography and the Great Lakes. *Rhode Island Oil Spill: Hearing, July 5, 1989, on Events Leading Up to the June 23, 1989 Oil Spill of the Greek-Flagged Tanker, World Prodigy, Off the Coast of Rhode*

Island at Brenton Reef. 101st Cong., 1st sess., Washington, DC: GPO, 1989. 90 p.

U.S. Congress. House. Committee on Merchant Marine and Fisheries. Subcommittee on Oversight and Investigations. *Monongahela River Oil Spill: Hearing, May 26, 1988, on the Effects of the Oil Spill in Floreffe, Pennsylvania, on the Environment, the Lives of Those in the Area, and To Design a Policy Which Will Protect This Nation's Environmental Resources.* 100th Cong., 2d sess., Washington, DC: GPO, 1988. 132 p.

U.S. Congress. House. Committee on Merchant Marine and Fisheries. Subcommittee on Panama Canal/Outer Continental Shelf. *The Environmental Effects of OCS Development: Hearing, September 12, 1989, on the Environmental Effects of Offshore Oil and Gas Development and Associated Land Based Activities on Our Coastal Wetlands and Benthic Ecosystems.* 101st Cong., 1st sess., Washington, DC: GPO, 1990. 130 p.

U.S. Congress. House. Committee on Public Works and Transportation. Subcommittee on Water Resources. *Contaminated Sediments in Our Nation's Rivers and Harbors, Particularly in the Great Lakes: Hearing, June 21, 1989.* 101st Cong., 1st sess., Washington, DC: GPO, 1990. 197 p.

U.S. Congress. Senate. Committee on Environment and Public Works. *Eliminating Floatable Refuse in the New York Area: Hearing, June 19, 1989, Before the Subcommittee on Water Resources, Transportation, and Infrastructure and the Subcommittee on Superfund, Ocean, and Water Protection, on S.506, a Bill To Authorize the Corps of Engineers To Collect and Remove All Floating Material Wherever It Is Collecting and Removing from New York Harbor Debris Which Is an Obstruction to Navigation.* 101st Cong., 1st sess., Washington, DC: GPO, 1989. 272 p.

Ocean Pollution

General

Baur, Donald C., and Suzanne Iudicello. "Stemming the Tide of Marine Debris Pollution: Putting Domestic and International Control Authorities to Work." *Ecology Law Quarterly* 17, no. 1 (1990): 71–142.

Boczek, Boleslaw A. "Global and Regional Approaches to the Protection and Preservation of the Marine Environment." *Case Western Reserve Journal of International Law* 16, no. 1 (1984): 39–70.

Borrelli, Peter. "To Dredge or Not To Dredge: A Hudson River Saga." *Amicus Journal* 6 (Spring 1985): 14–26.

Cycon, Dean E. "Calming Troubled Waters: The Developing International Regime To Control Operational Pollution." *Journal of Maritime Law and Commerce* 13 (October 1981): 35– 51.

Hirvonen, H., and R. P. Coté. "Control Strategies for the Protection of the Marine Environment." *Marine Policy* 10 (January 1986): 19–28.

Joyner, Christopher C. "Oceanic Pollution and the Southern Ocean: Rethinking the International Legal Implications for Antarctica." *Natural Resources Journal* 24 (January 1984): 1–40.

———. "The Southern Ocean and Marine Pollution: Problems and Prospects." *Case Western Reserve Journal of International Law* 17 (Spring 1985): 165–194.

Kocasoy, Gunay. "The Relationship between Coastal Tourism, Sea Pollution, and Public Health: A Case Study from Turkey." *Environmentalist* 9 (Winter 1989): 245–251.

Saliba, Louis J. "Making the Mediterranean Safer: The Coastal States of the Mediterranean Are Striving to Reduce Marine Pollution, a Significant Threat to Health." *World Health Forum* 11, no. 3 (1990): 274–281.

Song, Yann-Huei Billy. "Marine Scientific Research and Marine Pollution in China." *Ocean Development and International Law* 20, no. 6 (1989): 601–621.

United Nations. Office for Ocean Affairs and the Law of the Sea. *The Law of the Sea: Protection and Preservation of the Marine Environment: Repertory of International Agreements Relating to Sections 5 and 6 of Part XII of the United Nations Convention on the Law of the Sea.* New York: UN Agent, 1990. 95 p.

U.S. Congress. House. Committee on Merchant Marine and Fisheries. *Coastal Water Quality: Hearing, July 13, 1988, Before the Subcommittee on Fisheries and Wildlife Conservation and the Environment and the Subcommittee on Oceanography, on Impact and the Subcommittee on Oceanography, on Impact of Nonpoint Source Pollution.* 100th Cong., 2d sess., Washington, DC: GPO, 1988. 161 p.

———. *Coastal Waters in Jeopardy: Reversing the Decline and Protecting America's Coastal Resources: Oversight Report, December 1988.* 100th Cong., 2d sess., Washington, DC: GPO, 1989. 48 p.

———. *Coastal Zone Management and Resource Protection: Hearing, September 27, 1988, before the Subcommittee on Fisheries and Wildlife Conservation and the Environment and the Subcommittee on Oceanography, on the Problem of Coastal Pollution; Examines the Development and Land Use in the Coastal Zone and Effective, Predictable Plans That Should Not Be Permitted.* 100th Cong., 2d sess., Washington, DC: GPO, 1988. 105 p.

U.S. Congress. Office of Technology Assessment. *Wastes in Marine Environments.* Washington, DC: GPO, 1987. 313 p.

U.S. Environmental Protection Agency. Office of Research and Development. *Bimonitoring for Control of Toxicity in Effluent Discharges to the Marine Environment.* Cincinnati, OH: EPA, 1989. 58 p.

Environment

Davidson, John H. "Little Waters: The Relationship between Water Pollution and Agricultural Drainage." *Environmental Law Reporter* 17 (3): 10074–10081 (1987).

Doneski, David. "Cleaning Up Boston Harbor: Fact or Fiction?" *Boston College Environmental Affairs Law Review* 12 (Spring 1985): 559–625.

Eichbaum, William. "The Chesapeake Bay: Major Research Program Leads to Innovative Implementation." *Environmental Law Reporter* 14 (June 1984): 10237–10245.

"Has the Time Come to Reserve Chesapeake Bay?" *Conservation Foundation Letter* (March–April 1984): 1–6.

Helsinki Commission. Baltic Marine Environment Protection Commission. *Activities of the Commission, 1987: Report on the Activities of the Baltic Marine Environment Protection Commission During 1987 Including the Ninth Meeting of the Commission Held in Helsinki 15–19 February 1988: HELCOM Recommendations Passed During 1988.* Baltic Sea Environment Proceedings no. 26. Helsinki, Finland: Mannerheimintie, 1988. 170 p.

Horton, Tom. "The People's Bay: Maryland Unveils a Tough Strategy To Save the Chesapeake." *Amicus Journal* 6 (Fall 1984): 12–19.

Lahey, William L. "Economic Changes for Environmental Protection: Ocean Dumping Fees." *Ecology Law Quarterly* 11, no. 3 (1984): 305–342.

Ramakrishna, K. "Environmental Concerns and the New Law of the Sea [Marine Pollution Problems]." *Journal of Marine Law and Commerce* 16 (January 1985): 1–19.

Russell, Dick. "Fisheries Management on the Chesapeake from Fiasco to a Future After All." *Amicus Journal* 6 (Fall 1984): 20–25.

Shaw, Bill, et al. "The Global Environment: A Proposal To Eliminate Marine Oil Pollution." *Natural Resources Journal* 27 (Winter 1987): 157–185.

U.S. Congress. House. Committee on Merchant Marine and Fisheries. Subcommittee on Fisheries and Wildlife Conservation and the Environment. *Coastal Estuarine Pollution: Hearing, May 20, 1987, on the Declining Environmental Quality of Our Coastal Areas, the Future of Our Coasts, and the Major Findings of the Office of Technology Assessment's Report Titled 'Waste in Marine Environments'.* 100th Cong., 1st sess., Washington, DC: GPO, 1987. 185 p.

U.S. Congress. House. Committee on Science and Technology. Subcommittee on Natural Resources, Agriculture Research and Environment. *Environmental Effects of Sewage Sludge Disposal: Hearing, May 27, 1981.* 97th Cong., 1st sess., Washington, DC: GPO, 1981. 109 p.

U.S. Congress. Senate. Committee on Environment and Public Works. Subcommittee on Environmental Pollution. *Ocean Incineration: Hearings, June 19 and July 17, 1985.* 99th Cong., 1st sess., Washington, DC: GPO, 1985. 593 p.

U.S. Congress. Senate. Committee on Governmental Affairs. Subcommittee on Governmental Efficiency and the District of Columbia. *Chesapeake Bay Cleanup Program: Hearing, June 24, 1986.* 99th Cong., 2d sess., Washington, DC: GPO, 1986. 457 p.

———. *EPA Chesapeake Bay Program: Hearing, March 1, 1983.* 98th Cong., 1st sess., Washington, DC: GPO, 1983. 152 p.

Voigt, Klaus. "The Baltic Sea: Pollution Problems and Natural Environmental Changes." *Impact of Science on Society*, nos. 3–4 (1983): 413–420.

Ocean Dumping

Bakalian, Allan. "Regulation and Control of United States Ocean Dumping: A Decade of Progress, an Appraisal for the Future." *Harvard Environmental Law Review* 8, no. 1 (1984): 193–256.

Bewers, J. N., and C. J. R. Garrett. "Analysis of the Issues Related to Sea Dumping of Radioactive Wastes." *Marine Policy* 11 (April 1987): 105–124.

Boxer, Baruch. "Mediterranean Pollution; Problem and Response." *Ocean Development and International Law* 10, no. 3-4 (1982): 315–356.

Guarascio, John A. "The Regulation of Ocean Dumping After City of New York v. Environmental Protection Agency." *Boston College Environmental Affairs Law Review* 12 (Summer 1985): 701–741.

Lahey, William L. "Ocean Dumping of Sewage Sludge: The Tide Turns from Protection to Management [Regulation Efforts by EPA]." *Harvard Environmental Law Review* 6, no. 2 (1982): 395–431.

Lenssen, Nicholas. "The Ocean Blues: They Are the Source of Food, Livelihood and Meteorological Balance, But the Oceans Are Reeling from Human Assault." *World Watch* 2 (July–August 1989): 26–35.

New Jersey. General Assembly. Energy and Natural Resources Committee. *Public Hearing on AR-45 (Directs the Energy and Natural Resources Committee To Study the Dumping of Untreated Sewerage in the Coastal Waters): Held: Toms, New Jersey, August 20, 1984.* Trenton, NJ: 1984. 151 p.

New Jersey. Senate. Special Committee To Study Coastal and Ocean Pollution. *Public Hearing; Testimony Concerning Problems of Water Pollution in the New York/New Jersey Metropolitan Region: Middletown, New Jersey, September 29, 1987.* Trenton, NJ: 1987. 215 p.

Smart, Tim, and Emily T. Smith. "Troubled Waters: The World's Oceans Can't Take Much More Abuse." *Business Week* (October 12, 1987): 88–91+.

Spirer, Julian H. "The Ocean Dumping Deadline: Easing the Mandate Millstone [Burden of Federal Rules and Regulations Directed at State and Local Governments]." *Fordham Urban Law Journal* 11, no. 1 (1982–1983): 1–49.

U.S. Congress. House. Committee on Merchant Marine and Fisheries. *Ocean Dumping: Hearings, March 5, 1979–February 20, 1980, Before the Subcommittee on Oceanography and the Subcommittee on Fisheries and Wildlife Conservation and the Environment on Ocean Dumping Authorization (Fiscal Year 1980) and Oversight—H.R. 1963 [and Other Bills].* 96th Cong. Washington, DC: GPO, 1980. 404 p.

———. *Waste Dumping: Hearings, May 1–November 5, 1981, Before the Subcommittee on Oceanography and the Subcommittee on Fisheries and Wildlife Conservation and the Environment, on Title I, Marine Protection, Research, and Sanctuaries Act; Ocean Dumping and Dumping Deadline; Radioactive Waste Dumping; Land Based Alternatives to Ocean Dumping.* 97th Cong., 1st sess., Washington, DC: GPO, 1982. 544 p.

U.S. Congress. House. Committee on Merchant Marine and Fisheries. Subcommittee on Oceanography. *Coastal Pollution in the New York Bight and Mid-Atlantic: Hearing, September 8, 1987, on H.R. 562 To Amend Title I of the Marine Protection, Research, and Sanctuaries Act of 1972 and H.R. 2791, To Amend Title I of the Marine Protection, Research, and Sanctuaries Act of 1972, To Provide for Restoration of the New York Bight, and for Other Purposes.* 100th Cong., 1st sess., Washington, DC: GPO, 1988. 405 p.

U.S. Congress. House. Committee on Public Works and Transportation. Subcommittee on Water Resources. *Ocean Dumping of Municipal Sludge: Hearing, August 2, 1988, on H.R. 4338 To Amend the Marine Protection, Research, and Sanctuaries Act of 1972 To Impose Special Fees on the Ocean Disposal of Sewage Sludge, and for Other Purposes.* 100th Cong., 2d sess., Washington, DC: GPO, 1989. 369 p.

U.S. Congress. House. Committee on Science, Space, and Technology. Subcommittee on Natural Resources, Agriculture Research, and Environment. *Medical Waste and Ocean Pollution Research and Monitoring: Hearing, October 3, 1988.* 100th Cong., 2d sess., Washington, DC: GPO, 1989. 111 p.

Welsch, Hubertus. "The London Dumping Convention and Sub-Seabed Disposal of Radioactive Waste." In *German Yearbook of International Law, 1985,* edited by Jost Delbrück, et al., 322–354. Berlin: Duncker and Humblot, 1986.

Wilcoxen, Peter J. "Coastal Zone Erosion and Sea Level Rise: Implications for Ocean Beach and San Francisco's Westside Transport Project." *Coastal Zone Management Journal* 14, no. 3 (1986): 173–191.

Zeppetello, Marc A. "National and International Regulation of Ocean Dumping: The Mandate To Terminate Marine Disposal of Contaminated Sewage Sludge." *Ecology Law Quarterly* 12, no. 3 (1985): 619–664.

Incineration

Bailey, Conner, and Charles E. Faupel. "Out of Sight Is Not Out of Mind: Public Opposition to Ocean Incineration." *Coastal Management* 17, no. 2 (1989): 89–102.

Ditz, Daryl W. "The Phase Out of North Sea Incineration." *International Environmental Affairs* 1 (Summer 1989): 175–202.

Reitze, Arnold W., Jr., and Andrew N. Davis. "Reconsidering Ocean Incineration as Part of a U.S. Hazardous Waste Management Program: Separating the Rhetoric from the Reality." *Boston College Environmental Affairs Law Review* 17 (Summer 1990): 687–798.

Walker, Christopher A. "The United States Environmental Protection Agency's Proposal for At-Sea Incineration of Hazardous Wastes: A Transnational Perspective." *Vanderbilt Journal of Transnational Law* 21, no. 1 (1988): 157–189.

Nuclear Wastes

Boehmer-Christiansen, Soñja. "An End to Radioactive Waste Disposal 'at Sea'?" *Marine Policy* 10 (April 1986): 119–131.

Finn, Daniel P. "Nuclear Waste Management Activities in the Pacific Basin and Regional Cooperation on the Nuclear Fuel Cycle." *Ocean Development and International Law* 13 (2): 213–246 (1983).

Van Dyke, Jon M. "Ocean Disposal of Nuclear Wastes." *Marine Policy* 12 (April 1988): 82–95.

Laws and Regulations

Boczek, Boleslaw A. "The Baltic Sea: A Study in Marine Regionalism [Focuses on Regional Cooperation in the Management of Fisheries and Marine Pollution under the Gdansk and Helsinki Convention,

Signed 1973 and 1974]." In *German Yearbook of International Law, 1980,* edited by Jost Delbrück, et al., 196–230. Berlin: Duncker and Humblot, 1981.

Burroughs, R.H. "OCS Oil and Gas; Relationships between Resource Management and Environmental Research [Reviews the Outer Continental Shelf Environmental Studies Program]." *Coastal Zone Management Journal* 9, no. 1 (1981): 77–88.

Clingan, Thomas A., Jr. "Environmental Problems and the New Order of the Oceans [Proposals Made at the Third Law of the Sea Conference for the Conservation of Marine Resources and the Control of Ocean Pollution]." *Columbia Journal of World Business* 15 (Winter 1980): 45–51.

Kitsos, T. R., and J. M. Bondareff. "Congress and Waste Disposal at Sea." *Oceanus* 33 (Summer 1990): 23–28.

Ma, Xiangcong, and Chen Zhenguo. "On the Marine Environmental Protection Law." *Chinese Law and Government* 19 (Spring 1986): 64–78.

Mizukami, Chiyuki, "The Law and Practice Relating to Marine Pollution in Japan." In *Japanese Annual of International Law, 1986,* edited by Soji Yamamoto, 65–73. Tokyo, Japan: International Law Association of Japan, University of Tokyo, 1986.

New Jersey. General Assembly. Coordinating Panel on Ocean Pollution. *Public Hearing: Review of Legislative Bills Addressing the Problem of Ocean Pollution, Toms River, New Jersey, December 1, 1987.* Trenton, NJ: 1987. 221 p.

New Jersey. General Assembly. Select Committee on Ocean and Beach Protection. *Public Meeting To Receive a Briefing from Appropriate State Officials on the Progress, Current Status, and Future Schedule of Implementation of the 'Clean Ocean' Legislative Package and Related Legislation Enacted in 1988: Trenton, New Jersey, February 14, 1989.* Trenton, NJ: 1989. 89 p.

New Jersey. Senate. Energy and Environment Committee. *Public Hearing: Senate Bill no. 2787 (The Clean Water Enforcement Act): Brick, New Jersey, October 12, 1988.* Trenton, NJ: 1988. 98 p.

New Jersey. Senate. Special Committee To Study Coastal and Ocean Pollution. *Public Hearing: Testimony Concerning the Proper Disposal of Hospital Waste, the Monitoring of Garbage Vessels and Other Shipping Traffic in Coastal Waters, and the Dispute between New York and New Jersey Regarding the Fresh Kills Landfill. Trenton, New Jersey, August 26, 1987.* Trenton, NJ: 1987. 170 p.

———. *Public Hearing; Testimony Concerning the Various Sources of Ocean Pollution, Including Sludge and Dredge Spoil Dumping, Vessel Refuse, and Other Ocean Dumping Practices: Long Beach, New Jersey, January 7, 1987.* Trenton, NJ: 1987. 325 p.

———. *Public Hearing: The Pretreatment of Industrial Waste Waters Prior to Discharge Into Publicly Owned Treatment Works: Trenton, New Jersey, September 15, 1987.* Trenton, NJ: 1987. 138 p.

Sparrow, Glen, and Dana Brown. "Black Water, Red Tape: Anatomy of a Border Problem." *National Civic Review* 75 (July–August 1986): 214–218.

Spirer, Julian H. "The Ocean Dumping Deadline: Easing the Mandate Millstone [Burden of Federal Rules and Regulations Directed at State and Local Governments]." *Fordham Urban Law Journal* 11, no. 1 (1982–1983): 1–49.

Tulokas, Mikko. "The Baltic Sea and Pollution [Regulations and Principles Governing Pollution: Emphasis on the Helsinki Convention of 1974]." In *Scandinavian Studies in Law,* edited by Anders Victorin, 205– 221. Stockholm, Sweden: Almqvist and Wikesell International, 1981.

U.S. Congress. House. Committee on Government Operations. Government Activities and Transportation Subcommitteee. *Dumping of Human Waste from Amtrak Trains; Hearing, September 27, 1988.* 100th Cong., 2d sess., Washington, DC: GPO, 1988. 176 p.

U.S. Congress. House. Committee on Merchant Marine and Fisheries. *The Coastal Defense Initiative of 1989: Hearings, Pts. 1–3, June 27, 1989–March 6, 1990, on H.R. 2647, Before the Subcommittee on Fisheries and Wildlife Conservation and the Environment and the Subcommittee on Oceanography and the Great Lakes.* 101st Cong., 1st and 2d sess., Washington, DC: GPO, 1989–1990. 3 pts.

———. *Dredge Spoil Disposal and PGB Contamination: Hearings, March 14 and May 21, 1980, on Exploring the Various Aspects Related to the Dumping of Dredged Spoil Material in the Ocean and the PCB Contamination Issue.* 96th Cong., 2d sess., Washington, DC: GPO, 1980. 698 p.

———. *Ocean Dumping: Hearing, February 23, 1988, Before the Subcommittee on Oceanography and the Subcommittee on Fisheries and Wildlife Conservation and the Environment.* 100th Cong., 2d sess., Washington, DC: GPO, 1988. 362 p.

———. *Ocean Dumping: Hearings, March 5, 1979–February 20, 1980, Before the Subcommittee on Oceanography and the Subcommittee on Fisheries and Wildlife Conservation and the Environment, on Ocean Dumping Authorization (Fiscal Year 1980) and Oversight—H.R. 1963 [and Other Bills].* 96th Cong., 1st and 2d sess., Washington, DC: GPO, 1980. 404 p.

———. *Plastic Pollution in the Marine Environment: Hearings, June 17 and July 23, 1987, Before the Subcommittee on Coast Guard and Navigation and the Subcommittee on Fisheries and Wildlife Conservation and the Environment, on H.R. 940, a Bill To Provide for the Regulation of the Disposal of Plastic Materials and other Garbage at Sea; To Provide for Negotiation, Regulation, and Research Regarding Fishing with Plastic Draftnets.* 100th Cong., 1st sess., Washington, DC: GPO, 1987. 498 p.

———. *Title I of the Ocean Dumping Ban Act of 1988: Hearings, May 16–17, 1989, Before the Subcommittee on Oversight and Investigations, Subcommittee on Fisheries and Wildlife Conservation and the Environment, and the Subcommittee on Oceanography and the Great Lakes, on Implementation of Title I of Public Law 100-66.* 101st Cong., 1st sess., Washington, DC: GPO, 1989. 419 p.

———. *Waste Dumping: Hearings, May 1–November 5, 1981, Before the Subcommittee on Oceanography and the Subcommittee on Fisheries and Wildlife Conservation and the Environment, on Title I, Marine Protection, Research, and Sanctuaries Act; Ocean Dumping and Dumping Deadline; Radioactive Waste Dumping; Land Based Alternatives to Ocean Dumping.* 97th Cong., 1st sess., Washington, DC: GPO, 1982. 544 p.

U.S. Congress. House. Committee on Merchant Marine and Fisheries. Subcommittee on Coast Guard and Navigation. *Coast Guard Enforcement of Environmental Laws: Hearing, October 16, 1989, on H.R. 3394 To Provide for a Comprehensive Compensation and Liability Scheme for*

Discharges of Oil, and for Other Purposes. 101st Cong., 1st sess., Washington, DC: GPO, 1990. 320 p.

———. *Plastic Pollution in the Marine Environment: Hearing, August 12, 1986, on the Problem of Nonbiodegradable Plastic Refuse in the Marine Environment, and To Examine the Options That Exist On All Levels for Responding to It.* 99th Cong., 2d sess., Washington, DC: GPO, 1986. 210 p.

U.S. Congress. House. Committee on Merchant Marine and Fisheries. Subcommittee on Fisheries and Wildlife Conservation and the Environment. *Medical Waste and Sewage Contamination: Hearing, September 6, 1988 on the Recent Closure of Coastal Areas in New York, New Jersey, and Southern New England Due to Medical Waste and Sewage Contamination and H.R. 5231, a Bill To Amend the Marine Protection, Research and Control Act of 1972.* 100th Cong., 2d sess., Washington, DC: GPO, 1988. 231 p.

U.S. Congress. House. Committee on Merchant Marine and Fisheries. Subcommittee on Oceanography. *Coastal Pollution in the New York Bight and Mid-Atlantic: Hearing, September 8, 1987, on H.R. 562, To Amend Title I of the Marine Protection, Research, and Sanctuaries Act of 1972 and H.R. 2791, To Amend Title I of the Marine Protection, Research and Sanctuaries Act of 1972, To Provide for Restoration of the New York Bight, and for other Purposes.* 100th Cong., 1st sess., Washington, DC: GPO, 1988. 405 p.

———. *Incineration of Hazardous Waste at Sea: Hearing, November 17, 1989, on H.R. 737.* 100th Cong., 1st sess., Washington, DC: GPO, 1988. 261 p.

———. *Medical Waste Disposal: Hearing, August 8, 1988, on H.R. 3478.* 100th Cong., 2d sess., Washington, DC: GPO, 1988. 123 p.

U.S. Congress. House. Committee on Public Works and Transportation. Subcommittee on Investigations and Oversight. *Implementation of the Clean Water Act Concerning Ocean Discharge Waivers (A Case Study of Lawmaking by Rulemakers): Report, December 1982.* 97th Cong., 2d sess., Washington, DC: GPO, 1983. 198 p.

U.S. Congress. House. Committee on Public Works and Transportation. Subcommittee on Water Resources. *Water Resources Needs: Hearing, January 29, 1982. (Mid-Winter Meeting of the U.S. Conference of Mayors, Washington, DC.)* 97th Cong., 2d sess., Washington, DC: GPO, 1982. 106 p.

U.S. Congress. House. Committee on Science, Space, and Technology. Subcommittee on Natural Resources, Agriculture Research, and Environment. *Medical Waste and Ocean Pollution Research and Monitoring: Hearing, October 3, 1988.* 100th Cong., 2d sess., Washington, DC: GPO, 1989. 111 p.

U.S. Congress. House. Committee on the Judiciary. Subcommittee on Crime. *Health Waste Anti-Dumping Act of 1988: Hearing, August 4, 1988, on H.R. 5130.* 100th Cong., 2d sess., Washington, DC: GPO, 1989. 98 p.

U.S. Congress. Office of Technology Assessment. *Wastes in Marine Environments.* Washington, DC: GPO, 1987. 313 p.

U.S. Congress. Senate. Committee on Commerce, Science and Transportation. *Oceans Pollution: Hearing, April 11, 1979, on Ocean Pollution Research and Development and Monitoring Planning Act of 1978, and Title II of the Marine Protection, Research and Planning Act of 1972.* 96th Cong., 1st sess., Washington, DC: GPO, 1979. 44 p.

U.S. Congress. Senate. Committee on Commerce, Science, and Transportation. National Ocean Policy Study. *Plastic Pollution in the Marine Environment: Hearing, July 29, 1987.* 100th Cong., 1st sess., Washington, DC: GPO, 1987. 130 p.

U.S. Congress. Senate. Committee on Environment and Public Works. *Implementation of Title I of the Marine Protection, Research, and Sanctuaries Act: Joint Hearing, February 18, 1988, Before the Subcommittees on Superfund and Environmental Oversight and Environmental Protection.* 100th Cong., 2d sess., Washington, DC: GPO, 1988. 242 p.

———. *Protection of Marine and Coastal Waters: Joint Hearings, July 12–August 11, 1989, on S. 587, [and other] Bills Concerning Research and Protection of the Marine and Coastal Waters, Before the Subcommittee on Superfund, Ocean, and Water Protection and the Subcommittee on Environmental Protection.* 101st Cong., 1st sess., Washington, DC: GPO, 1990. 579 p.

U.S. Congress. Senate. Committee on Environment and Public Works. Subcommittee on Environmental Protection. *Controlling and Reducing Pollution from Plastic Waste: Hearings, June 1–September 17, 1987, on S.559, S.560, and S.633.* 100th Cong., 1st sess., Washington, DC: GPO, 1987. 393 p.

———. *Environmental Trends and Conditions in Marine and Coastal Waters and the Marine Research Act of 1988: Hearings, April 20 and 28, 1988, on S. 2608, a Bill To Amend the Marine Protection Research, and Sanctuaries Act To Protect Marine and Near-Shore Coastal Waters through Establishment of Regional Marine Centers.* 100th Cong., 2d sess., Washington, DC: GPO, 1988. 302 p.

Zaleski, Alexander V. "A New Authority for Massachusetts: Best Solution for a Difficult Task." *National Civic Review* 74 (December 1985): 531–537.

Regional—United States

Western United States

Agthe, Donald E., and Kambiz Raffice. "Pricing as a Conservative Practice of Nevada Water Companies: A Survey." *Nevada Review of Business and Economics* 8 (Fall 1984): 19– 23.

Arrandale, Tom. "Western Water." *Editorial Research Reports* (January 30, 1987): 42–51.

Barloon, Marvin J. "The Diversion of Great Lakes Water to the Arid West." *Illinois Business Review* 45 (April 1988): 5–10.

Bates, Karl L. "Water Diversion from the Great Lakes [to Western States]." *Illinois Issues* 10 (December 1984): 17–20.

Becker, Nir, and K. William Easter. *Diversions from the Great Lakes: Opportunities and Dangers.* Staff Paper P89-28. St. Paul: University of Minnesota, Institute of Agriculture, Forestry, and Home Economics, 1989. 13 leaves.

Brajer, Victor, and Wade E. Martin. "Allocating a 'Scarce' Resource, Water in the West: More Market-Like Incentives Can Extend Supply, but Constraints Demand Equitable Policies." *American Journal of Economics and Sociology* 48 (July 1989): 259–271.

Brown, Lee. "Conflicting Claims to Southwestern Water: The Equity and Management Issues." *Southwestern Review of Management and Economics* 1 (Spring 1981): 35–60.

Cannon, James S. "The Big Splash Over Western Water: Vast Amounts of Water Are Needed To Exploit the West's Natural Resources [Especially Coal Production]." *Business and Society Review* (Summer 1984): 34–38.

"Clash over the American." *Western Water* (November–December 1985): 4–10.

DuMars, Charles, and Helen Ingram. "Congressional Quantification of Indian Reserved Water Rights: A Definite Solution or a Mirage?" *Natural Resources Journal* 20 (January 1980): 17–43.

Duncan, Marvin, and Ann Laing. "Western Water Resources: Coming Problems and the Policy Alternatives." *Federal Reserve Bank of Kansas City Economic Review* 65 (February 1980): 14–22.

Durant, Robert F., and Michelle Deany Homes. "Thou Shalt Not Covet Thy Neighbor's Water: The Rio Grande Basin Regulatory Experience." *Public Administration Review* 45 (November–December 1985): 821–831.

Gross, Sharon P. "The Galloway Project and the Colorado River Compact: Will the Compacts Bar Transbasin Water Diversions?" *Natural Resources Journal* 25 (October 1985): 935– 960.

Hartshorn, J. K. "Xeriscape: A New Word for Saving Water." *Western Water* (May–June 1986): 4–11.

"The Impact of Recent Court Decisions Concerning Water and Interstate Commerce on Water Resources of the State of New Mexico [Report of the Water Law Study Committee]." *Natural Resources Journal* 24 (July 1984): 689–744.

Landry, Stephanie. "The Galloway Proposal and Colorado Water Law: The Limits of the Doctrine of Prior Appropriation." *Natural Resources Journal* 25 (October 1985): 961–983.

Langfur, Hal. "Colorado's Water Fight." *New Leader* 71 (July 25, 1988): 11–14.

Lemarquand, David G. "Precondition to Cooperation in Canada-United States Boundary Waters." *Natural Resources Journal* 26 (Spring 1986): 221–242.

McCool, Daniel. "Marketing Water From Indian Lands." *Forum for Applied Research and Public Policy* 5 (Spring 1990): 73–78.

McDonald, Briand, et al. "An Evolutionary History of the Middle Rio Grande Conservancy District." *New Mexico Business* 33 (April 1980): 3–29.

McKinnon, Janet E. "Water to Waste: Irrational Decisionmaking in the American West." *Harvard Environmental Law Review* 10, no. 2 (1986): 503–532.

Stevens, Anastasia S. "Public Water Rights in New Mexico." *Natural Resources Journal* 28 (Summer 1988): 535–583.

U.S. Congress. Senate. Committee on Energy and Natural Resources. Subcommittee on Water and Power. *Western Water Policy Review Act: Hearing, August 27, 1990, on S. 1996 To Provide for a Comprehensive Review by the Secretary of the Interior of Western Water Resource Problems and Programs Administered by the Geological Survey, the Bureau of Reclamation, and other Operations of the Department of the Interior, and for other Purposes.* 101st Cong., 2d sess., Washington, DC: GPO, 1990. 84 p.

U.S. General Accounting Office. *Colorado River Basin Water Problems: How To Reduce Their Impact: Report to the Comptroller General of the United States.* Washington, DC: GPO, 1979. 137 p.

Wescoat, James L., Jr., "On Water Conservation and Reform of the Prior Appropriation Doctrine in Colorado." *Economic Geography* 61 (January 1985): 3–24.

California

Argent, Gala. "Banking for the Future: Conjunctive Use of California's Surface and Ground Water." *Western Water* (March–April 1990): 4–11.

"California's Growing Urban Thirst." *Western Water* (January–February 1990): 4–11.

Brickson, Betty. "The Endangered Species Act and California Water." *Western Water* (November–December 1990): 4–11.

California. Resources Agency. Department of Water Resources. *The California State Water Project: Current Activities and Future Management Plans.* Sacramento, CA: 1981. 311 p.

"California Water Interests: Working Toward Consensus." *Western Water* (July–August 1986): 3–14.

"Can Farm Water Conservation Save Californians from Building More Water Projects?" *Western Water* (November–December 1984): 4–12.

Carson, Dan. "Water Wars '87: New Bills Revive North-South Struggle." *California Journal* 18 (August 1987): 378–382.

Clifton, Merritt. "Water from the North? Last Summer's Drought Focused New Attention on Plans To Channel Water from Canada's Wild North to the Thirsty Neighbor Down South." *Environmental Action* 20 (January–February 1989): 25–27.

de Lambert, Deborah. "District Management for California's Groundwater [Recommends the Creation of a Statewide Program of Local Groundwater Management Authorities]." *Ecology Law Quarterly* 11, no. 3 (1984): 373–400.

Goldman, George E., et al. "California's Water Quality, Quantity, and the [Sacramento-San Joaquin] Delta." *Public Affairs Report* 23 (April 1982): 1–12.

Parker, Richard. "Water Supply for Urban Southern California: An Historical and Legal Perspective." *Glendale Law Review* 8, no. 1–2 (1988): 1–66.

Tucker, William. "Billions Down the Drain: Why Californians are Squandering Their Most Precious Resource [Emphasis on the Proposed Peripheral Canal, Intended To Bring Water from Northern to Southern California]." *Reason* 14 (June 1982): 24–32.

U.S. Congress. Senate. Committee on Energy and Natural Resources. Subcommittee on Water and Power. *California Water Controversies and the U.S. Bureau of Reclamation: Hearing, August 29, 1989.* 101st Cong., 1st sess., Washington, DC: GPO, 1989. 416 p.

Vogel, Nancy. "Tapped Out in California: Five-Year Drought Exposes Deficiencies in State's Water System." *California Journal* 22 (April 1991): 154–163.

Walker, Richard, and Michael Storper. "The Expanding California Water System." *Northern California Review of Business and Economics* (Winter 1982): 2–9.

Zeigler, Richard. "Water, Water: Does the State Need a New Way of Thinking About a Vital Resource?" *California Journal* 19 (March 1988): 104–109.

Arizona

Balchin, W. G. V. "Land and Water Use Policy in Arizona: Limits to Desert Development." *Land Use Policy* 5 (April 1988): 197–206.

Billings, R. B., and W. M. Day. "Demand Management Factors in Residential Water Use: The Southern Arizona Experience." *American Water Works Association Journal* 81 (March 1989): 58–64.

Cuthbert, R. W. "Effectiveness of Conservation-Oriented Water Rates in Tucson." *American Water Works Association Journal* 81 (March 1989): 65–73.

Davis, Tony. "Trouble in a Thirsty City: Tucson's Water Budget Is Badly Out Of Balance: Technological Advances and Cultural Changes Are in the Works—But Will They Be Enough?" *Technology Review* 87 (August–September 1984): 66–71.

Leshy, John D., and James Belanger. "Arizona Law Where Ground and Surface Water Meet." *Arizona State Law Journal* 20 (Fall 1988): 657–748.

McNulty, Michael F., and Gary C. Woodard. "Arizona Water Issues: Contrasting Economic and Legal Perspectives." *Arizona Review* 32 (Fall 1984): 1–13.

"Water for Arizona: The Central Arizona Project." *Western Water* (May–June 1985): 4–11.

Woodard, Gary C. "Arizona's New Environmental Quality Act." *Arizona's Economy* (September 1986): 1–4.

Eastern and Southern States

Cox, William E. "Water Supply Management in Virginia." *University of Virginia News Letter* 61 (June 1985): 57–61.

Gellis, Ann J. "Water Supply in the Northeast: A Study in Regulatory Failure." *Ecology Law Quarterly* 12, no. 3 (1985): 429–479.

"Managing Florida's Water Resources." *Business and Economic Dimensions* 16, no. 1 (1980): 1–28.

Marshall, Arthur R. "The Crisis in Water Management in South Florida." *Business and Economic Dimensions* 18, no. 2 (1982): 22–27.

New Jersey. General Assembly. Agriculture and Environment Committee. *Briefing On Water (Supplies and Drought Conditions): Held: Trenton, New Jersey, July 31, 1985.* Trenton, NJ: 1985. 88 p.

———. *Public Housing on Toms River Experience—Water Pollution and Related Issues: Held: Toms River, New Jersey, March 25, 1982.* Trenton, NJ: 1982. 67 p.

Schwartz, Larry N., and Mary F. Smallwood. "Regulation of Wastewater Discharge to Florida Wetlands." *National Wetlands Newsletter* 8 (November–December 1986): 5–9.

U.S. Congress. Senate. Committee on Governmental Affairs. Subcommittee on Governmental Efficiency and the District of Columbia. *A 1980's View of Water Management in the Potomac River Basin: A Report, November 2, 1981.* 97th Cong., 2d sess., Washington, DC: GPO, 1982. 82 p.

Zaleski, Alexander, V. "A New Authority for Massachusetts: Best Solution for a Difficult Task." *National Civic Review* 74 (December 1985): 531–537.

Midwest

Caldwell, Lynton K. "Garrison Diversion: Constraints on Conflict Resolution [Transboundary Environmental Dispute, Involving Canada, the U.S., Manitoba, and North Dakota, Stemming from Diversion of Missouri River Water from the Garrison Dam to East-Central North Dakota: Case for Authorizing the International Joint Commission of United States and Canada, or another Bi-National Body, To Deal with International Environmental Controversies]." *Natural Resources Journal* 24 (October 1984): 839–863.

Easterbrook, Gregg. "'Deep Tunnel: How Our Money Flows Into Chicago's Sewers: The World According to TARP [Critical of the Tunnel Reservoir Plan, Intended To Control Flooding and Water Pollution and Rehabilitate Chicago's Unborn Canal System]." *Washington Monthly* 11 (November 1979): 30–36.

Hanson, Mark E., and Harvey M. Jacobs. "Private Sewage System Impacts in Wisconsin: Implications for Planning and Policy." *Journal of the American Planning Association* 55 (Spring 1989): 169–180.

Krohe, James, Jr., "Illinois Water: Cleaner But Not Clean." *Illinois Issues* 8 (October 1982): 23– 28.

"Water Policy in Illinois: The Search for Balance." *Illinois Issues* 8 (November 1982): 12–17.

Kromm, David E., and Stephen E. White. "Public Preferences for Recommendations Made by the High Plains–Ogallala Aquifer Study." *Social Science Quarterly* 67 (December 1986): 841–854.

Merritt, Raymond H. *The Corps, the Environment, and the Upper Mississippi River Basin [1866–1980]*. Environmental History Series. Washington, DC: U.S. Army Corps of Engineers. Office of the Chief of Engineers, 1984. 119 p.

"Protecting Minnesota's Waters: The Land-Use Connection." *Minnesota Cities* 72 (June 1987): 4–6+.

U.S. Army. Corps of Engineers. North Central Division. *Water Resources Development in Illinois.* Chicago, IL: 1981. 139 p.

U.S. Congress. Senate. Committee on Energy and Natural Resources. Subcommittee on Water and Power. *Future of Water Resource Development in South Dakota: Hearing, September 12, 1981, on S.1553, a Bill To Authorize the Secretary of the Interior To Proceed with Development of the WEB Pipeline, To Provide for the Study of South Dakota Water Projects To Be Developed in Lieu of the Oahe and Pollock Herreid Irrigation Projects, and To Make available Missouri Basin Pumping Power to Projects Authorized by the Flood Control Act of 1944 To Receive Such Power.* 97th Cong., 1st sess., Washington, DC: GPO, 1981. 1004 p.

Regional—Foreign

Canada

Alper, Donald K., and Robert L. Monahan. "Regional Transboundary Negotiations Leading to the Skagit River Treaty: Analysis and Future Application." *Canadian Public Policy* (Guelph) 23 (March 1986): 163–174.

Canada. International Development Research Centre. *Fresh Water: The Human Imperative.* Ottawa, Ontario: The Research Center, 1989. 40 p.

Jayal, N. D. "Research Priorities for Planning Water Resource Development." *Canadian Journal of Development Studies* 7, no. 1 (1986): 37–46.

Sadler, Barry. "The Management of Canada-U.S. Boundary Waters: Retrospect and Prospect." *Natural Resources Journal* 26 (Spring 1986): 359–376.

Asia

Coleby, David. "Isotopes Aid Asian Water Management." *Australian Foreign Affairs Record* 54 (February 1983): 64–69.

Cooley, John K. "The War Over Water [Competition for Water Rights and Supplies in the Near East]." *Foreign Policy* (Spring 1984): 3–26.

Grieves, Robert. "China Focus: Chongqing and the Three Gorges." *Asian Business* 22 (September 1986): 56–59.

Hafner, James. "View from the Village: Participatory Rural Development in North East Thailand." *Community Development Journal* 22 (April 1987): 87–97.

Skutel, H. J. "Water in the Arab-Israel Conflict: Why Israel Won't Budge; Aquifers Are Not Enough." *International Perspectives* (Canada) (July–August 1986): 22–24.

Stork, Joe. "Water and Israel's Occupation Strategy [Attempts to Control Water Resources in the Occupied Territories to Encourage Israeli Settlements and Development in Israel]." *MERIP (Middle East Research and Info Project) Reports* 13 (July–August 1983): 19–24.

"The Three Gorges Project." *Chinese Geography and Environment* 1 (Fall 1988): 1–102 (Winter 1988): 3–119.

United Nations Economic and Social Commission for Asia and the Pacific. *Development and Conservation of Ground-Water Resources and Water-Related Natural Disasters and Their Mitigation: in Selected Least Developed Countries and Developing Island Countries in the ESCAP Region.* New York: UN Agent, 1989. 102 p.

———. *Water Resources Development in Asia and the Pacific: Dam Safety Evaluation and Monitoring, Water Tariffs and Rain-Water Harvesting.* New York: UN Agent, 1989. 118 p.

———. *Water Resources Development in Asia and the Pacific: Some Issues and Concerns.* Water Resources Series no. 62. New York: UN Agent, 1987. 202 p.

Other Countries

Darst, Robert G., Jr. "Environmentalism in the USSR: The Opposition to the River Diversion Projects." *Soviet Economy* 4 (July–September 1988): 223–252.

Day, Diana G. "Australian Natural Resources Policy: Water and Land." *Resources Policy* 13 (September 1987): 228–248.

———. "Resources Development or Instream Protection? The Case of Queensland, Australia." *Environmentalist* 9 (Spring 1989): 7–23.

Dourojeanni, A., and M. Molina. "The Andean Peasant, Water and the Role of the State." *CEPAL Review* (April 1983): 145–166.

Eldridge, Roger Lee. "A Comprehensive Approach to U.S.-Mexico Border Area Water Management." *Southwestern Review of Management and Economics* 4 (Summer 1985): 89–101.

Fung, Chung-Yue. "Water Resources Activities in Taiwan, 1968–1978." *Industry of Free China (Taipei)* 52 (July 1979): 10–34.

Heilbroon, S. G. "Water Laws, Prior Rights and Government Apportionment of Water in Swaziland, Southern Africa." *Journal of African Law* 25 (Autumn 1981): 136–149.

Mercer, David. "Australia's Constitution, Federalism, and the 'Tasmanian Dam Case'." *Political Geography Quarterly* 4 (April 1985): 91–110.

Mohanty, R. P. "Multi-Purpose Reservoir System Management [Hirakud Dam Project in the State of Orissa, India]." *Impact of Science on Society* no. 1 (1983): 83–95.

Parker, Dennis J., and W. R. Derrick Sewell. "Evolving Water Institutions in England and Wales: An Assessment of Two Decades of Experience." *Natural Resources Journal* 28 (October 1988): 751–785.

Thomi, Walter. "Man-Made Lakes as Human Environments: The Formation of New Socio-Economic Structures in the Region of the Volta Lake in Ghana/West Africa [Volta River Project]." *Applied Geography and Development* 23 (1984): 109–127.

"Water Resources." *Banque Morocaine du Commerce Extérieur Information Review* (September 1986): 11–20.

Williams, D. A. R., and M. C. Holm. "Recent Developments in New Zealand Water Law [Quality Allocation, Water Rights, River Protection, and Related Matters]." *New Zealand Law Journal* (August 1983): 245–251.

Selected Journal Titles

The journals listed here publish articles on many aspects of water. Because of increasing demand for water and problems of pollution, water availability has become an important issue in recent years. New journals are continually appearing. For other journals and additional information please consult:

Ulrich's International Periodicals Directory 1991–1992. 30th ed., New York: R. R. Bowker, 1991. 3 vols.

Information on the journals listed is arranged in the following sample entry:

Journal Title

1. Editor
2. Year First Published
3. Frequency of Publication
4. Code
5. Special Features
6. Address of Publisher

American Journal of Economics and Sociology

1. Frank Genovese
2. 1941
3. Quarterly
4. ISSN 0002-9246
5. Stat., index
6. American Journal of Economics and Sociology, Inc.
   41 E. 72nd Street
   New York, NY 10021

American Planning Association Journal

1. —
2. 1925
3. Quarterly
4. ISSN 0194-4363
5. Adv., bk. rev., abstr., bibl., charts, illus., maps, cum. index
6. American Planning Association
   1313 E. 60th Street
   Chicago, IL 60637

American Water Works Association Journal

1. Nancy M. Zeilig
2. 1914
3. Monthly
4. ISSN 0003-150X
5. Adv., bk. rev., abstr., bibl., charts, illus., stat., tr. lit., index, cum. index 1946–1980 (5 vols)
6. American Water Works Association
   6666 W. Quincy Avenue
   Denver, CO 80235

American Water Works Association Proceedings

1. —
2. —
3. Annual
4. ISSN 0360-814X
5. Illus.

6. American Water Works Association
   6666 W. Quincy Avenue
   Denver, CO 80235

Amicus Journal

1. Peter Borrelli
2. 1979
3. Quarterly
4. ISSN 0276-7201
5. Bk. rev., illus., index, cum. index
6. Natural Resources Defense Council Inc.
   40 W. 20th Street
   New York, NY 10011

Arizona Review

1. Lynne Schwartz and Jo Marie Gellerman
2. 1952
3. Twice annually
4. ISSN 0004-1629
5. Charts, illus.
6. University of Arizona
   College of Business and Public Administration
   Division of Economic and Business Research
   Tucson, AZ 85721

Arizona State Law Journal

1. Editorial Board
2. 1969
3. Quarterly
4. ISSN 0164-4297
5. Adv., bk. rev., bibl.
6. Arizona State University
   College of Law
   Tempe, AZ 85287

Boston College Environmental Affairs Law Review

1. Editorial Board
2. 1971
3. Quarterly

4. ISSN 0190-7034
5. Adv., bk rev., bibl., charts
6. Boston College
   School of Law
   885 Centre Street
   Newton, MA 02159

California Business

1. Christopher Bergonzi
2. 1965
3. Monthly
4. ISSN 0008-0926
5. Adv., illus., stat.
6. California Business News, Inc.
   4221 Wilshire Boulevard
   Suite 400
   Los Angeles, CA 90010-3503

California Journal

1. Richard Zeiger
2. 1970
3. Monthly
4. ISSN 0008-1205
5. Adv., charts, stat., index
6. California Journal, Inc.
   1714 Capitol Avenue
   Sacramento, CA 95814

Coastal Management

1. Marc J. Hershman
2. 1973
3. Quarterly
4. ISSN 0892-0753
5. Adv., bk rev., abstr., bibl., charts, illus., index, cum. index
6. Taylor & Francis
   1900 Frost Road
   Suite 101
   Bristol, PA 19007

Congressional Digest

1. —
2. 1921
3. 10 times a month
4. ISSN 0010-5899
5. Index
6. Congressional Digest Corp
3231 P Street NW
Washington, DC 20007

EPA Journal

1. John Heritage
2. 1975
3. Bimonthly
4. —
5. Adv.
6. Environmental Protection Agency
Office of Public Affairs
Waterside Mall
401 M Street SW
Washington, DC 20460

Ecology Law Quarterly

1. —
2. 1971
3. Quarterly
4. ISSN 0046-1121
5. Adv., bk. rev., bibl., index
6. University of California Press
Journals Division
2120 Berkeley Way
Berkeley, CA 94720

Economic Development Quarterly

1. Editorial Board
2. 1987
3. Quarterly
4. ISSN 0891-2424
5. Adv., bk. rev.

6. Sage Publications Inc.
   2111 West Hillcrest Drive
   Newbury Park, CA 91320

Environmental Action

1. Editorial Board
2. 1970
3. Bimonthly
4. ISSN 0013-922X
5. Adv., bk. rev., film rev., illus., index
6. Environmental Action Inc.
   1525 New Hampshire Avenue NW
   Washington, DC 20036

Environmental Forum

1. Carole Parker
2. 1981–1986; resumed 1988
3. Monthly
4. ISSN 0731-5732
5. —
6. Environmental Law Institute
   1616 P Street NW
   Suite 200
   Washington, DC 20036

Environmental Law Reporter

1. Barry Breen
2. 1971
3. Monthly
4. ISSN 0046-2284
5. Index
6. Environmental Law Institute
   1616 P Street NW
   Washington, DC 20036

Environmental Science & Technology

1. William H. Glaze
2. 1967
3. Monthly

4. ISSN 0013-936X
5. Abstr., adv., bk. rev., bibl., charts, illus., stat., tr. lit., index
6. American Chemical Society
   1155 16th Street NW
   Washington, DC 20036

Environmentalist

1. John F. Potter
2. 1980
3. Quarterly
4. ISSN 0251-1088
5. Adv.
6. Science and Technology Letters
   P.O. Box 81
   Northwood, Middlesex HA6 3DN
   Great Britain

Glendale Law Review

1. Juan Dominquez
2. 1976
3. Irregular
4. ISSN 0363-2423
5. —
6. Glendale University
   College of Law
   220 N. Glendale Avenue
   Glendale, CA 91206

Ground Water

1. Jay H. Lehr
2. 1963
3. Bimonthly
4. ISSN 0017-467X
5. Abst., bibl., charts., illus., index
6. Water Well Journal Publishing Co.
   6375 Riverside Drive
   Dublin, OH 43017

Harvard Environmental Law Review

1. —
2. 1976
3. 2 times a year
4. ISSN 0147-8257
5. —
6. Harvard University
   Law School
   Publications Center
   Holmes Hall
   Cambridge, MA 02138

Illinois Issues

1. Caroline Gherardini
2. 1975
3. Monthly (except Aug. & Sept. combined)
4. ISSN 0738-9663
5. Bk. rev.
6. Sangamon State University
   Springfield, IL 62794-9243

Impact of Science on Society

1. Howard J. Moore
2. 1950
3. Quarterly
4. ISSN 0019-2872
5. Illus.
6. UNESCO
7. 7 Place de Fontenoy
   75700 Paris
   France

Journal of Environmental Engineering

1. Chin-Pao Huang
2. 1956
3. Bimonthly
4. ISSN 0733-9372
5. —

6. American Society of Civil Engineers
   Environmental Engineering Division
   345 E. 47th Street
   New York, NY 10017-2398

Journal of Irrigation and Drainage Engineering

1. Otto J. Helweg
2. 1956
3. Bimonthly
4. ISSN 0733-9437
5. —
6. American Society of Civil Engineers
   Irrigation and Drainage Division
   345 E. 47th Street
   New York, NY 10017-2398

Journal of Maritime Law and Commerce

1. Nicholas J. Healy
2. 1969
3. Quarterly
4. ISSN 0022-2410
5. Adv., bk. rev., bibl.
6. Anderson Publishing Co.
   2035 Reading Road
   Cincinnati, OH 45202

Journal of Planning and Environment Law

1. Victor Moore
2. 1948
3. Monthly
4. ISSN 0307-4870
5. Adv., bk. rev., index
6. Sweet & Maxwell
   South Quay Plaza
   8th Floor
   183 Marsh Wall
   London E14 9FT
   England

Journal of Water Resources Planning and Management

1. William W. G. Yeh
2. 1982
3. Bimonthly
4. ISSN 0733-9496
5. —
6. American Society of Civil Engineers
   Water Resources Planning and Management Division
   345 E. 47th Street
   New York, NY 10017-2398

Kentucky Law Journal

1. Brian A. Cromer
2. 1912
3. Quarterly
4. ISSN 0023-026X
5. Adv., bk. rev., index
6. University of Kentucky
   College of Law
   Lexington, KY 40506

Land Economics

1. Daniel Bromley
2. 1925
3. Quarterly
4. ISSN 0023-7639
5. Adv., bk. rev., bibl., index
6. University of Wisconsin Press
   Journal Division
   114 N. Murray Street
   Madison, WI 53715

Land Use Policy

1. Penny Street
2. 1984
3. Quarterly
4. ISSN 0264-8377
5. Index

6. Butterworth-Heinemann Ltd.
   P.O. Box 63
   Westbury House, Bury Street
   Guilford, Surrey GU2 5BH
   England

Management Review

1. A. J. Rutigliano
2. 1973
3. Monthly
4. ISSN 0025-1895
5. Bk. rev., charts, illus., index
6. American Management Association
   135 W. 50th Street
   New York, NY 10020

National Forum

1. Stephen W. White
2. 1913
3. Quarterly
4. ISSN 0162-1831
5. Bk. rev.
6. Honor Society of Phi Kappa Phi (Auburn)
   c/o Stephen W. White
   129 Quad Center
   Mall Street
   Auburn, AL 36849-5306

National Wetlands Newsletter

1. Nicole Veilleux
2. 1979
3. Bimonthly
4. ISSN 0164-0712
5. Bk. rev., bibl., film rev.
6. Environmental Law Institute
   1616 P Street NW
   Suite 200
   Washington, DC 20036

National Resources Journal

1. Albert E. Utton
2. 1961
3. Quarterly
4. ISSN 0028-0739
5. Adv., bk. rev., charts, index, cum. index every 10 years
6. University of New Mexico
   School of Law
   1117 Stanford NE
   Albuquerque, NM 87131

Notre Dame Law Review

1. —
2. 1925
3. 5 times a year
4. ISSN 0029-4535
5. Adv., bk. rev.
6. University of Notre Dame
   School of Law
   Box 988
   Notre Dame, IN 46556

Ocean and Shoreline Management

1. Editorial Board
2. 1973
3. 8 times a year
4. ISSN 0951-8312
5. Bk. rev., charts
6. Elsevier Science Publishers Ltd.
   Crown House Linton Road
   Barking, Essex IG11 8JU
   England

Ocean Development and International Law

1. Daniel S. Cheever
2. 1973
3. Bimonthly
4. ISSN 0090-8320
5. Adv., bk. rev., abstr., charts, stat., index, cum. index

6. Taylor & Francis
   1900 Frost Road
   Suite 101
   Bristol, PA 19007

Population and Environment

1. Burton Mindick and Ralph Taylor
2. 1978
3. Quarterly
4. ISSN 0199-0039
5. Adv., bk. rev., charts, index
6. Human Sciences Press, Inc.
   72 Fifth Avenue
   New York, NY 10011

Public Health Reports

1. Marian Priest Tebben
2. 1878
3. Bimonthly
4. ISSN 0090-2918
5. Bibl., charts, illus., stat., index
6. U.S. Public Health Service
   Department of Health and Human Services
   Parklawn Building, Room 13C-26
   5600 Fishers Lane
   Rockville, MD 20857

Rural Development Perspectives

1. Sara Mills Mazie
2. 1984
3. 3 times a year
4. —
5. Bk. rev.
6. U.S. Department of Agriculture
   Economic Research Service
   1301 New York Avenue NW
   Room 324
   Washington, DC 20005-4788

Southwestern Review of Management and Economics

1. Roger Norton and William Peters
2. 1981
3. Quarterly
4. —
5. —
6. University of New Mexico
   Albuquerque, NM 87131

Technology Review

1. Jonathan Schlefer
2. 1899
3. 8 times a year
4. ISSN 0040-1692
5. Adv., bk. rev., illus.
6. Massachusetts Institute of Technology
   Association of Alumni and Alumnae
   W59-200
   Cambridge, MA 02139

UCLA Journal of Environmental Law and Policy

1. —
2. 1980
3. Twice annually
4. ISSN 0733-401X
5. Adv.
6. University of California, Los Angeles
   School of Law
   405 Hilgard Avenue
   Los Angeles, CA 90024

Water Pollution Control Federation Journal

1. Peter J. Piecuch
2. 1928
3. Monthly
4. ISSN 0043-1303
5. Adv., bk. rev., illus., index., cum. index: vols. 1–42, 1928–1970

6. Water Pollution Control Federation
   601 Wythe Street
   Alexandria, VA 22314-1994

Water Research

1. John Andrews
2. 1967
3. Monthly
4. ISSN 0043-1354
5. Adv., bk. rev., charts, illus., stat., index
6. Pergamon Press Inc.
   Journals Division
   Maxwell House
   Fairview Park
   Elmsford, NY 10523

Western Water

1. Rita Schmidt Sudam
2. 1973
3. Bimonthly
4. —
5. Bk. rev., illus., tr. lit.
6. Water Education Foundation
   717 K Street
   No. 517
   Sacramento, CA 95814-3406

Yale Law Journal

1. —
2. 1891
3. 8 times a year
4. ISSN 0044-0094
5. Adv., bk. rev., index
6. Yale University
   School of Law
   Yale Law Journal Co., Inc.
   401-A Yale Station
   New Haven, CT 06520

6

Films, Filmstrips, and Videocassettes

FILMS AND VIDEOCASSETTES PROVIDE a wide range of information on the environmental issues of water. This selection begins with a series of general films followed by films covering such specific topics as physical properties of water, groundwater, water supplies, quality, and others. The list concludes with case studies of how important water is to a variety of peoples in the world. These graphic presentations provide information about the importance of water more vividly than the written word. Water is so fundamental to the well- being of people that it may become the most critical issue faced by many of the world's people in the immediate future. The public must reach a critical decision about providing not only an adequate supply of water, but a quality product that will protect the health of the total population. A film presentation may aid in reaching a satisfactory solution. Although the films vary greatly in date of production, all films included here remain technically and scientifically accurate.

The following sources list films and videos in English:

AAAS Science Film Catalog, Washington, DC: American Association for the Advancement of Science, 1975. 398 p.

Educational Film & Video Locator of the Consortium of College and University Media Centers and R. R. Bowker. 4th ed. New York: R. R. Bowker Company, 1990–1991. 2 vols. 3361 p.

Film & Video Finder. 2d ed. Medford, NJ: Plexus Publishing Company, 1989. 3 vols. 1424 p.

Films in the Sciences: Reviews and Recommendations. Washington, DC: American Association for the Advancement of Science, 1980. 172 p.

Index to Environmental Studies Multimedia. University Park, Los Angeles, CA: University of Southern California, National Information Center for Educational Media (NICEM), 1977. 1113 p.

Video Rating Guide for Libraries. Santa Barbara, CA: ABC-CLIO, 1990–.

The Video Source Book. 12th ed. Syosset, NY: National Video Clearing House, Inc., 1991. 2 vols. 2745 p.

The following data are provided for each film:
Title of film
Distributor
Phone number if available
Data on film
Description

General

Can We Survive?
Stuart Finley Inc.
3428 Mansfield Road
Falls Church, VA 22041
Color, 20 minutes, sound, 16mm, 1975.

Abel Wolman shows problems faced in developing countries in water supply management and population control. His approach to the problem is humanistic.

The Desert (2d ed.)
Barr Films
12801 Schabarum Avenue
P.O. Box 7878
Irwindale, CA 91107
Phone: 818-338-7878
Color, 10 minutes, optical sound, 16mm, 1965.

Shows the desert as a place that gets little rain and how plants and animals adapt to the region.

Water
Children's Television International
8000 Forbes Place
Suite 201
Springfield, VA 22151
Phone: 703-321-8445
Color, 15 minutes, sound, 3/4 or 1/2 video, n.d.

Shows the relationship between water, the Earth's surface, and living things.

Water and Life: A Delicate Balance
Films for the Humanities and Sciences
P.O. Box 2053
Princeton, NJ 08543-2053
Phone: 800-257-5126
609-452-1128
Color, 13 minutes, VHS, Beta, U-matic, 1990.

There is no life without water. This program shows the role of water in the human body, the cycle of water, industrial water consumption, and pollution, as well as ways of increasing water supply and a reminder that water is not necessarily a renewable resource.

The Water Crisis
Time-Life Video
1271 Avenue of the Americas
New York, NY 10020
Phone: 212-484-5940
Color, 57 minutes, sound, Beta, VHS, 3/4 U-matic, 1981.

This *Nova* film shows how water could become the next national issue. Water problems from the Adirondack Mountains to the Mississippi River to the West Coast are examined.

Water from Another Time
Documentary Educational Resources Inc.
101 Morse Street
Watertown, MA 02172
Phone: 617-926-0491
Color, 29 minutes, sound, 3/4 U-matic, 1986.

A documentary of the life-styles of the U.S. past through the experiences of three elderly residents of Orange County, Indiana.

Water for Californians
California Department of Water Resources
P.O. Box 388
Sacramento, CA 95802
Color, 26 minutes, optical sound, 16mm, n.d.

Shows the role water plays in the life of people in the state. Stresses drinking water, flood control, clean power for homes and industry, recreation, and wildlife.

Water, The Plain Wonder
Great Plains Instructional TV Library
University of Nebraska
P.O. Box 80669
Lincoln, NE 68501
Color, 9 minutes, optical sound, 16mm, n.d.

Shows how important a natural resource water is. Americans take water for granted even though our supplies are threatened.

Water—What Happens to It
Agency for Instructional Technology
Box A
Bloomington, IN 47402
Phone: 800-457-4509
Color, 15 minutes, sound, 3/4 or 1/2 video, 1975.

This film gives a graphic picture of the different uses of water and the importance of water to all aspects of life.

Water—Why It Is, What It Is
Moody Institute of Science
Educational Film Division
12000 E. Washington Boulevard
Whittier, CA 90606
Phone: 213-698-8256
Color, 11 minutes, sound, 16mm, ½ video, n.d.

Shows that water is one of the Earth's most valuable resources, from enabling life to affecting the weather.

Water's Way
Phoenix/BFA Films and Video, Inc.
468 Park Avenue South
New York, NY 10016
Phone: 212-684-5910
800-221-1274
Color, 7 minutes, sound, 16mm, ¾ or ½ video, 1983.

A water droplet guides a little boy through the story of water—our greatest natural resource.

Science and Physical Properties

Earth: Its Water Cycle
Coronet Instructional Media
65 East South Water Street
Chicago, IL 60601
Color, 11 minutes, sound, 16mm, 1974.

Concentrates on the movement of water in and out of the atmosphere by showing the effect of temperature and pressure on the rate of water change from a liquid state to gaseous stage and vice versa.

Water—A Cutting Edge with Time
PBS Video/Public Broadcasting Services
1320 Braddock Place
Alexandria, VA 22314-1698
Phone: 202-488-5220

Color, 28 minutes, sound, 3/4 or 1/2 video, n.d.

Deals with the science of hydrology on a field trip to the canyon area of Utah and a raft ride down the Arkansas River.

Water Affects the Weather
Agency for Instructional Technology
Box A
Bloomington, IN 47402
Phone: 800-457-4509

Color, 15 minutes, sound, 3/4 or 1/2 video, n.d.

Shows how water influences weather through changes in water content of the atmosphere. Shows how the amount of moisture influences precipitation in a region.

Water and Energy
Visual Aids Service
1325 South Oak Street
University of Illinois
Champaign, IL 61820
Phone: 800-367-3456

Color, 17 minutes, sound, 3/4 Beta, VHS, n.d.

Describes principal properties of water in terms of hydrogen bonding. Energy changes of water are described.

Water and Plant Life
Films for the Humanities and Sciences
P.O. Box 2053
Princeton, NJ 08543-2053
Phone: 800-257-5126
609-452-1128
Color, 28 minutes, VHS, Beta, U-matic, 1990.

Covers the water cycle in plants, examining different adaptations to insufficiency of available water as problems of water shortage, including shallow roots, freezing temperatures, adaptation to heat, night and day blooming.

Water and Rain
Agency for Instructional Technology
Box A
Bloomington, IN 47402
Phone: 800-457-4509
Color, 15 minutes, 3/4 or 1/2 video, n.d.

Shows how water recycles, with examples of evaporation and cloud formation.

Water and the Weather
Stanton Films
2417 Artesia Boulevard
Redondo Beach, CA 90278
Phone: 213-542-6573
Color, 11 minutes, sound, 16mm, 1983.

Presents the role of water in determining weather. Explains meteorological terms, such as evaporation and condensation, high and low humidity, and the hydrologic cycle.

Water, Birth, Planet Earth—The Land
Coronet Instructional Films
Distributed by Simon and Schuster Film & Videos
108 Wilmot Road
Deerfield, IL 60015
Color, 27 minutes, sound, 16mm, 3/4 or 1/2 video, 1986.

Shows how plants and animals came from the sea and developed on land.

Water, Birth, Planet Earth—The Sea
Coronet Instructional Films
Distributed by Simon and Schuster Film & Videos
108 Wilmot Road
Deerfield, IL 60015
Color, 22 minutes, sound, 16mm, 3/4 or 1/2 video, 1986.

Shows how the Earth's atmosphere was formed, minerals were deposited in the sea, and life developed from a molecule.

Water—Coming and Going
Sterling Educational Films
241 East 34th Street
New York, NY 10016
Color, 20 minutes, optical sound, 16mm, n.d.

Portrays experiments showing evaporation, condensation, and precipitation of water. Explains stages of the water cycle and shows examples of streams, clouds, fog, snow, rain, and oceans. Includes multiple choice questions for students.

The Water Cycle (2d Edition)
Britannica Films
310 South Michigan Avenue
Chicago, IL 60604
Phone: 312-347-7958
Color, 14 minutes, sound, Beta, VHS, 3/4 U-matic, 1981.

Examines evaporation, condensation, and the effects of the water cycle on climate and the land.

Water in the Air
Educational Images
P.O. Box 3456, West Side
Elmira, NY 14905
Phone: 607-732-1090
Color, 30 minutes, sound, Beta, VHS, 3/4 U-matic, 1987.

Discusses the water cycle from evaporation to final precipitation.

Water Journey
Norman Burger Productions
3217 South Arville Street
Las Vegas, NV 89102-7612
Phone: 702-876-2328
Color, 60 minutes, sound, Beta, VHS, 1988.

Follows the movement of water beginning in the Colorado Rockies to the Atlantic Coast.

The Water Planet
Coast District Telecourses
11460 Warner Avenue
Fountain Valley, CA 92708-2597
Phone: 714-241-6109
Color, 30 minutes, 3/4 or 1/2 video, n.d.

Discusses the Earth as a water planet, with water covering more than 71 percent of the surface. Explores the origins of life.

Water Resources
Dallas County Community College District
Center for Telecommunications
4343 North Highway 67
Mesquite, TX 75150-2095
Phone: 214-324-7784
Color, 29 minutes, sound, 3/4 U-matic, 1977.

Shows the importance of water in monitoring different types of ecosystems from areas of swamp to dryland vegetation.

Water Runs Downhill
Journal Films, Inc.
930 Pitner Avenue
Evanston, IL 60202
Phone: 800-323-5448
312-328-6700
Color, 13 minutes, sound, 16mm, 3/4 or 1/2 video, n.d.

Observes actions of rain, clouds, and running streams. Defines erosion, water cycle, and evaporation.

Water: The Timeless Compound
American Educational Films
3807 Dickerson Road
Nashville, TN 37207
Color, 42 minutes, sound, 16mm, 1986.

Treats all aspects of water: physical properties, ocean waves and tides, laminar and helicoid flow, rapids and dams, frozen water, and more.

The Waters of the Earth
Coast District Telecourses
11460 Warner Avenue
Fountain Valley, CA 92708-2597
Phone: 714-241-6109
Color, 30 minutes, 3/4 or 1/2 video, n.d.

Describes the properties of water and physical conditions of the sea in respect to salinity, density, and temperature. Looks at water-sampling instruments.

Who Stole the Water?
Macmillan Films
34 MacQuesten Parkway South
Mt. Vernon, NY 10550
Color, 10 minutes, sound, 16mm, 1975.

This is an animated film with music to explain evaporation, condensation, and precipitation.

Groundwater

Ground Water
Dallas County Community College District
Center for Telecommunications
4343 North Highway 67
Mesquite, TX 75150-2095
Phone: 214-324-7784
Color, 29 minutes, sound, Beta, VHS, 3/4 U-matic, 1973.

Illustrates how the action of water on soluble rock produces characteristic soluble and depositional features.

Ground Water
Dallas County Community College District
Center for Telecommunications
4343 North Highway 67
Mesquite, TX 75150-2095
Phone: 214-324-7784

Color, 30 minutes, sound, 3/4 or 1/2 video, n.d.

Shows uses and problems associated with groundwater. In many areas, groundwater is being used faster than nature is replacing it, demonstrating the need for groundwater management.

Ground Water
Media Guild
11722 Sorrento Valley Road
Suite E
San Diego, CA 92121
Phone: 619-755-9191

Color, 19 minutes, sound, Beta, VHS, 3/4 U-matic, 1973.

Discusses the origin of groundwater as surface water sinks into the ground and accumulates in underground aquifers and defines each step in the utilization of groundwater through the wells that penetrate the water zone.

Ground Water—A Physician's Perspective
Marshfield Regional Video Network
1000 North Oak Avenue
Marshfield, WI 54449-5777
Phone: 715-387-5127
800-782-8581

Color, 59 minutes, sound, Beta, VHS, 3/4 U-matic, 1983.

Discusses chemicals found in water in central Wisconsin, together with health effects, acceptable channel levels, and medical recommendations about nitrates and pesticides.

Ground Water: California's Sunken Treasure
California Department of Water Resources
P.O. Box 388
Sacramento, CA 95802
Phone: 916-445-8569
800-952-5530

Color, 14 minutes, sound, 3/4 U-matic, 1977.

Demonstrates the importance of groundwater development to California's water users. The statewide distribution is shown, together with animated sequences illustrating the physical characteristics of groundwater reserves and the effects of degradation and depletion.

Ground Water Processes
Gulf Publishing Co. Video
P.O. Box 2608
Houston, TX 77001
Phone: 713-529-4301

Color, 58 minutes, sound, 3/4 or 1/2 video, n.d.

Demonstrates how water moves underground by gravity and the overlying pressure of water in a gently sloping aquifer—by these processes water can move hundreds of miles in an aquifer.

Ground Water—The Hidden Reservoir
Media Guild
11722 Sorrento Valley Road
Suite E
San Diego, CA 92121
Phone: 619-755-9191

Color, 19 minutes, sound, 16mm, 3/4 or 1/2 video, 1971.

Narrates the role of ground water in the hydrologic cycle.

Supply

Down to the Last Drop
Films for the Humanities and Sciences
P.O. Box 2053
Princeton, NJ 08543-2053
Phone: 800-257-5126
609-452-1128
Color, 26 minutes, VHS, Beta, U-matic, 1990.

Societies do not miss water until the wells dry up. Only during droughts is the public conscious of the need for water conservation. Yet, water is not necessarily a renewable resource. The time to plan for a water shortage is at the time when there is a water surplus. This program examines techniques of water management in Israel and the United States.

Drought and Flood: Two Faces of One Coin
Films for the Humanities and Sciences
P.O. Box 2053
Princeton, NJ 08543-2053
Phone: 800-257-5126
609-452-1128
Color, 18 minutes, VHS, Beta, U-matic, 1990.

The effects of global warming appear to be paradoxical. Even as hotter air will increase the size of desert areas, ocean heating will increase water temperatures resulting in massive storms in other areas.

Drought on the Land
British Broadcasting Co-TV
630 Fifth Avenue
New York, NY 10020
Color, 20 minutes, sound, 16mm, ¾ or ½ video, n.d.

Shows suffering in a small village in central Brazil due to drought. Describes how a family lost all their crops and some of their children died from malnutrition. Many of the family moved to São Paulo for better conditions.

The Flowing Oasis
KD Enterprises
P.O. Box 8321
Incline Village, NV 89450
Phone: 702-831-8178
702-827-3821
Color, 55 minutes, sound, video, 1987.

Presents a desert river, the East Walker in Nevada, and the impact it has on the surrounding environment. The unusual regime of the river is shown, with universal application to dry regions.

Water
Journal Films, Inc.
930 Pitner Avenue
Evanston, IL 60202
Phone: 800-323-5448
312-328-6700

Color, 14 minutes, sound, 16mm, Beta, VHS, 3/4 U-matic, 1977.

Examines the world's need for water from the point of view of doctors, meteorologists, agronomists, hydrologists, and engineers.

Water—A Precious Resource
National Geographic Society
Educational Services
Department 86
Washington, DC 20036
Phone: 800-368-2728

Color, 23 minutes, sound, 16mm, 3/4 or 1/2 video, 1979.

Examines many aspects of water, such as where it comes from and how it is used and considers whether there will be enough clean water for the future.

Water & Hunger
Maryknoll World Productions
Media Relations
Maryknoll, NY 10545
Phone: 800-227-8523
Color, 24 minutes, sound, VHS, 1988.

Shows the importance of water in the poverty-stricken areas of the world.

Water and Life
CTV Television Network
48 Charles Street East
Toronto, Ontario M4Y 1T4
Canada
Black and white, 25 minutes, 16mm, 3/4 or 1/2 video, n.d.

Shows how important water is to animals in East Africa and how they struggle to survive during droughts when water holes dry up.

Water Decisions
Idaho Water Resources Board
Pocatello, Idaho, 83209
Color, 28 minutes, sound, 16mm, 1976.

Problems of water shortage in Idaho are revealed by water resource planners, together with recommendations for economic development, agriculture, and power generation.

Water for a City
Australian Information Service
Australian Consulate General
636 Fifth Avenue
New York, NY 10010
Phone: 212-245-4000
Color, 13 minutes, optical sound, 16mm, 1972.

Shows the problems of obtaining and storing enough water for a large Australian city.

Water for Jordan
Media Guild
11722 Sorrento Valley Road
Suite E
San Diego, CA 92121
Phone: 619-755-9191
Color, 26 minutes, sound, 16mm, 3/4 or 1/2 video, 1985.

Shows how water can be located and stored in the desert. Traces geology and rainfall pattern in Jordan Valley and explains the Jordan Valley Water Project.

Water: It's What We Make It
PBS Video
1320 Braddock Place
Alexandria, VA 22314-1698
Phone: 202-739-5380

Color, 18 minutes, sound, VHS, 3/4 U-matic, 1986.

Examines the Earth's water supply, the pollution over the years, and efforts to maintain water purity.

Water Wars: The Battle of Mono Lake
University of California at Berkeley
Extension Media Center
2176 Shattuck Avenue
Berkeley, CA 94704
Phone: 415-642-0460

Color, 39 minutes, sound, 3/4 U-matic, special orders formats, 1984.

Covers the controversial diversion of freshwater from California's Mono Lake to the city of Los Angeles.

Quality

Clean Water
Films for the Humanities and Sciences
P.O. Box 2053
Princeton, NJ 08543-2053
Phone: 800-257-5126
609-452-1128
Color, 20 minutes, VHS, Beta, U-matic, 1990.

Reviews the unsuspected environmental and health problems people create at home and offers suggestions on purchasing, storing, and using common household products and nontoxic alternatives.

The Effects of Water Pollution
Films for the Humanities and Sciences
P.O. Box 2053
Princeton, NJ 08543-2053
Phone: 800-257-5126
609-452-1128
Color, 19 minutes, VHS, Beta, U-matic, 1990.

Uses the seal disaster of spring 1988 to show how pollution destroys the ecosystem of oceans. The message is that pollutants affect the food chain and can ultimately affect all life on the Earth.

Fit to Drink
Films for the Humanities and Sciences
P.O. Box 2053
Princeton, NJ 08543-2053
Phone: 800-257-5126
609-452-1128
Color, 20 minutes, VHS, Beta, U-matic, 1990.

Traces the water cycle, beginning with the collection of rainwater in rivers and lakes, passage through a water treatment plant to some of the places where water is used, and finally back to the atmosphere.

Pointless Pollution: America's Water Crisis
Bullfrog Films
Oley, PA 19547
Phone: 800-543-FROG
Color, 28 minutes, sound, video, 1989.

Presents the viewpoint that water pollution is a problem not only for industry but for all persons in the United States. Demonstrates that nonpoint sources of pollution are major causes of water contamination. These include oil and street trash, fertilizer, pesticides, herbicides, and all the wastes created by humans.

Safe Water: The High Cost of Drinking
Universal Education and Visual Arts
1200 Universal Plaza
Universal City, CA 91608
Color, 15 minutes, sound, 16mm, 1975.

On-the-spot interviews showing the problems of obtaining safe water, what must be done to ensure conservation and purification, and the cost involved. Also shows the dangers if water supplies are not cleaned up. The film gives a sense of urgency to do something about clean water before it is too late.

Seas under Siege
Films for the Humanities and Sciences
P.O. Box 2053
Princeton, NJ 08543-2053
Phone: 800-257-5126
609-452-1128
Color, 56 minutes, VHS, Beta, U-matic, 1990.

Shows the pollution of toxic wastes flowing or being dumped into the oceans, with a look at some of the sources of this pollution: industrial effluent, by-products and wastes, storm runoff, herbicides, and motor oil spills.

Ten Miles To Fetch Water: A Crisis in the West Virginia Coalfield
Asymmetry Productions
P.O. Box 5657
Athens, OH 45701
Phone: 614-592-3456
Color, 29 minutes, sound, video, 1989.

Shows the inefficiencies of a water system in the coal region of West Virginia. Polluted water is an example of corporate neglect of the human side of resource extraction.

Turning the Tide: Into Deep Water
Bullfrog Films
Oley, PA 19547
Phone: 800-543-FROG
Color, 26 minutes, sound, video, 1986.

Examines water pollution in Great Britain due to pesticides, nitrates, human sewage, and increasing acid precipitation. These problems are presented with the message that the next generation will inherit an unlivable world.

Water: A Clear and Present Danger
MTI Teleprograms Inc.
108 Wilmot Road
Deerfield, IL 60015-9990
Phone: 800-621-2131
312-940-1260

Color, 26 minutes, sound, Beta, VHS, ¾ U-matic, 16mm, 1983.

Examines the growing problems of groundwater contamination from polluted surface water containing pesticides, chemicals, organic materials, and other substances. Methods are given to control pollutants.

Water Analysis
Audio Visual Medical Marketing Inc.
404 Park Avenue South
9th Floor
New York, NY 10016
Phone: 800-221-3995

Color, 32 minutes, sound, ¾ or ½ video, n.d.

Gives steps in purifying water for drinking by determining *coliform* in the water.

Water and Wastewater Treatment
University of Wisconsin Extension
Department of Engineering & Applied Science
432 North Lake Street
Madison, WI 53706
Phone: 608-262-2061

Color, 30 minutes, sound, ¾ U-matic, 6 programs, 1977.

A course treating chemistry and methods of water supply and treatment, including wastewater treatment and disposal. Programs include chemical basics, water supply, water treatment plants, and wastewater treatment and disposal.

Water Chemistry and Chemical Handling Practices
Industrial Training Corp.
13515 Dulles Technology Drive
Herndon, VA 22070
Phone: 800-638-6757

Color, 60 minutes, sound, ¾ or ½ video, n.d.

Identifies chemicals used in water treatment and safety gear to be used with them.

Water for a Thirsty World
Counselor Films, Inc./Career Futures Inc.
1728 Cherry Street
Philadelphia, PA 19103
Phone: 215-568-7904
Color, 15 minutes, optical sound, 16mm, n.d.

Shows how pollution and demand threaten the freshwater supply. Explains operation of desalting. Shows new ways to map groundwater deposits.

Water for the City
Phoenix/BFA Films and Video, Inc.
468 Park Avenue South
New York, NY 10016
Phone: 212-684-5910
800-221-1274

Color, 11 minutes, sound, Beta VHS, ¾ U-matic.

Demonstrates the need to purify water regardless of its origin.

Water Means Life
CARE, Inc.
National Field Director
660 First Avenue
New York, NY 10016
Color, 5 minutes, optical sound, 16mm, 1979.

Shows that people often die from drinking contaminated water.

Water Means Life
New York State Education Department
Center for Learning Technologies
Media Distribution Network
Room C-7, Concourse Level
Cultural Education Center
Albany, NY 12230
Phone: 518-474-1265

Color, 20 minutes, sound, Beta, VHS, ¾ U-matic, 1979.

Shows how water is polluted by lack of sanitation facilities such as sewage plants or chlorination of water in Tanzania, Guatemala, and Mexico.

Water Pollution
Dallas County Community College District
Center for Telecommunications
4343 North Highway 67
Mesquite, TX 75150
Phone: 214-324-7784

Color, 30 minutes, sound, 3/4 or 1/2 video, n.d.

Deals with sources of water pollution and effects on people and wildlife. Also looks at technological and economic solutions.

Water Pollution—A First Film
Phoenix/BFA Films and Video, Inc.
468 Park Avenue South
New York, NY 10016
Phone: 212-684-5910
800-221-1274

Color, 8 minutes, sound, 16mm, 3/4 or 1/2 video, 1971.

Studies a stream from beginning to end to see the sources of pollution. Also shows how people can help clean up our water sources.

Water Pollution—A First Film (Revised)
Phoenix/BFA Films and Video, Inc.
468 Park Avenue South
New York, NY 10016
Phone: 212-684-5910
800-221-1274

Color, 12 minutes, sound, Beta, VHS, 3/4 U-matic, 1985.

A study designed to help viewers understand water pollution with information on aquifers to aid in the learning process.

Water Pollution—Can We Keep Our Water Clean
Journal Films, Inc.
930 Pitner Avenue
Evanston, IL 60202
Phone: 800-323-5448
312-328-6700

Color, 16 minutes, sound, 16mm, 3/4 or 1/2 video, 1971.

Shows sources of pollution in trash and garbage, fertilizers, chemicals, oil from boats, industrial waste from factories, and thermal pollution from nuclear plants. Available as a captioned film.

Water Resources Videos
Educational Images
P.O. Box 3456, West Side
Elmira, NY 14905
Phone: 607-732-1090

Color, 28 minutes, sound, Beta, VHS, 3/4 U-matic, 3 programs, 1987.

A series portraying water usage and quality in California: 1. Water, Water, Everywhere, 2. California's Coastline, 3. Tahoe at the Turning Point.

The Water Treatment Engineer
Access Network
295 Midparkway SE
Calgary, Alberta T2X 2A8
Canada

Black and white, 15 minutes, sound, 3/4 Beta/VHS, n.d.

Treatment of water, such as by chlorination, to make it safe for drinking. The techniques and chemicals are graphically illustrated.

Management

The Desert Doesn't Bloom Here Anymore
Coronet/MTI Film and Video
108 Wilmot Road
Deerfield, IL 60015
Phone: 800-323-5343
Color, 58 minutes, sound, VHS, 1987.

Documents the continuing problems of water in the Sahel. The management of land and water resources in an areas of climatic change, expanding population, local traditions, and shortsighted governmental policies. Compares the irrigation system of the Sahel and southwestern United States. A *Nova* production.

Water for Development
University of Illinois
Visual Aids Service
1325 South Oak Street
Champaign, IL 61820
Color, 12 minutes, sound, Beta, VHS, 3/4 U-matic, 16mm, 1976.

Investigates difficult problems of water management in the South African republic of Lesotho. Considers problem of exploration, storage, transportation, contamination, irrigation, and consumption.

Water—More Precious Than Oil
PBS Video/Public Broadcasting Service
1320 Braddock Place
Alexandria, VA 22314-1698
Phone: 202-488-5220
Color, 58 minutes, sound, 3/4 or 1/2 video, 1980.

Shows the use and abuse of water around the world, and the necessity for proper management of water resources.

Water Policy Debate
National Federation of State High Schools Assn.
P.O. Box 20626
11726 Plaza Circle
Kansas City, MO 64195
Color, 60 minutes, sound, VHS, 1985.

As demands for water increase there is growing awareness it must not be wasted. This film presents viewpoints as to how water is conserved as a precious resource.

Water Roundup
California Department of Water Resources
P.O. Box 388
Sacramento, CA 95802
Phone: 916-445-8569
800-952-5530

Color, 15 minutes, sound, 3/4 U-matic, 1982.

Describes how the California State Water Project supplies water for agricultural irrigation and for such major urban centers as Los Angeles and San Diego.

Conservation

Problems of Conservation: Water
Encyclopedia Britannica Educational Corp.
425 North Michigan Avenue
Chicago, IL 60611
Phone: 800-554-9862
Color, 16 minutes, sound, 16mm, n.d.

Shows the problem of water supply in southern California and the ways to conserve water and develop new supplies through desalination of saltwater and artificial weather control. It also recognizes the importance of pollution control.

Water California Style
California Department of Water Resources
P.O. Box 388
Sacramento, CA 95802
Phone: 916-445-8569
800-952-5530

Color, 16 minutes, sound, 3/4 U-matic, 1976.

Traces California water development for the past 200 years via paintings, historical pictures, and live-action scenes. Develops current water management policies, emphasizing water conservation and reclamation.

Water Crisis?
Self Reliance Foundation
Box 1
Las Trampas, NM 87576
Phone: 505-689-2250

Color, 45 minutes, sound, Beta, VHS, 3/4 U-matic, 1982.

Examines water conservation techniques for small farmers.

Water for Industry
California Department of Water Resources
P.O. Box 388
Sacramento, CA 95802
Phone: 916-445-8569
800-952-5530

Color, 5 minutes, sound, 3/4 U-matic, 1978.

Cartoon telling the story of water in food processing. The cleaning and recycling of water is shown, demonstrating how water can be used more efficiently.

Water—Its Many Voices
Ideal Pictures Film Library
44 W. North Avenue
Milwaukee, WI 53208
Phone: 414-873-4616
Color, 20 minutes, optical sound, 16mm, n.d.

Shows how to conserve water by use of water conservation devices and a change in the attitude of consumers that water is available in unlimited quantities.

Wise Use of Water Resources
Universal Education and Visual Arts
100 Universal City Plaza
Universal City, CA 91608
Color, 14 minutes, sound, 16mm, n.d.

Develops the concept of properties of water, its abundance, its value as a natural resource, and its use for consumer supply. Conservation methods emphasized throughout the film.

Acid Precipitation

Acid from Heaven
National Film Board of Canada
1251 Avenue of the Americas
New York, NY 10020-1173
Phone: 212-586-5131
Color, 30 minutes, sound, 16mm, 1982.

Acid rain results from coal-burning industries, cars, and trucks producing sulfur and nitrogen oxides. These oxides undergo changes to produce acid-causing sulfates and nitrates. Acid rain transforms the ecosystems.

Acid Precipitation: Particles and Rain
Film Fair Communications
10900 Ventura Boulevard
P.O. Box 1728
Studio City, CA 91604
Phone: 818-985-0244
Color, 16 minutes, sound, Beta, VHS, ¾ U-matic, 1984.

Examines the origin of acid rain, the problem of its control, and its effect on the ecosystems of streams and lakes.

Acid Rain
American School Publishers
Princeton Road
P.O. Box 408
Highstown, NJ 08520-9377
Phone: 800-843-8855
Color, 36 minutes, sound, video, 1982.

This film provides background on how acid rain is formed and the worldwide research to find ways to control, reduce, and eventually end its destructive impact.

Acid Rain
Film Fair Communications
10900 Ventura Boulevard
P.O. Box 1728
Studio City, CA 91604
Phone: 818-985-0244

Color, 17 minutes, sound, Beta, VHS, 3/4 U-matic, 1984.

A basic study showing how sulfur dioxide and nitric oxide are converted into acids and how these acids are distributed over hundreds of thousands of square miles; concludes with the disastrous effect of acid rain on ecosystems.

Acid Rain
Films for the Humanities
743 Alexander Road
Princeton, NJ 08540
Phone:800-257-5126
609-452-1128
Color, 20 minutes, sound, VHS, Beta, U-matic, 1989.

History and explanation of acid rain, including chemical definitions, geological and meteorological interactions, and sources of acid precipitation. Problems of the ecology and some possible solutions.

Acid Rain
Films for the Humanities and Sciences
P.O. Box 2053
Princeton, NJ 08543-2053
Phone: 800-257-5126
609-452-1128
Color, 20 minutes, VHS, Beta, U-matic, 1990.

A history and exploration of acid rain and the dangers posed by industrial carbon and sulfur that interacts with airborne moisture to produce acid rain. Demonstrates chemical reactions and geological and meteorological interactions and traces the sources of acid precipitation.

Acid Rain
Time-Life Video
1271 Avenue of the Americas
New York, NY 10020
Phone: 212-484-5940

Color, 57 minutes, sound, Beta, VHS, ¾ U-matic, 1984.

A comprehensive view of the origin of acid rain and the effect on the environment. A *Nova* film.

Acid Rain: A North America Challenge
National Film Board of Canada
1251 Avenue of the Americas
New York, NY 10020-1173
Phone: 212-586-5131
Color, 16 minutes, sound, video, 1988.

Shows effects of acid rain on vegetation with emphasis on the need for a U.S.–Canadian agreement on control of acid rain.

Acid Rain: Requiem or Recovery
National Film Board of Canada
1251 Avenue of the Americas
New York, NY 10020-1173
Phone: 212-586-5131

Color, 27 minutes, sound, Beta, VHS, ¾ U-matic, 1983.

Examines the origin of acid rain and impacts on the environment and wildlife.

Acid Rain: The Choice Is Ours
TV Sports Scene Inc.
5804 Ayrshire Boulevard
Minneapolis, MN 55436
Phone: 612-925-9661

Color, 20 minutes, sound, Beta, VHS, ¾ U-matic, 1982.

A study of the causes of acid rain from the use of fossil fuels in North America and Europe. The negative aspects are stressed.

Acid Reign
Kinetic Film Enterprises
255 Delaware Avenue
Buffalo, NY 14202
Color, 10 minutes, sound, 3/4 U-matic, 1983.

This film, narrated by Ralph Nader, presents the problem of acid rain as another part of the ongoing battle to preserve the environment.

Oil Spills

Between the Devil and the Deep Blue Sea
Landmark Films
3450 Slade Run Drive
Falls Church, VA 22042
Phone: 800-342-4336
703-241-2030
Color, 32 minutes, video, 1989.

A study of the *Exxon Valdez* Alaska oil spill in March 1989, with emphasis on new legislation and improved technologies to protect against future ecological destruction in ocean waters. The portrayal of the rehabilitation of the ecosystem in Prince William Sound is excellent.

The Great Oil Disaster
Coronet Films & Video
108 Wilmot Road
Deerfield, IL 60015
Color, 30 minutes, sound, 16mm, video, 1987.

A documentary on the *Amoco Cadiz* oil spill, beginning with how oil spills occur. Shows cleanup efforts and the continuing adverse effects years later. Concludes with ways to prevent future spills and new procedures for cleanup use.

Oil Pollution Prevention Regulations
University of Texas at Austin
Petroleum Extension Services
10100 Burnet Road
Austin, TX 78758

Color, 18 minutes, sound, Beta, VHS, 3/4 U-matic, 1975.

Gives orientation to the Federal Water Pollution Control Act of 1972 and the Spill Prevention Control and Countermeasure (SPCC) plans required to comply with the act.

Oil Spill: Patterns in Pollution
University of Michigan
Ann Arbor, MI 48109
Color, 17 minutes, sound, 16mm, 1974.

The search for oil climaxed by a spill. The attempt to control the spill with pitifully inadequate bundles of straw. Environmental problems include sea birds, their plumage matted, running in bewildered panic, unable to fly.

Oil Spill Report
Kluge Motion Pictures
5350 W. Clinton Avenue
Milwaukee, WI 53223
Color, 15 minutes, sound, 16mm, 1974.

Shows how fast action and good emergency techniques contained a spill of 250,000 gallons of crude oil at a pumping station near Lake Ripley. The effort was so successful that 90 percent of the crude was recovered and sent to a refinery.

Oil Spills
University of Toronto
Media Centre/Distribution Office
121 St. George Street
Toronto, Ontario
Canada M5S 1A1

Color, 7 minutes, sound, Beta, VHS, 3/4 U-matic, 1977.

Deals with research being conducted on oil properties in the Arctic regions.

Floods

Flood
MTI Teleprograms/Coronet International
Distributed by Simon and Schuster
108 Wilmot Road
Deerfield, IL 60015
Phone: 312-940-1260
Color, 14 minutes, sound, 16mm, 3/4 or 1/2 video, n.d.

Story of a canoeist who becomes adventurous on a flood-swollen river.

Flood!
Warner Home Video Inc.
4000 Warner Boulevard
Burbank, CA 91522
Phone: 818-954-6000
Color, 98 minutes, sound, Beta, VHS, 1976.

A made-for-TV disaster film, involving a dam that threatens to burst, which would cause downstream flooding. Actors include Robert Culp, Martin Milner, Barbara Hershey, Richard Basehart, Carol Lynley, Roddy McDowell, Cameron Mitchell, Teresa Wright, and Francine York.

Flood Below
Intercollegiate Video Clearing House
P.O. Drawer 33000 R
Miami, FL 33133
Phone: 305-443-3500
Color, sound, 3/4 or 1/2 video, n.d.

Shows the evolution and progress in monitoring floods by use of high-altitude LANDSAT photos.

Flood Forecasting
Britannica Films
310 South Michigan Avenue
Chicago, IL 60604
Phone: 312-347-7958

Color, 20 minutes, sound, Beta, VHS, 3/4 U-matic, 1987.

Modern meteorological technology used to predict a possible flood.

The Flooding River
Blackhawk Films
5959 Triumph Street
Commerce, CA 90040-1688
Phone: 213-888-2229
800-826-2295

Color, 34 minutes, sound, Beta, VHS, 3/4 U-matic, 1973.

A profile of the Connecticut River from its headwaters to the sea, showing the problem of flooding on the river and its effects on the valley.

Natural Waste Water Treatment
Bullfrog Films
Oley, PA 19547
Phone:800-543-FROG

Color, 29 minutes, sound, Beta, VHS, 3/4 U-matic, 1987.

The function of a small decentralized sewage treatment plant using the natural purifying characteristics of marsh plants. Considers the ecology and environment of the treatment plant.

Dams and Water Power

Water Power
Bullfrog Films
Oley, PA 19547
Phone: 800-543-FROG

Color, 25 minutes, sound, Beta, VHS, 3/4 U-matic, 1982.

Traces the history of water power and visits operating units. Manufacturers and consultants are interviewed.

Water Power
Education Development Corp.
10302 East 55th Place
Tulsa, OK 74146-6507
Phone: 918-622-4522

Color, 10 minutes, sound, 3/4 U-matic, Special order formats, 1973.

Shows the influence of the Volta Dam on the people of Ghana. Begins with traditional Ghana and continues with a montage of scenes showing the changes to the African landscape and way of life.

Water Power
Lucerne Media
37 Ground Pine Road
Morris Plains, NJ 07950
Phone: 800-341-2293

Color, 23 minutes, sound, 16mm, 3/4 or 1/2 video, 1981.

The technical steps in the movement of water through turbines and the conversion of energy to electric power.

Water Power
Viewfinders, Inc.
P.O. Box 1665
Evanston, IL 60204

Color, 24 minutes, sound, 16mm, 3/4 or 1/2 video, n.d.

Shows use of water wheels in the past and examines new ways of getting water power from tides, waves, and torrents.

Irrigation

Desert Garden
British Columbia Department of Agriculture
Information Center
Kelowna, British Columbia
Canada
Color, 25 minutes, 16mm, optical sound, 1974.

Shows how irrigation has changed the desert of Okanagan Valley into lush orchards.

Irrigation
Australian Information Service
Australian Consulate General
636 Fifth Avenue
New York, NY 10010
Phone: 212-245-4000
Color, 17 minutes, optical sound, 16mm, 1973.

Studies irrigation projects in Australia and asks if the benefits justify the cost.

Irrigation Farming (2d ed.)
Encyclopaedia Britannica Educational Corp.
425 North Michigan Avenue
Chicago, IL 60611
Phone: 800-554-9862
Color, 10 minutes, optical sound, 16mm, n.d.

Shows the needs for dams for irrigation in certain sections of the United States. Illustrates irrigation by several methods—furrow, flooding, and sprinkling.

Irrigation Specialist
Access Network
295 Midparkway SE
Calgary, Alberta, T2X 2A8
Canada
Black and white, 15 minutes, sound, 3/4 Beta/VHS, n.d.

Shows how water is distributed by canal systems to irrigate fields. Shows origin of water and the means of transporting it to irrigated areas.

Irrigation Systems
British Columbia Department of Agriculture
Information Center
Kelowna, British Columbia
Canada
Color, 48 minutes, optical sound, 16mm, 1977.

Shows types of irrigation systems found in British Columbia's Okanagan Valley.

Water
Walt Disney
Educational Media Co.
500 S. Buena Vista Street
Burbank, CA 91521
Phone: 800-621-2131
Color, 9 minutes, sound, 16mm, VHS, 1984.

Presents a proper method for management of water resources. Shows how improper irrigation can destroy soil for future use and shows how drip irrigation reduces water runoff.

Water Farmers
University of California at Berkeley
Extensions Media Center
2176 Shattuck Avenue
Berkeley, CA 94704
Phone: 415-642-2525
Color, 29 minutes, sound, VHS, ¾ U-matic, 16mm, 1984.

Explores how the farmers of Shaoxing, China, utilize their water environment.

Water for Farming
California Department of Water Resources
P.O. Box 388
Sacramento, CA 95802
Phone: 916-445-8569
800-952-5530

Color, 5 minutes, sound, $^3/_4$ *U-matic, 1978.*

An animated character tells the story of agriculture as the largest consumer of water in California. Explains where farmers get their water and how they get it to crops during the long hot summers.

Water for the Columbia Basin
Petite Film Company
708 North 62nd Street
Seattle, WA 98103
Black and white, 10 minutes, sound, 16mm, n.d.

Shows how water from the Grand Coulee Dam is used for irrigation of the desert. The flow is traced through the irrigation system until it reaches the fields. Also shows the crops produced through the work of the Bureau of Reclamation.

Water in the Desert: The Imperial Valley
California Biological Supply Company
2700 York Road
Burlington, NC 27215

Color, 15 minutes, sound, $^3/_4$ *or* $^1/_2$ *video, 35mm filmstrip, n.d.*

Color photographs show how water for irrigation was brought from the Colorado River to the Imperial Valley, flooding the Salton Sea.

Life and Health

Water: A Treasure in Trouble
Pyramid Film and Video
Box 1048
Santa Monica, CA 90406
Phone: 213-828-7577
800-421-2304

Color, 14 minutes, sound, Beta, VHS, $^3/_4$ *U-matic, 1989.*

Explores the role of water in life and some of the problems as it becomes scarcer and scarcer.

Water and Life: A Delicate Balance
Films for the Humanities
743 Alexander Road
Princeton, NJ 08540
Phone: 800-257-5126
609-452-1128

Color, 13 minutes, sound, VHS, Beta, 3/4 U-matic, 1989.

Examines the role of water in the human body, cycles of water, industrial water consumption and pollution, methods to increase water supply, and misconceptions about water as a renewable resource.

Water—Fluid for Life
Coronet Instructional Films
Distributed by Simon and Schuster Film and Videos
108 Wilmot Road
Deerfield, IL 60015
Phone: 312-940-1260

Color, 16 minutes, sound, 16mm, 3/4 or 1/2 video, 1978.

A broad interpretation of the importance of water to the existence of life on Earth. Reveals the many aspects of water as a fundamental Earth resource.

Water—The Common Necessity
Moody Institute of Science
Educational Film Division
12000 E. Washington Boulevard
Whittier, CA 90606
Phone: 213-698-8256

Color, 9 minutes, sound, 16mm, 1/2 video, 1975.

Shows the water cycle and ways that water is necessary for life. Examines conservation and uses of water.

Water: The Hazardous Necessity
National Film Board of Canada
1251 Avenue of the Americas
New York, NY 10020
Phone: 212-586-5131
Color, 27 minutes, sound, 16mm, 1977.

Investigates the public health problems resulting from unclean water supplies in several African countries and analyzes the causes and attempts at solutions.

Water, Water Everywhere
Area 16 Productions
917 N Highland Avenue
Hollywood, CA 90038
Color, 15 minutes, optical sound, 16mm, 1977.

Defines cross-connections in water systems and illnesses that have resulted. Describes remedial measures such as double check valves and vacuum breakers.

Water, Water, Water
Medfact Inc.
P.O. Box 418
Massillon, OH 44648
Phone: 213-837-9251
800-824-9225

Color, 39 minutes, sound, Beta, VHS, 3/4 U-matic, 1975

Demonstrates the importance of an adequate water intake plus a short discussion of the physiologic role of water.

Glossary

absorption The process by which substances are assimilated or taken up by other substances.

acid precipitation Precipitation that is acidic due to sulfur dioxide (SO_2) and/or nitric oxide (NO) combining with water in the atmosphere to form weak sulfuric acid (H_2SO_4) or nitric acid (HNO_3).

acid water Water with a pH of less than 7.0.

acre-foot The amount of water required to cover an acre of land one foot deep, about 325,851 gallons or 43,500 cubic feet.

advection The horizontal movement of a mass of air which causes changes in temperature or any other physical properties of air.

alkaline A solution having a high concentration of hydroxyl ions (OH) with a pH greater than 7.0

anion Negatively charged ion.

aquifer A geological formation, or group of formations, that yields a significant quantity of water to wells and springs.

aquifer system Intercalated materials that act as a water-yielding, hydraulic unit.

artesian aquifer An aquifer in which groundwater is confined under pressure that is greater than atmospheric pressure.

artesian well A well securing water from a confined aquifer. A flowing artesian well is one in which the water level is above the level of outlet.

brackish water Water containing dissolved minerals from 1,000 to 10,000 mg/l in excess of potable water standards.

brine A highly mineralized solution of more than 100,000 g/l, often produced as a waste product of desalination of seawater.

buffering capacity The ability of a large body of water, such as lake water, to resist change.

carrying capacity The upper limit of a water system capable of supporting all components with the resources available.

cation Positively charged ion.

chlorination Primary method of disinfection used to treat water in the United States.

coliform count Number of coliforms present in a sample of water (usually 100 ml).

commercial withdrawal Water obtained from a public supply or self-supplied for use by motels, hotels, office and commercial buildings, and civilian and military institutions.

confined groundwater Water in an aquifer that is bounded by confining beds. The water is under pressure greater than the atmosphere.

confining bed A rock layer having a low hydrologic conductivity that retards the movement of water into or out of an adjoining aquifer.

connate water Water entrapped in the pores of sedimentary rock at the time of its deposition.

consumptive use Use of water where a large proportion is lost by evapotranspiration as in irrigation.

contamination Water quality that is degraded.

correlative rights doctrine Extension of the reasonable use doctrine of water use allowing nonoverlying groundwater use by nonoverlying users.

cubic feet per second Measurement for the amount of water passing a given point.

current meter Device to measure water velocity.

Darcy's law The fundamental principle upon which the rate of groundwater flow is derived.

deep well injection The emplacement of a fluid under pressure into a deep aquifer.

degradation All processes that reduce the quality of water.

desalination Process of changing saline or brackish water to water suitable for human consumption.

discharge The rate of flow of surface or underground water, generally expressed in cubic feet per second.

discharge area An area in which subsurface water, including groundwater and water in unsaturated zones, is discharged to the land surface, to surface water, or to the atmosphere.

domestic withdrawal Water used for normal household purposes.

drainage divide The boundary between one drainage basin and another.

draw A tributary valley or coulee that discharges water only after a rainstorm.

drawdown A process of lowering the water level in reservoirs.

effluent A waste liquid that is discharged into the environment.

epidemiological Concerned with infectious diseases.

eutrophication An enriched condition of an aquatic ecosystem, characterized by the increased growth of certain plants and animals and reduction in the ability of many organisms to adjust to changes in the environment.

evaporation The process by which water is changed from a liquid to a gas or vapor.

evapotranspiration Water discharged to the atmosphere as a result of evaporation and plant transpiration.

flood A high stream flow that overtops the natural or artificial banks in a stream.

flume A structure built to transport ditch or canal water over a depression or a stream.

gage height Elevation of the water surface measured in feet and parts of feet.

gaging station Where records of flow, discharge, and water contents are taken.

groundwater Water that is held in underground reservoirs called aquifers.

groundwater reservoir Permeable rocks in the zone of saturation.

groundwater system A groundwater reservoir and the water it contains.

hardness (water) Ions of calcium and magnesium causing water to form an insoluble residue when used with soap.

hydraulic gradient The ratio of the difference of water height to the horizontal distance traveled—slope of the water table.

hydrograph A plot of discharge as a function of time.

hydrologic cycle Cyclical movement of water from the ocean to the atmosphere by evaporation, through rain to the surface of the Earth, through runoff and groundwater to streams, and back to the sea.

hydrology The scientific study of the water found on the Earth's surface and subsurface and in the atmosphere.

industrial withdrawal Water withdrawn from a public supply or self-supplied for industrial uses.

infiltration The movement of water into soil or porous rocks.

in-stream use Use of water within the stream channel.

irrigation Distribution of water over the ground surface by canals, pipes, or flooding or by overhead sprinkling and dripping, such as rain, to promote plant growth.

irrigation efficiency A measure of how much of the water applied to a field is actually used by plants.

irrigation return flow The part of applied water not consumed by evapotranspiration that migrates to an aquifer or surface water body.

irrigation withdrawal Water withdrawn for use on land to assist in the growing of crops and pastures or to maintain recreational areas.

laminar flow A flow pattern in which particle paths are straight or gently curved and parallel.

leaching The process of removing salts and alkali from soils that takes place when water percolates through the soil.

mean annual flood The highest flow a river normally reaches during the year with no overflow.

mean annual flow The amount of water that normally flows in a stream annually.

milligram per liter (mg/l) A measure of mass per volume. One one-thousandth of a gram per liter. 1mg/l of a substance equals one part per million (1 ppm) of the substance multiplied by its density.

mining of groundwater Withdrawal of groundwater in excess of recharge and therefore excessive.

mutagenic Capable of inducing a major change.

nonconsumptive use Water that performs a function, such as cooling, but in the process is not consumed.

nonpoint source of pollution A contaminant that cannot be traced to one source.

overdraft Withdrawal of groundwater in excess of recharge.

perched groundwater Unconfined groundwater separated by an unsaturated zone from the underlying main body of groundwater.

permeability The ability of a formation to transmit water or fluids through cracks or pores of the formation.

pH scale The scale that measures acidity or alkalinity in water. The scale runs from O to 14 with 7 indicating absolutely neutral substance. Zero is maximum acidity and 14 is maximum alkalinity.

point source of pollution A contaminant that can be traced to an individual source.

potable water Water that is safe for human use.

ppb Part(s) per billion.

ppm Part(s) per million.

precipitation Atmospheric discharge of water vapor in the form of rain, hail, sleet, or snow.

reasonable use doctrine Groundwater that is used without waste on overlying land.

recharge Water added to the zone of saturation.

recharge area An area where water infiltrates the ground and reaches the zone of saturation.

recharged groundwater Underground storage of water.

renewable water supply The proportion of the water supply available for use in a region on a permanent basis.

return flow That part of withdrawn water not consumed and available for further use.

riparian rights The right to use water in a stream based on ownership of the land adjacent to the stream.

runoff The part of snowmelt or precipitation that reaches streams or surface water bodies.

saline water Water not suitable for human use or for irrigation because it contains great quantities of dissolved solids and salts.

salinity The quality or state of being saline. A concentration of salt.

salt balance To maintain soil productivity, salt deposited on the surface as water evaporates must be dissolved by irrigation water to create a neutral balance.

salts Soluble compounds such as sodium chloride dissolved from earth and rocks by water that flows through them.

saturated zone The subsurface zone in which all the pore spaces or all openings are filled with water.

sedimentation Deposition of suspended materials in streams by gravity.

semiarid An area with annual precipitation of 8 to 16 inches.

spring A flow of groundwater emerging naturally at the ground surface.

toxicology The science of the nature and effects of poisons, and their detection and treatment.

transmissivity The rate at which water is transmitted through a unit width of an aquifer under a unit hydraulic gradient.

transpiration The process by which water passes through plants and living organisms and into the atmosphere.

troposphere The portion of the atmosphere below the stratosphere, extending 7 to 10 miles above the Earth's surface, in which the temperature generally rapidly decreases with altitude, clouds form, and convection is active.

turbidity Cloudiness caused by suspended matter in water.

unconfined aquifer An aquifer whose upper surface is free of the water table and can fluctuate under atmospheric pressure.

underground water Subsurface water in saturated and unsaturated zones.

unsaturated zone A subsurface zone in which spaces are not all filled with water.

water budget Inflow to, outflow from, and storage changes in a hydrologic unit.

water harvesting The capture of runoff water from depressional areas in fields or the floodplains of streams.

waterlogging Soil that is saturated with water that cannot be removed by the existing drainage system.

watershed An elevated line that forms the division of two areas drained by separate streams, systems, or bodies of water.

water table The top of the zone of saturation in a rock.

water year A continuous 12-month period used to study hydrologic or meteorologic phenomenon during which a complete annual hydrologic cycle occurs.

weathering The action of natural elements altering the shape, color, or texture of rocks or earth surfaces.

withdrawal Water removed from the ground or diverted from a surface water source for use.

Appendix

Precipitation

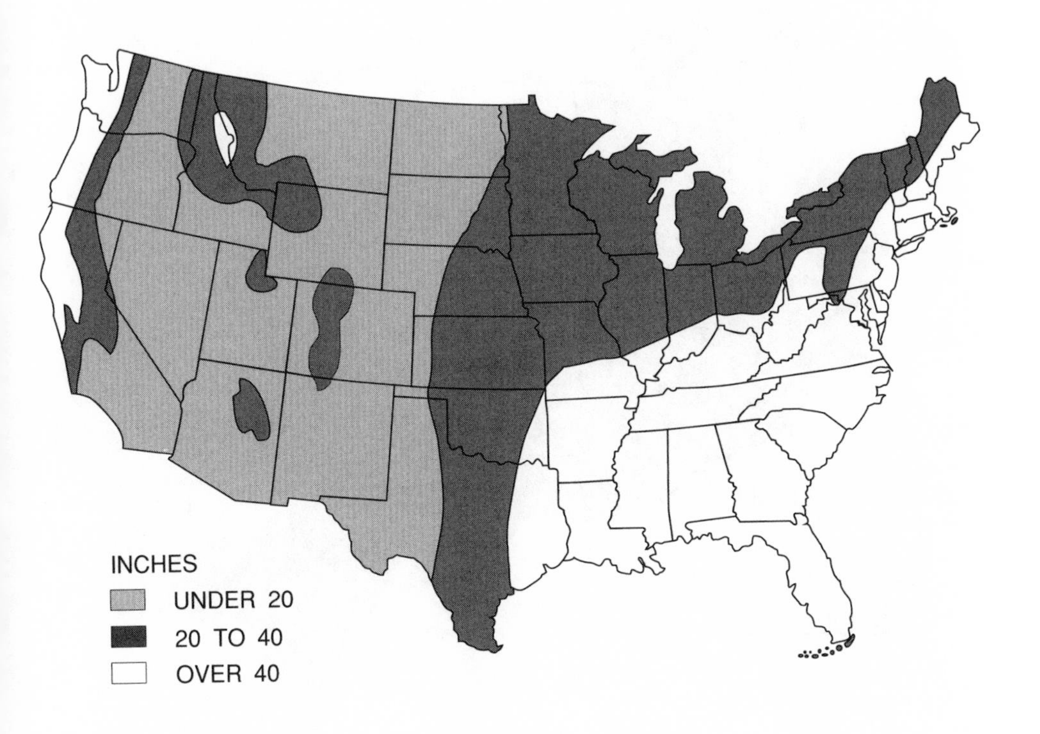

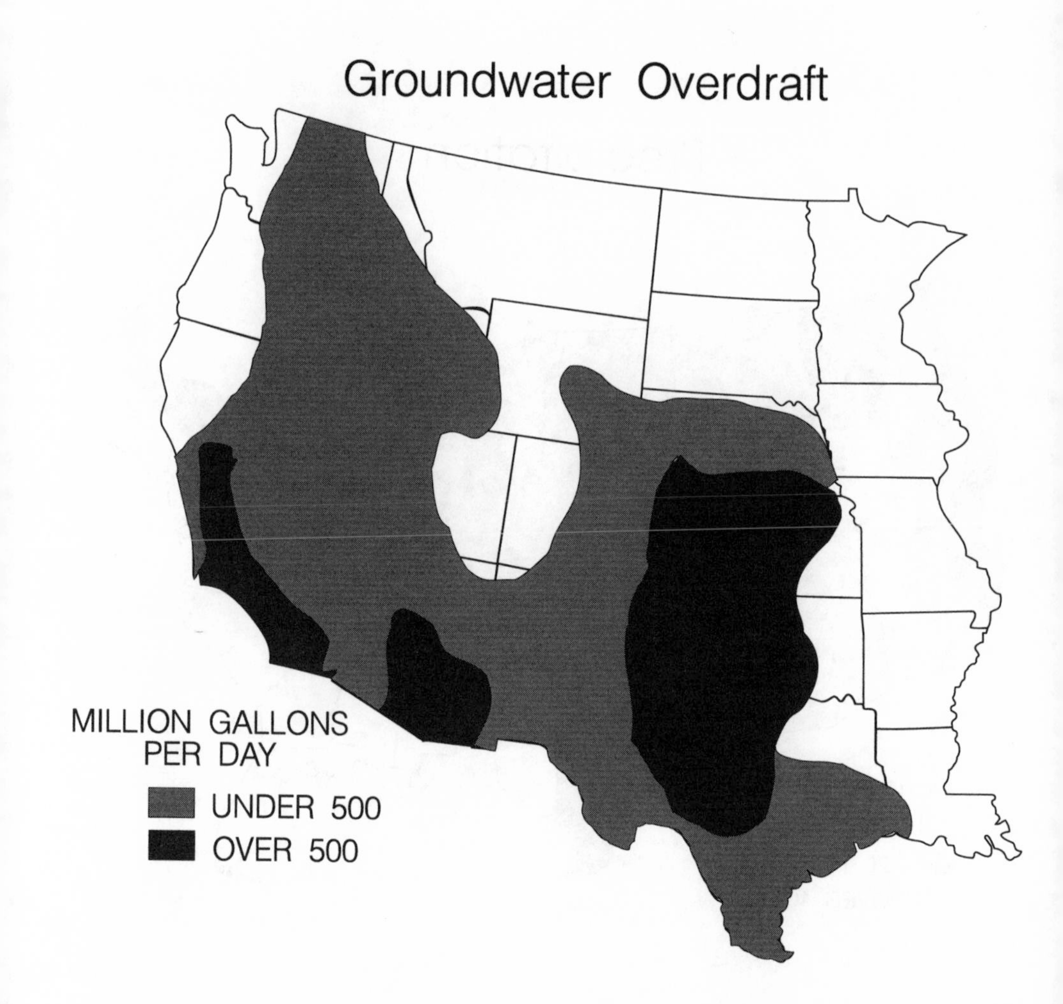
Groundwater Overdraft
MILLION GALLONS
PER DAY
UNDER 500
OVER 500

Major Saline-Water Areas of the West

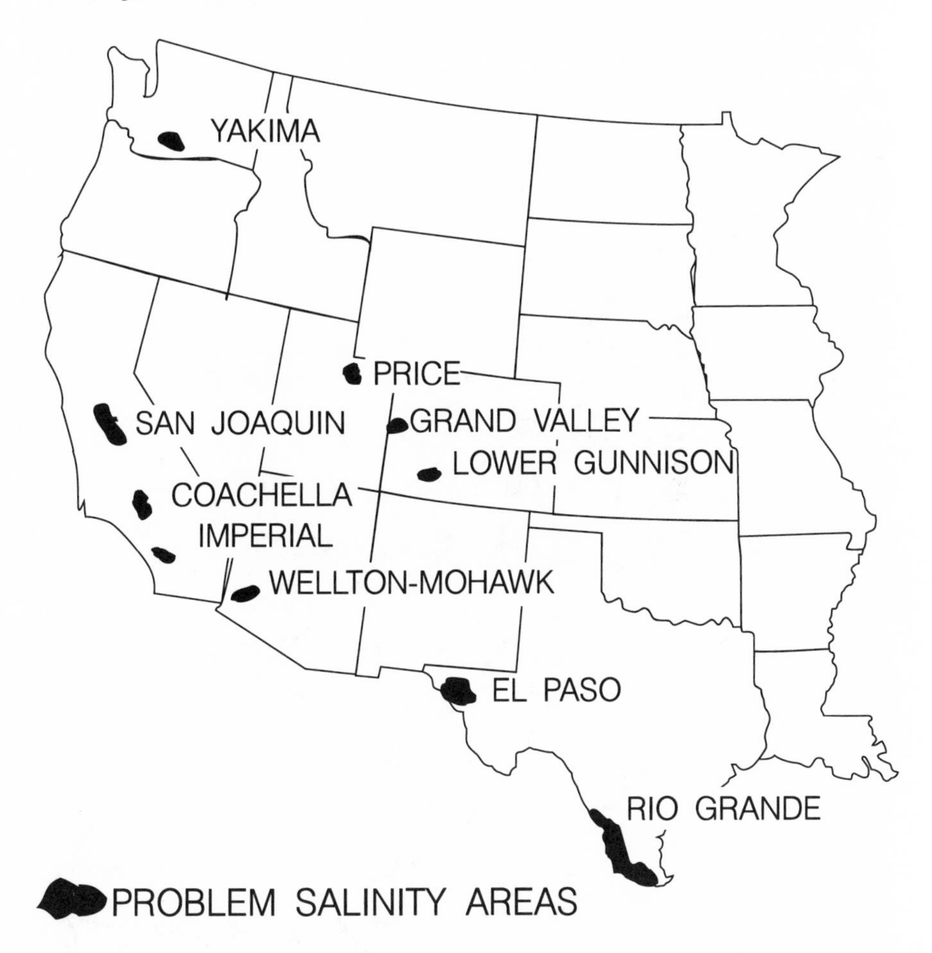

California's Principal Water-Supply System

Index